Structural Design of Low-Rise Buildings in Cold-Formed Steel, Reinforced Masonry, and Structural Timber

About the International Code Council

The International Code Council (ICC), a membership association dedicated to building safety, fire prevention, and energy efficiency, develops the codes and standards used to construct residential and commercial buildings, including homes and schools. The mission of ICC is to provide the highest quality codes, standards, products, and services for all concerned with the safety and performance of the built environment. Most United States cities, counties, and states choose the International Codes, building safety codes developed by the International Code Council. The International Codes also serve as the basis for construction of federal properties around the world, and as a reference for many nations outside the United States. The Code Council is also dedicated to innovation and sustainability, and a Code Council subsidiary, ICC Evaluation Service, issues Evaluation Reports for innovative products and reports of Sustainable Attributes Verification and Evaluation (SAVE).

Headquarters: 500 New Jersey Avenue NW, 6th Floor, Washington, DC 20001-2070

District Offices: Birmingham, AL; Chicago, IL; Los Angeles, CA

1-888-422-7233; www.iccsafe.org

Structural Design of Low-Rise Buildings in Cold-Formed Steel, Reinforced Masonry, and Structural Timber

J. R. Ubejd Mujagic

J. Daniel Dolan

Chukwuma G. Ekwueme

David A. Fanella

Roger A. LaBoube

New York Chicago San Francisco
Lisbon London Madrid Mexico City
Milan New Delhi San Juan
Seoul Singapore Sydney Toronto

The McGraw-Hill Companies

Cataloging-in-Publication Data is on file with the Library of Congress

McGraw-Hill books are available at special quantity discounts to use as premiums and sales promotions, or for use in corporate training programs. To contact a representative please e-mail us at bulksales@mcgraw-hill.com.

Structural Design of Low-Rise Buildings in Cold-Formed Steel, Reinforced Masonry, and Structural Timber

1 2 3 4 5 6 7 8 9 0   DOC/DOC   1 9 8 7 6 5 4 3 2

ISBN: 978-0-07-176792-7
MHID: 0-07-176792-4

The pages within this book were printed on acid-free paper.

Sponsoring Editor Joy Evangeline Bramble	**Proofreader** Mohana Shakthi, MEGAS
Acquisitions Coordinator Molly Wyand	**Indexer** Doreen McLaughlin
Editorial Supervisor David E. Fogarty	**Production Supervisor** Pamela A. Pelton
Project Manager Aloysius Raj, Newgen Publishing and Data Services	**Composition** Newgen Publishing and Data Services
Copy Editor James K. Madru	**Art Director, Cover** Jeff Weeks

Contents

Preface vii

1 Low-Rise Building Systems   **1**
  Introduction 1
  Definition of Low-Rise Buildings 2
  Applicable Design Documents and Design Process 2
  Low-Rise Building Types and Properties 13
  References 34

2 Loads and Load Paths in Low-Rise Buildings **37**
  Introduction 37
  Load Effects 37
  Load Combinations 81
  Load Paths 87
  Examples 92
  References 120

3 Structural Materials **121**
  Cold-Formed Steel Materials 121
  Reinforced Masonry 123
  Timber Mechanical Properties 133
  Open-Web Joist Systems 135
  References 143

4 Design of Cold-Formed Steel Structures **147**
  Introduction 147
  Cold-Formed Steel Framing 147
  Design Specifications and Materials 149
  Manufacturing Methods and Effects 149
  Design Methodology 150
  Section Property Calculations 150
  Effective-Width Concept 151
  Tension Members 153
  Flexural Members 153
  Concentrically Loaded Compression Members 162
  Combined Tensile Axial Load and Bending 163
  Combined Compressive Axial Load and Bending 164
  Welded Connections 167
  Bolted Connections 168
  Screw Connections 171

Other Resources 173
Example Problems 173
References 180

5 Structural Design of Reinforced Masonry **183**
Introduction 183
Load Paths and Analysis 183
Design of Masonry Members Subjected Primarily
   to Flexure (and Shear) 194
Design of Masonry Members Subjected to
   Axial Loads and Flexure 210
Shear Walls 236
Design of Walls Loaded Out of Plane (Slender Walls) 265
Connections in Masonry 280
References 295

6 Design of Structural Timber **297**
Introduction 297
Basic Design Philosophy 298
Flexural Members 303
Tension Members 307
Compression Members 308
Connections 313
Shear Walls 330
Diaphragms 345
Serviceability 351
Seismic Design and Detailing Requirements 354
International Building Code 354
References 355

7 Open-Web Steel Joist Systems **357**
Introduction 357
Non-composite Open-Web Steel Joists 361
Composite Open-Web Steel Joists 392
Cold-Formed Steel Deck 405
International Building Code 413
References 414

Index **417**

Preface

As the title suggests, this book focuses on the structural design of low-rise buildings constructed using cold-formed steel, reinforced masonry, or structural timber. The motivation to materialize this text has several sources.

While the vast majority, if not all, civil and structural engineering undergraduate curricula mandate one- or two-semester-long design courses, these almost always deal with reinforced concrete and structural steel, given their perception as the primary construction materials. A few programs periodically offer elective courses in reinforced-masonry and structural timber design, whereas only a handful of colleges and universities teach cold-formed steel design. Furthermore, some schools elect to offer single-semester combined courses providing instruction on the most important elements of reinforced-masonry and structural timber design. With this text, salient information on the design of structures of cold-formed steel, reinforced masonry, and structural timber are presented with the hope that it will facilitate a single-semester course covering all three materials. A course covering the three important construction materials that most structural engineers are likely to deal with throughout their career ideally would complement the existing curricula emphasizing structural steel and reinforced concrete, providing future structural engineers with an adequate theoretical background in all five main construction materials. A thorough coverage of load standard requirements and load paths in low-rise buildings should further enhance the understanding of each particular material and the associated type of construction.

It is anticipated that this book will be of significant use to practicing structural engineers as well, especially those with little or no educational background in cold-formed steel, reinforced masonry, or structural timber. Specifically, this text should provide the information and background in these three materials that are most critical to an understanding of member and system behavior found in common building structures. Those familiar with the three materials covered herein but seeking information on latest design code changes also should find this text useful.

Finally, candidates for structural engineering licensing examinations should find this text useful when reviewing design requirements for cold-formed steel, reinforced masonry, and structural timber.

Given their vast prevalence among all construction, low-rise buildings were selected as a benchmark for determining the content focus. Furthermore, low-rise buildings represent an excellent mean for instructing design of structural elements, load path, and overall behavior. In short, the intent of this book is to present all critical information pertinent to the design of a typical low-rise building constructed with cold-formed steel, reinforced masonry, or structural timber. Furthermore, many low-rise structures framed with structural steel or reinforced concrete often incorporate reinforced masonry, cold-formed steel, or wood components. The information presented in this book also will aid in the design of such elements.

Chapter 1 provides a general discussion of the low-rise building as a structural system and outlines the design process. Chapter 2 covers determination of structural loadings and load paths pertinent to low-rise buildings. Chapter 3 discusses the most important characteristics and properties of materials used in construction of cold-formed steel, reinforced-masonry, and structural timber buildings. The focus is on understanding of engineering properties critical to the behavior of members and the system. Chapters 4, 5, and 6 present the design requirements for cold-formed steel, reinforced masonry, and structural timber, respectively. Finally, Chapter 7 presents the design requirements and explains the behavior of non-composite and composite open-web joists. Open-web joist systems are common in low-rise buildings constructed with reinforced masonry. It is expected that the material discussing the design of open-web joist systems will enhance the text and facilitate overall understanding of the framing and design aspects of low-rise buildings.

The text uses the design code requirements referenced by 2009 *International Building Code*, the latest available building code at the time of completion of this manuscript. Therefore, it is anticipated that those familiar with general design requirements of cold-formed steel, reinforced masonry, and structural timber will find the text useful when seeking information on the latest load standard and design code updates.

The Authors

About the Authors

J. R. Ubejd Mujagic, Ph.D., P.E., S.E., is a Senior Associate at Uzun & Case Engineers, LLC, in Atlanta, Georgia. He has diverse experience in the design of metal building systems and low- and high-rise conventional buildings of most types and in all major building materials. Dr. Mujagic holds a Ph.D. in Structural Engineering from Virginia Tech. His research interests include structural reliability, composite structures, and cold-formed steel. Dr. Mujagic is a member of the American Iron and Steel Institute (AISI) Committee on Specifications and its subcommittees on connections, diaphragms, and seismic design. He is also a member of the American Institute of Steel Construction (AISC) Task Committees on Composite Members and Loads and Analysis. Dr. Mujagic is a licensed Structural Engineer in California, Illinois, and Washington, and a registered Professional Engineer in several states.

J. Daniel Dolan, Ph.D., P.E., is a Professor of Civil and Environmental Engineering and Director of Codes and Standards for the Composite Materials and Engineering Center at Washington State University. He holds a Ph.D. in Civil Engineering from the University of British Columbia. Dr. Dolan has been involved in the development of many of the building codes and design standards used in the United States. He has conducted extensive research in the area of the dynamic response of timber structures, especially their response to earthquakes and hurricanes. Dr. Dolan has published over 300 technical publications and has given over 500 technical presentations dealing with these topics. He holds professional engineering licenses in Wisconsin and Virginia.

Chukwuma G. Ekwueme, Ph.D., P.E., S.E., LEED AP, is an Associate Principal with Weidlinger Associates, Inc., in Marina del Rey, California. He holds a Ph.D. in Civil Engineering from the University of California, Los Angeles. Dr. Ekwueme has an extensive background in the design and analysis of a wide variety of structures, including concrete and masonry construction, steel and aluminum structures, and light-framed wood buildings. He is a registered Structural Engineer in California and Nevada and is an active member of the main committee, seismic subcommittee, and axial flexural loads and shear subcommittee of the Masonry Standards Joint Committee (MSJC).

David A. Fanella, Ph.D., S.E., P.E., F.ASCE, is Principal and Vice President at Klein and Hoffman Inc., in Chicago, Illinois. He holds a Ph.D. in Structural Engineering from the University of Illinois at Chicago and is a licensed Structural Engineer in the State of Illinois and a licensed Professional Engineer in numerous states. Dr. Fanella is past-president and current board member of the Structural Engineers Association of Illinois (SEAoI). He is an active member of a number of American Concrete Institute (ACI) Committees and is an Associate Member of the ASCE 7 Committee. He has authored or coauthored many structural publications, including a textbook on reinforced concrete design.

Roger A. LaBoube, Ph.D., P.E., is Curator's Teaching Professor Emeritus of Civil Engineering, Director of the Wei-Wen Yu Center for Cold-Formed Steel Structures, and Director of the Student Design and Experiential Learning Center at the Missouri University of Science and Technology (formerly University of Missouri—Rolla). He holds a Ph.D. in Civil Engineering from the University of Missouri—Rolla. Dr. LaBoube has an extensive background in the design and behavior of cold-formed steel structures. His research and design activities have touched on many facets of cold-formed steel construction to include cold-formed steel beams, panels, trusses, headers, wall studs, and bolt, weld, and screw connections. Dr. LaBoube is active in several professional organizations and societies, and is a member of the American Iron and Steel Institute (AISI) Committee on Specifications and a member of the AISI Committee on Framing Standards. He is a registered Professional Engineer in Missouri.

Structural Design of Low-Rise Buildings in Cold-Formed Steel, Reinforced Masonry, and Structural Timber

CHAPTER 1

Low-Rise Building Systems

1.1 Introduction

With respect to their height, building structures typically are categorized as low rise, mid-rise, and high rise. Furthermore, the high-rise classification contains a special subcategory of supertall buildings, which, according to the Council on Tall Buildings and Urban Habitat (CTBUH),[16] comprises buildings in excess of 300 m (984 ft) in height. According to the same source, as of mid-2011, only 51 such buildings have been completed and occupied in the whole world. While the precise breakdown of low-rise, mid-rise, and high-rise categories depends on many parameters and is somewhat subjective, low-rise buildings represent over 90 percent of all constructed buildings by almost any criteria. It is thought that 85 to 90 percent of all buildings subjected to significant seismic and wind hazards are low-rise structures.[19] Nearly 90 percent of all nonresidential building construction by floor area constructed from 1994 to 1996 belongs to buildings of three or fewer stories in height.[17] Furthermore, in the United States, over 90 percent of all nonresidential construction value in 2008 was attributed to low-rise buildings.[28] Per square foot of constructed floor space, low-rise buildings typically are significantly cheaper than their high-rise counterparts. Given such a significant prevalence of low-rise buildings in the overall construction market, it is clear that such structures accommodate a significant portion of the nation's industrial, commercial, institutional, residential, educational, and other activities. Furthermore, in some instances, height limitations imposed on certain structural systems in regions of high seismic activity, as well as zoning requirements, in some situations may lead to low-rise buildings as the sole economical choice. It is therefore of paramount importance that the behavior of such structures be thoroughly understood to ensure reliable designs and performance to protect the originally intended use of the building.

A significant portion of low-rise structures is constructed with reinforced masonry, cold-formed steel, and structural timber. The use of each material is highly dependent on the intended function of the building. For instance, while over half of all low-rise hotels in the United States are constructed with structural timber, its use in industrial applications is negligible.[28] The three materials also can be used in combination. For instance, masonry bearing walls generally are used in combination with cold-formed steel or plywood deck diaphragms, and many cold-formed steel and wood-framed buildings can include subgrade and lateral elements constructed with reinforced masonry. Furthermore, cold-formed steel, reinforced masonry, and structural timber often represent significant cladding and framing components of buildings of other types of construction. For instance, reinforced concrete buildings can have exterior cladding and interior partitions constructed with cold-formed steel, nearly all structural steel buildings contain

1

diaphragms in the form of cold-formed steel decking, and cladding and shear-wall elements in preengineered metal buildings in most cases are constructed with cold-formed steel sheathing and members. Similarly, reinforced-masonry shear walls, exterior cladding, and bearing walls often are used in conjunction with structural steel framing. Structural timber permits significant architectural expression when used for monumental roofs supported by either structural steel or reinforced concrete. The recent trend in both low-rise and mid-rise constructions involves construction of a reinforced-concrete or structural steel single- or multistory platform supporting a cold-formed steel or a wood stud-framed building of up to five or six stories in height. This particular type of construction permits increased heights or combinations of different occupancies, such as retail and residential or parking and residential, in the footprint of a single building while still permitting the use of economical light frames in the upper floors, thus reducing the overall building weight and cost.

It is certain that most structural engineers will encounter the need to design or specify low-rise structures of cold-formed steel, reinforced masonry, and structural timber at some point in or throughout their careers. Understanding of such structures and the ways to go about their design and assess their behavior therefore is critical to the practice of structural engineering.

1.2 Definition of Low-Rise Buildings

An official or formal definition of low-rise buildings does not exist. An actual terminology used to define a low-rise building can be based on its height, number of stories, presence of a mechanical elevator, or its aspect ratio, and it all depends on the context, the purpose for which the definition is needed, a particular industry, or even geography. While a specific building may be considered low-rise in Manhattan, NY, that the same building may be considered mid-rise or even high rise in a rural or suburban midwestern surrounding. Occasionally, a low-rise building is used solely as a vague colloquial expression to define a building with relatively few stories without a particular need for any particular quantifiable detail.

The most regularly used definition states that *low-rise buildings* are structures that are four or fewer stories in height. The definition of perhaps the most used to structural engineers comes from the wind-load requirements of American Society of Civil Engineers (ASCE) *Minimum Design Loads for Buildings and Other Structures* (ASCE/ SEI 7–05).[41] For the purposes of stipulating wind provisions in this particular standard, ASCE/SEI 7–05 defines *low-rise buildings* as structures of 60 or fewer feet in height in which the height is not larger than the least horizontal dimension.

1.3 Applicable Design Documents and Design Process

With respect to applicable design criteria, low-rise buildings do not differ appreciably from other types of buildings. The applicable design requirements are contained in four types of sources:

- Building code
- Loading standard
- Design standards
- Test reports

The role of these sources in the design process as it relates to the design of typical cold-formed steel, reinforced masonry, and structural timber buildings is discussed in this section.

1.3.1 Building Code

The building code applicable to a specific project is stipulated by the governing jurisdiction in which the building to be constructed will be located. A *building code* is a legal document governing all facets of design and construction of a building within the jurisdiction adopting it. It stipulates, either by reference or by direct requirement, all loading, design, and construction provisions critical to the proper execution of a building system. Regulation of construction constitutionally belongs to the states, although the jurisdictions adopting a specific building code vary from state to state. In some instances, the state itself represents a jurisdiction specifying the building code for all or certain types of structures within its borders. Examples of such states are Arkansas, California, Texas, and Tennessee. Where a statewide building code exists, the state often permits local governments, such as counties, townships, and cities, to augment the statewide building criteria within their jurisdiction. For example, the Texas state law allows the local jurisdictions to adopt more stringent provisions than stipulated by the state building code or to adopt a newer edition of the building code. The building code enforcement in such states is typically at the local level. There are also states such as Mississippi and Illinois that do not stipulate a statewide building code, leaving most responsibilities for code adoption and enforcement to local jurisdictions.

The primary model building code in the United States is the *International Building Code* (IBC).[23] A new edition of this code is issued every three years, although the International Code Council (ICC) issues critical code updates in the form of an IBC Supplement between regular three-year edition updates. Its current edition is 2009 IBC. The model building code itself is not a building code in a jurisdiction unless and until it is adopted by that jurisdiction. It simply represents a template the jurisdiction can use to create its building code. Models for adoption and implementation of building codes vary from state to state. The state of Tennessee, for example, adopted the 2006 IBC[22] in its entirety. Other states adopt a model building code, such as the IBC, with moderate state amendments. An example of this is the state of Georgia, whose *Georgia State Minimum Standard Building Code—International Building Code with Georgia State Amendments*[18] represents the 2006 IBC with relatively minor state-specific amendments. Finally, certain states elect to publish their own building codes comprising most IBC provisions with extensive state revisions. A very prominent example of this is the state of California and its *California Building Code (2010 CBC).*[14] It incorporates the 2009 IBC with numerous state amendments and special provisions aimed at specific classes of buildings. Most pronounced are the additional requirements imposed by the Office of Statewide Health Planning and Development and the Division of State Architect regulating construction of health-care facilities and public schools, respectively. As can be expected, the latest available version of the model code is not necessarily the one adopted in a given jurisdiction, often owing to administrative reasons. For instance, the state of Georgia typically omits the adoption of every other IBC model code. Specifically, it will not adopt the currently available 2009 IBC but is expected to implement 2012 IBC when it is published. Consequently, the user must use the edition of the building code adopted in the jurisdiction governing a particular project, not necessarily the latest available edition thereof.

The following 2009 IBC sections are relevant to the structural design of cold-formed steel, reinforced masonry, and structural timber:

- Chapter 16, Structural Design
- Chapter 17, Structural Tests and Special Inspections
- Chapter 18, Soils and Foundations
- Chapter 19, Concrete
- Chapter 21, Masonry
- Chapter 22, Steel
- Chapter 23, Wood
- Chapter 35, Referenced Standards

Chapter 16 of 2009 IBC stipulates the loading criteria for building design either directly or by reference to the loading standard. Chapter 2 of this book covers the details of load determination for low-rise buildings. Furthermore, Chap. 16 of 2009 IBC stipulates the loading information that must be shown on construction documents. Chapter 17 of 2009 IBC stipulates requirements for fabrication, welding requirements for reinforcement and cold-formed steel, testing, and inspection during and after construction. Chapter 18 of 2009 IBC stipulates requirements for the design of various types of foundation systems, including special requirements for seismic design categories C and above. Chapter 19 of 2009 IBC governs the design and construction of concrete elements and systems. It stipulates the concrete design standard, as well as significant exceptions thereto, especially as it relates to the design of concrete elements in seismic design categories D and above. Chapter 19 is relevant to cold-formed steel, reinforced-masonry, and structural timber construction when performing the foundation design and anchorage of structural elements to concrete. Design requirements for reinforced masonry are contained in Chap. 21 of 2009 IBC. This chapter stipulates the reinforced-masonry design standard, along with a number of exceptions to the design standard. Design provisions for cold-formed steel are stipulated by Chap. 22 of 2009 IBC, Secs. 2209 and 2210. Section 2209 references the design standards for and deals with generic aspects of cold-formed steel design, including stainless steel, steel decks, and non-light cold-formed steel framing. Section 2210 defines the design requirements and references specific requirements for cold-formed steel light framing. Finally, Sec. 2206 deals with open-web joist systems, including joist girders and composite and noncomposite joists. Aside from referencing specific design standards, a significant portion of this section deals with the division of responsibility for the specification, design, and construction aspects of open-web joist systems between the system manufacturer and the registered design professional responsible for the particular project. Chapter 23 of 2009 IBC references the design standards for wood framing and diaphragms. It also offers stand-alone alternative provisions, as well as prescriptive design provisions, for conventionally framed light-frame buildings. The role of Chap. 35 of 2009 IBC is to specify the edition of loading, design, and other standards referenced throughout the IBC. It is important to recognize that the edition of a standard specified in Chap. 35 is not necessarily the latest available edition. Consequently, the user must apply the edition of a standard stipulated in Chap. 35, not necessarily the latest one. When published, each new edition of a design or a loading standard is reviewed by the

ICC technical committees responsible for assessing the suitability of that particular document for adoption into the subsequent edition of IBC. Therefore, even though a standard may represent state-of-art provisions pertaining to its area of applicability, its eventual adoption into the model code may be associated with significant exceptions and additional requirements that typically cannot be foreseen in advance.

Other sections of 2009 IBC are mainly nonstructural, although they can have an impact on the design of structural members. For example, Chap. 7 of 2009 IBC deals with fire-protection requirements. While structural engineers do not deal with specific requirements pertaining to configuration of fire protection, the minimum required fire rating of various floor and wall structural elements could have an impact on the minimum required thickness of such elements. Furthermore, Chap. 34 of 2009 IBC can have significant structural implications with respect to alteration, repair, and change of occupancy of existing buildings.

When implementing various building code provisions in design, it is of utmost importance to understand and conscientiously apply the code intent. Specifically, the building code is a finite document providing general criteria for a virtually infinite number of scenarios that can be experienced during the design and construction process. To that end, it is possible to encounter situations where a particular provision can be interpreted in two or more ways. Further, owing to a certain lag that often inevitably exists between the instances of various code-referenced load and design standards and the ICC code renewal cycle, and given the robustness and complexity of the process, it is possible occasionally to experience either a conflict or an incompatibility between various provisions in the model code and the referenced standard or between certain referenced standards. In such cases, the correct interpretation of the code intent is critically important and must not be lost in an attempt to interpret finite points of a particular sentence or a paragraph in a black-or-white manner. ICC members, including design professionals, may submit a code interpretation request by either contacting the ICC staff by phone or by submitting an inquiry online. For more complex issues, code interpretation inquiries also can be submitted to the standing ICC Interpretation Committee. Official interpretations are issued as a publicly available written document. As an example, Table 1607.1 in 2006 IBC[22] stipulates a live load of 40 psf for hotels and multifamily dwellings such as condos, as well as for private rooms and corridors serving them. At the same time, it stipulates a 100-psf live load for public rooms and corridors serving them. A structural engineer designing a multistory condo building might reasonably infer that the corridors on a typical residential level not containing public areas connect only individual dwelling units and therefore can be designed for a live load of 40 psf. Alternatively, another designer might conclude that since the residential-level corridors are publicly accessible, 100-psf live load applies. The IBC Interpretation No. 37–07[24] confirms that the former of the two conclusions is correct because such corridors serve no public spaces, only individual living spaces. While the final authority to interpret or accept interpretation regarding a particular code provision rests with the building code authority having jurisdiction over the project, the ICC interpretations are commonly accepted as dependable.

1.3.2 Loading Standard

One of the most critical ingredients of the structural design process is the proper determination of load effects. Sections 1606 through 1613 of 2009 IBC stipulate the requirements for determination of various load effects applicable to building structures.

In the case of many loads, this is done by partially or completely referencing the loading standard ASCE/SEI 7–05. Section 1605 of 2009 IBC specifies load combinations for establishing the design-load effects for *load and resistance factor design* (LRFD) and *allowable stress design* (ASD). As is the case with all IBC-referenced standards and for the reasons outlined in Sec. 1.3.1, it is important to use the edition of the standard specified in Chap. 35 of 2009 IBC. The 2012 IBC[23a] is expected to reference ASCE 7–10.[42]

Although ASCE/SEI 7–05 has an appearance of a complete and stand-alone load standard incorporating provisions for determination of all load effects, load combinations for combining those effects into design forces, and even quality assurance and seismic detailing requirements, it is important to use this standard as prescribed by the building code. Therefore, when determining the load effects applicable to a building structure, the first point of reference always must be the building code. In the case of 2009 IBC, this point of reference is Chap. 16. In some instances, Chap. 16 of 2009 IBC provides the loading criteria directly. This is the case for live load, whose requirements are prescribed entirely by Sec. 1607 of 2009 IBC. Other load criteria are prescribed completely by reference to ASCE/SEI 7–05, as is the case with snow in Sec. 1608 of 2009 IBC. Similarly, Chap. 16 of 2009 IBC references ASCE/SEI 7–05 for the specific criteria, although it provides a number of optional alternatives to ASCE/SEI 7–05. For example, wind loads on a building can be computed entirely using only ASCE/SEI 7–05, but as an alternative, they also can be calculated using the alternate all-heights method provided in Sec. 1609.6 of 2009 IBC. Chapter 2 of this book provides a detailed discussion on determination of structural loadings according to 2009 IBC and therein referenced ASCE/SEI 7–05.

1.3.3   Design Standards

A *design standard* or a *design specification* is a document guiding the user in determining the strength of structural systems, members, and connections. Additionally, design standards prescribe permissible materials, detailing requirements, and certain aspects of the material and member fabrication and construction and also can lay down requirements pertaining to analysis, serviceability, seismic design, and standard practice. In short, while the role of a loading standard is to define the effects acting on a structural system and its members, the role of a design standard is to predict the behavioral response to such effects while providing prescriptive requirements pertaining to analysis, detailing, and construction to ensure that the predicted behavior matches the actual response of a structural system and individual members and connections within that system.

For cold-formed steel construction, the design standard referenced by Chap. 35 of 2009 IBC is the 2007 American Iron Steel Institute (AISI) *North American Specification for the Design of Cold-Formed Steel Members* (ANSI/AISI S100-07).[8] For specific light-frame applications, Chap. 35 of 2009 IBC further references AISI framing standards. These can be viewed as design standards containing specific prescriptive requirements for specific cold-formed steel light-frame construction elements, such as stud, header, or lateral framing design. These framing standards are listed in Chap. 4 of this book. In some instances, such as is the case with header design, 2009 IBC provides an option for the user to use either the framing standard or the ANSI/AISI S100-07 design specification without a reference to the framing standard. Also, framing standards often refer to ANSI/AISI S100-07 for certain specific provisions.

For the design of structural masonry, Chap. 35 of 2009 IBC references the 2008 Masonry Standards Joint Committee (MSJC) *Specification for Masonry Structures* (TMS

602–08/ACI 530.1–08/ASCE 6–08)[31] and MSJC *Building Code Requirements for Masonry Structures* (TMS 402–08/ACI 530–08/ASCE 5–08).[30] The MSJC is a joint body sponsored by three organizations with interest in structural masonry design—The Masonry Society (TMS), the American Concrete Institute (ACI), and the American Society of Civil Engineers (ASCE). Consequently, these two standards each bear the three separate designations, each of which is associated with one of the three organizations involved with its development and maintenance. The purpose of TMS 402–08/ACI 530–08/ASCE 5–08 is to provide specific design criteria for determining the design strength of structural masonry elements in bearing, bending, compression, and shear, as well as shear and tension strength of anchorage to masonry. Furthermore, this document sets out rules for seismic detailing and evaluating serviceability, creep, and shrinkage. TMS 402–08/ACI 530–08/ASCE 5–08 deals with properties of constructed masonry assemblies. TMS 602–08/ACI 530.1–08/ASCE 6–08, on the other hand, can be viewed as a background document to TMS 402–08/ACI 530–08/ASCE 5–08. It regulates the materials for the individual components of masonry assemblage—masonry units, grout, mortar, and reinforcing steel—and it regulates various aspects of construction, such as storage and handling of materials, quality assurance, submittal of construction documents, and construction conditions. In short, TMS 602–08/ACI 530.1–08/ASCE 6–08 is charged with ensuring the material quality and construction conditions to justify the design strength and behavior predicted by TMS 402–08/ACI 530–08/ASCE 5–08. Some provisions of TMS 402–08/ACI 530–08/ASCE 5–08 and TMS 602–08/ACI 530.1–08/ASCE 6–08 are further amended or altered by the requirements of Chap. 21 of 2009 IBC.

For structural wood, Chap. 35 of 2009 IBC primarily references the 2005 American Forest and Paper Association (AF&PA) *National Design Specification for Engineered Wood Construction with 2005 Supplement* (ANSI/AF&PA NDS-2005)[2] for the design of wood framing and *Special Design Provisions for Wind and Seismic* (ANSI/AF&PA SDPWS-08)[6] for the design of diaphragm and shear walls. For allowable stress design, Chap. 23 of 2009 IBC further supplements by or provides as an alternative to the ANSI/AF&PA NDS-2005 requirements a number of other standards, including American Institute of Timber Construction (AITC), American Society of Agricultural Engineers (ASAE), American Plywood Association (APA), and Truss Plate Institute (TPI) standards for the design of specific wood construction methods and elements such as trusses, laminated members, foundations, and inspection and construction details. The user should refer to Sec. 2306 of 2009 IBC for the complete list of these additional standards and Chap. 35 of 2009 IBC for the required edition. In many cases, Chap. 23 itself of 2009 IBC provides supplemental or alternative design criteria to those provided by ANSI/AF&PA NDS-2005 and ANSI/AF&PA SDPWS-08.

For the design of noncomposite joists, long-span joists, and joist girders, Chap. 35 of 2009 IBC references the 2005 Steel Joist Institute (SJI) *Standard Specification for Open Web Steel Joists, K-series* (SJI-K-1.1–05),[38] *Standard Specification for Long-Span Steel Joists, LH-Series, and Deep Longspan Steel Joists, DLH Series* (SJI-LH/DLH-1.1–05),[39] and *Standard Specification for Joist Girders* (SJI-JG-1.1–05),[37] respectively. The IBC-referenced design standards for the design of noncomposite floor and roof decks are the 2006 Steel Deck Institute (SDI) *Standard for Noncomposite Steel Floor Deck* (ANSI/SDI-NC1.0–06)[34] and *Standard for Steel Roof Deck* (ANSI/SDI-RD1.0–06),[35] respectively. The SDI *Standard for Composite Steel Floor Deck* (ANSI/SDI-C1.0–06)[33] is used for deck applications in which the deck serves as a permanent support and steel reinforcement for the supported concrete slab. Although this document is not an IBC-referenced standard, it is generally used to supplement the requirements of 2009 IBC referenced *Standard for the Structural Design of Composite Slabs* (ANSI/ASCE 3–91),[13a] which is often viewed as outdated. A newer SDI standard or a joint ASCE/SDI consensus standard is expected to replace ANSI/ASCE 3–91 and ANSI/SDI-RD1.0–06 as the new consensus standard for composite slabs.

Other design standards typically associated with the construction of cold-formed steel, structural masonry, and structural wood low-rise buildings include the IBC-referenced 2008 American Concrete Institute (ACI) *Building Code Requirements for Structural Concrete* (ACI 318–08)[1] and 2005 American Institute of Steel Construction (AISC) *Specification for Structural Steel Buildings* (ANSI/AISC 360–05).[7] ACI 318–08 is required for the design of foundations, development of masonry reinforcement into concrete, bearing on concrete, and anchorage into concrete. ANSI/AISC 360–05 is required for the design of elements of the open-web joist systems not otherwise covered by the preceding SJI specifications. Furthermore, it is required for the design of connections in cold-formed steel framing where the thicknesses of both the connected members exceed 3/16 in. Also, ANSI/AISC 360–05 is used in the design of steel connection and anchorage elements in wood and masonry construction.

The user should take note of the common manner in which the design specifications are designated. Commonly, the name of the parent organization is followed by the name of the parent organization publishing and maintaining the standard, followed by the number designating a particular standard and the year of the standard edition. For instance, ACI 318–08 indicates a 2008 edition of the ACI-published *Building Code Requirements for Structural Concrete* that ACI designates by the number 318. Also, when the designation ANSI precedes the name of the standard, this indicated that the American National Standards Institute (ANSI) accredited the organization publishing the standard and consequently approved the standard itself. ANSI approval guarantees a certain measure of quality in the committee process associated with publishing and maintaining a design specification. An example of this is ANSI/AISC 360–05.

There are two types of documents associated with design standards that complement the design specifications—specification commentaries and design manuals. These are sometimes published as completely separate documents, as is the case with the 2007 AISI *Commentary to the North American Specification for the Design of Cold-Formed Steel Structural Members*[9] and the 2005 *National Design Specification for Engineered Wood Construction Commentary*.[11] In other instances, such is the case with as ANSI/AISC 360–05, TMS 602–08/ACI 530.1–08/ASCE 6–08, and TMS 402–08/ACI 530–08/ASCE 5–08, the commentary is separate from the specification but issued with it in a single document. In the case of the SDI steel deck standards, the commentary is embedded in the specification itself. The SJI open-web joist specifications do not contain a commentary. The role of a commentary is multifold. It explains the rationale behind and cites the research work leading to the specific design specification provisions and provides a narrative of historical basis for various provisions while chronologically tracking their development and modification. Furthermore, it provides nonmandatory design advice complementing specification provisions. The reason for inclusion of such items into a commentary rather than in the specification or for including them in the commentary in the first place is the fact that they represent significant design considerations that cannot be omitted, yet the particular methods of taking into account these design considerations are subject to significant engineering judgment or come in many different forms. One example of this is the alignment charts for computation of column-slenderness K-factors found in most design specifications dealing with consideration of structural stability via the effective length method. As is the case with column K-factor, there exist many methods for computing their values, heavily depending on the configuration of the structure, as well as significant engineering judgment in assessing end-column fixity and applicability of different methods. Therefore, this particular issue is too complex for a consensus-based specification provision. Instead, the specification would mandate the consideration of column slenderness while outlining in its commentary one of the methods covering various basic conditions. Finally, possibly the most important role of the commentary is that it explains the specification provisions and their intended uses, limitations, and proper interpretation. It is important to note, however, that although the commentary text has a useful advisory role, it does not represent mandatory language stipulating building code–referenced requirements.

Design manuals, much like commentaries, have an instructive role. They typically contain numerical examples illustrating design applications of various specification provisions. Furthermore, they also contain useful design advice, based on standard practice, for the aspects of the design not explicitly addressed by the specification, including construction details. Finally, they often offer design aids in the form of tabular or graphic solutions implementing various specification provisions and other useful documents such as testing standards and codes of standard practice. For cold-formed steel construction, the AISI publishes the *Cold-Formed Steel Design Manual* (AISI D100–08).[12] Similarly, TMS puts forth the *Masonry Designers' Guide* (MDG-6),[29] whereas AF&PA publishes the *ASD/LRFD Manual for Engineered Wood Construction*[4] and *Structural Wood Design Solved Example Problems*[5] as a part of a publication package also containing ANSI/AF&PA NDS-2005, ANSI/AF&PA NDS-2005 *Commentary*, and ANSI/AF&PA SDPWS-08. As with design-specification commentaries, design manuals do not represent building code referenced mandatory provisions. Instead, they can be viewed as authoritative interpretative textbooks.

1.3.4 Test Reports

There are instances where design specifications and building codes do not provide design criteria to determine the strength and performance of a structural component or a system or acceptability of a construction method. These components and systems typically come in the form of proprietary or recently developed systems, connectors, and complex assemblies for which it is not possible to develop closed-form criteria in the form of a design equation or a prescriptive requirement. Consequently, testing is used to establish the strength and performance of such items or their equivalence to another component prescribed by the code. The building code basis for such a course of action lays in Sec. 104.11 of 2009 IBC, which states the following:

> The provisions of this code are not intended to prevent the installation of any material or to prohibit any design or method of construction not specifically prescribed by this code, provided that any such alternative has been approved. An alternative material, design or method of construction shall be approved where the building official finds that the proposed design is satisfactory and complies with the intent of the provisions of this code, and that the material, method or work offered is, for the purpose intended, at least the equivalent of that prescribed in this code in quality, strength, effectiveness, fire resistance, durability and safety.

The key role of the evaluation process belongs to two ICC limited-liability companies—the International Code Council Evaluation Service (ICC-ES) and the International Accreditation Service (IAS). The ICC-ES and the interested party, usually the manufacturer of a structural component, will create through a collaborative effort an *Acceptance Criteria* (AC) document for the component in question. If the AC document was developed previously for the same type of product, it can be used as the benchmark for product acceptance instead of creating another AC document. The AC document so created is reviewed by the ICC-ES Evaluation Committee and, if approved, is assigned a number and made available to both the manufacturer and the public through the ICC-ES database. This database can be accessed at www.icc-es.org. For example, AC01 designates *Acceptance Criteria for Expansion Anchors in Masonry Elements*.[25] Each AC document bears the approval date, effective date, and dates of previous approvals. Since the role of an AC document is to establish acceptability of a particular component under the provisions of a specific building code or multiple building codes, each AC document is building code–specific. For example, with respect to the IBC, the AC01 approved in December 2009 and effective January 1, 2010, is applicable to both 2006 IBC and 2009 IBC. An AC document must be reapproved for each desired edition of each desired building code, even if no revision to the AC document in question is necessary. However, this approach ensures that any new requirements or revised requirements for a particular component are captured in the AC document. Among other things, an AC document also typically lists the applicable materials, permitted testing procedures and equipment, required elements of a test report, and statistical methods of evaluating data, including establishment of a proper factor of safety.

The next step in the evaluation process is to perform testing in accordance with the applicable AC document. The IAS certifies laboratories in accordance with its *Accreditation Criteria for Testing Laboratories* (AC89).[20] The ICC-ES uses the list of certified laboratories provided by the IAS to determine whether a particular laboratory is qualified to evaluate the component. Under certain circumstances, IAS certification may not be required. Based on the review of a laboratory data report in accordance with the applicable ICC-ES AC document by the ICC-ES technical staff, ICC-ES ultimately issues an *Evaluation Service Report* (ESR), which indicates the conditions of applicability of a specific product under specific building codes, along with the tested design values established through testing. Additionally, an ESR will stipulate the component's limitations, installation details, and any required inspection procedures to ensure the validity of the tested strength values. Typically, an ESR will indicate allowable design values (i.e., allowable shear or allowable tension) of a component suitable for ASD, reflecting tested strength values divided by the appropriate safety factor, determined as stipulated by the applicable AC document. The ICC-ES ESRs for numerous products are publicly available at www.icc-es.org. For example, the acceptability and design values under the 2009 IBC and the 2006 IBC for the masonry anchor Kwik Bolt 3 manufactured by Hilti are established by ESR-1385.[27] The component manufacturer typically is indicated as the report holder.

It should be noted that the role of ICC-ES and an ESR is advisory rather than mandatory. In short, the decision to permit implementation of a particular component's ESR on a specific project rests solely with the authority having jurisdiction over the project—in other words, the local building official. The ICC-ES nonetheless is well established as an independent and authoritative third-party service, and its ESRs are commonly accepted. To this end, building officials from various regions of the country make up the membership of the ICC-ES Evaluation Committee.

In stating the evaluation criteria for a specific product, an AC document in many cases will refer to other standards for establishing proper testing techniques or for determination of resistance factors or factors of safety. As an example, *Acceptance Criteria for Tapping Screw Fasteners* (AC118)[26] refers to AISI *Standard Test Method for Determining the Tensile and Shear Strength of Screws* (AISI S904–08)[12] and *Test Methods for Mechanically Fastened Cold-Formed Steel Connections* (AISI S905–08)[12] for testing procedures used for determining the strength of screw connections joining cold-formed steel members, while also refer to Chap. F of ANSI/AISI S100–07 for determination of safety and resistance factors.

1.3.5 Design Process

The primary role of structural design is to protect the public safety through the design of structures for adequate strength, stability, and ductility while also ensuring that the structure remains serviceable during and following the application of loads. *Serviceability* can be viewed as the set of building performance characteristics that are not critical to the strength of the structure and consequently the protection of life. These characteristics affect comfort, appearance, porosity, and performance of nonstructural components. Poor serviceability performance can be manifested through excessive deflections, cracking, damage to nonstructural components, excessive vibrations, noise, and so on. Unlike strength, serviceability does not depend on material strength properties. Instead, depending on a specific serviceability issue, it is affected by material modulus of elasticity, stiffness of structural elements, relationship of stiffness and mass, and proper detailing.

The only exception to the preceding design goals is the design for seismic loads. The role of seismic design is to prevent loss of life and collapse of the structure during seismic events. It would be economically prohibitive to design structures to remain serviceable during and following a seismic event of significant magnitude, considering the low probability of occurrence of such an event. In other words, while protection of life remains a primary design objective, the lack of usability of a structure following a sizable seismic event is economically leveraged by the low probability of occurrence of such an event. Consequently, the primary goal of seismic design is to establish a mechanism within the structure that will dissipate the energy successfully that the structure absorbs during a seismic event. The most common way of achieving this is to allow for inelastic deformations in a designated part of a structural system, such as a member within a lateral force resisting system, while ensuring that the other parts of the same system, such as connections, diaphragms, and other members, remain essentially elastic. The ability of a member or a structure to undergo significant inelastic deformations without fracturing or collapsing is known as its *ductility*, representing one of the most critical properties in seismic design. Operational continuity of certain critical buildings, such as fire stations and hospitals, is critical during and after a significant seismic event. Consequently, while a certain degree of inelasticity in such structures may be tolerated, they are designed for significantly higher seismic forces than other buildings with similar seismic characteristics.

The critical aspect of structural design for strength is to establish a valid load path that can absorb the externally applied forces and transfer them through the framing into foundations. It should be noted that a structural system can have more than one valid load path, although one path may be more obvious, efficient, or economical than another. In addition to the need to establish a valid load path, seismic design requires a sensible selection of an energy-dissipating mechanism, known as a *fuse*, that can undergo significant inelastic deformations. Chapter 2 provides a detailed discussion of loads and load paths in low-rise buildings.

A structural design is only as useful as its ability to properly capture and accommodate the architectural design, site conditions, and mechanical, electrical, and plumbing features. Also, it must ensure that the resulting design is constructible and economical. Finally, it is important to clearly communicate the structural design intent in the construction documents. To this end, the continuous coordination of the structural engineer with the project architect is crucial throughout the design process. Initially, this coordination ensures that the chosen architectural concept design is structurally feasible, both economically and functionally, whereas continuous coordination throughout the project is aimed at structural accommodation of specific architectural functional features. Following the initial definition of an architectural conceptual design, the geotechnical engineer will develop seismic characteristics of the project site and foundation recommendations in the form of a comprehensive geotechnical report. This information will assist the structural engineer in computing seismic design forces and determining seismic design category. Additionally, the identification of an appropriate type of foundation system typically will be accompanied by specific design values the structural engineer can use in foundation design. Design of the substructure also will entail coordination with the civil engineer to determine the need for and subsequent design of any retaining walls based on the site grading plan generated by the civil engineer. Similarly, the coordination with the civil engineer also will result in a decision on how to deal with high groundwater, if present. Specifically, absent groundwater mitigation procedures designed by the civil engineer, various parts of substructure and structure may be subjected to buoyancy and increased lateral pressure. Structural design inevitably involves coordination with mechanical, plumbing, and electrical designers so that the issues stemming from additional loads imposed on the structure and disruption of the structural system stemming from ductwork and other equipment can be avoided or accommodated by the structural design. The portion of the construction documents generated by the structural engineer typically includes structural drawings and structural specifications. The role of these documents is to communicate the requirements of the design documents presented in Secs. 1.3.1 through 1.3.4 in the form of project-specific advice. The purpose of construction documents is twofold. First, they are used in obtaining permit documents from the local authority having jurisdiction. Second, they are used for the project construction and construction inspection.

Section 1603 of 2009 IBC stipulates the required contents of construction documents. In short, the construction documents must indicate structural design criteria and provide a complete picture of the framing of the entire structural system, including individual member sizes, material specifications, their details, and any required instructions. Other than this, the specific contents and organization of construction documents are to a large extent a matter of individual preference, the nature of the project, or the standard practice common to a specific type of construction. Typically, structural drawings will contain general notes summarizing the basic design criteria required by Sec. 1603 of 2009 IBC, material properties, and any requirements that can be expressed in a narrative form and applied to the entire project (i.e., development lengths for various bar sizes used in a structural masonry project). The presentation of structural framing and associated member and component sizes will depend greatly on the type of construction. Typically, framing members and their locations and sizes will be shown in plans or on elevations. Accompanying plans and elevations are framing details depicting framing connections, attachment of structural framing to foundations, and any other details required to fabricate and construct a structural system.

Project specifications typically are narrative instructions specifying acceptable materials, quality-assurance criteria, and any other criteria relevant to construction of the structural system not otherwise stipulated by the drawings. One of the crucial aspects of structural specification is its role as a performance specification. A *performance specification* is a document outlining minimum performance or design criteria for structural components and systems designed by a third party. Cold-formed steel trusses represent a common example of a third-party designed system. Compliance with a performance specification is established by the third party's submittal consisting of drawings and calculations supporting the adequacy of the system by its in-house structural engineer.

Finally, the process of construction administration concludes the structural design of a building. The role of this process is to verify that the design intent communicated through the construction document is implemented in the construction. It consists of a review of third-party design submittals, a review of fabrication submittals, and supplemental structural design aimed at resolving any errors or conflicts arising during construction. For example, in structural masonry construction, fabrication submittals include grout mix reports and reinforcement shop drawings. Fabrication submittals are generated by the contractor or its subcontractor and are created from the information communicated in construction documents.

1.4 Low-Rise Building Types and Properties

This section provides a brief review of typical structural systems employed in the construction of low-rise buildings in cold-formed steel, structural masonry, and structural timber and their properties.

1.4.1 Structural Systems

According to the manner in which they support vertical loads and lateral loads, low-rise building structural systems can be classified into the following general categories:

- Moment frames
- Braced frames
- Nonbearing walls
- Bearing walls
- Cantilever columns

A *moment frame*, depicted schematically in Fig. 1.1a, is a structural system that derives its stability under gravity loads and lateral strength from rigidly connected beams and columns. Moment frames depend on the flexural strength and stiffness of rigidly connected beams and columns to maintain the lateral strength and stability of the system. Gravity load in moment frames is supported in beams, which, in turn, transmit the loads into columns and consequently foundations. The wall framing is nonstructural and serves as cladding rather than for the support of gravity loads. As shown in Fig. 1.1a, the column bases may be either fixed or pinned, and beams, whose end fixity is not required to maintain the structure's strength, stability, and stiffness, may remain pinned.

Moment frames are not a type of structural system commonly employed in low-rise buildings constructed of cold-formed steel, structural masonry, or structural timber. However, there are some notable exceptions. In cold-formed steel construction, these exceptions are mostly related to the material handling industry. Figure 1.2 shows an all-bolted cold-formed steel moment frame commonly employed in the construction of nonbuilding structures such as industrial mezzanines. Although their primary use is seen in industrial applications, cold-formed steel bolted moment frames are conceptually similar to building-structure framing methods and have recently seen an increased use in building applications. The use of this system for building purposes in circumstances of higher seismic design categories is facilitated by the recently published 2007 AISI *Standard for Seismic Design of Cold-Formed Steel Structural Systems—Special Bolted Moment Frames with Supplement No. 1* (ANSI/AISI S110–07/S1–09)[8] and its recognition as a seismic lateral force resisting system in ASCE/SEI 7–10. The moment frame comprised by this system is created by side-bolting a cold-formed steel channel to one or both faces of a cold-formed steel hollow structural shape. Other variations of this system not yet included in ANSI/AISI S110–07/S1–09 exist mainly to reflect different configurations of beam-to-column bolted connections. Storage racks fabricated by the materials-handling industry also employ moment frames. Although these are nonbuilding structures, the framing, analysis, and design concepts employed in their

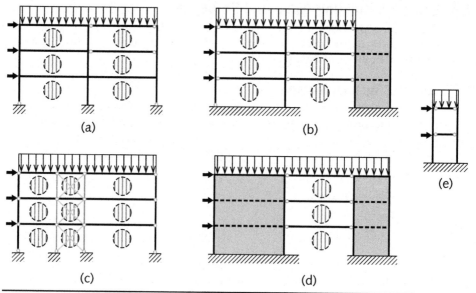

FIGURE 1.1 Structural systems: (a) moment frames; (b) building frame walls; (c) braced frames; (d) bearing frame walls; (e) cantilever columns.

design and construction can be useful in certain building applications. As can be seen in Fig. 1.3, prefabricated storage rack frames are highly modular and are manufactured in a manner that facilitates the ease of frame erection. Moment connections in storage rack frames are commonly accomplished by bolting of a cold-formed steel beam, typically a C-shape, to the side of a cold-formed steel column, typically a perforated C-shape or a tube. As shown in Fig. 1.4, beam ends usually feature shop-welded brackets to facilitate bolting. Although capable of transmitting moment, such joints are usually fairly flexible and undergo significant joint rotations. Therefore, storage rack frames typically require a rigorous semirigid frame analysis to accurately assess stability, drift, and member forces. Column perforations allow for field adjustments of beam elevations and, consequently, storage levels. The primary design standard for moment frames in cold-formed steel construction is ANSI/AISI S100–07. For the design of storage racks, Sec. 2208.1 of 2009 IBC refers to the 2008 Rack Manufacturers Institute (RMI) *Specification for Design, Testing and Utilization of Industrial Steel Storage Racks* (RMI ANSI/MH16.1–08).[32] However, RMI ANSI/MH 16.1–08 references ANSI/AISI S100–07 for most cold-formed steel design requirements.

A moment frame structural system can appear in low-rise buildings featuring structural timber post-and-beam construction. The main characteristic of post-and-beam wood construction is that the gravity loads are supported by distinct beam elements that transfer the applied load into distinct column elements and consequently into column foundations. The infill wall elements, when present, act as nonstructural

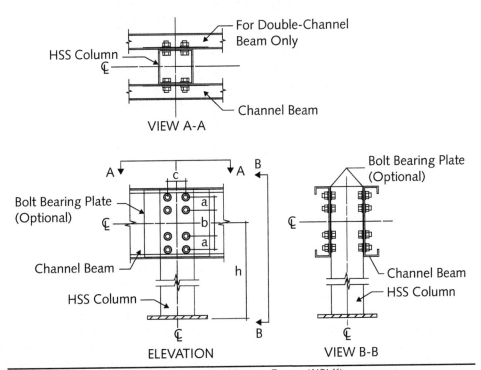

Figure 1.2 Cold-formed steel Special Bolted Moment Frame. (AISI.[11])

Figure 1.3   Storage rack moment frames. (*James Crews, P.E., Unarco Material Handling, Inc.*)

Figure 1.4   Storage rack moment connection. (*James Crews, P.E., Unarco Material Handling, Inc.*)

cladding and partition elements and have no role in supporting the gravity load. The most common manner for establishing a moment connection in post-and-beam construction is by using knee braces near the intersections of beams and columns, as shown in Fig. 1.5. With this arrangement, the beam remains pin-connected to the column, but the knee brace mobilizes the flexural strength of the beams and columns, resulting in a moment frame restrained against sway. The knee-braced post-and-beam moment frames typically are used in creative architectural applications, residential construction, and agricultural utility buildings. Figure 1.6 shows a variation of this type

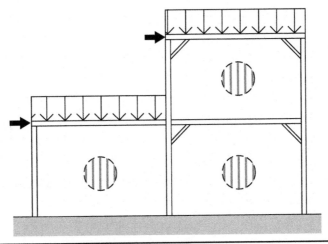

Figure 1.5   Knee-braced moment post-and-beam frame.

Figure 1.6   Knee-braced moment post-and-beam application. (Joe Briglevich, Briglevich Engineering, Inc.)

of system in an architecturally expressive residential application. Another form of structural wood is achieved through application of plywood members. The challenge of creating moment joints in wood construction stems from the granular structure of wood material, whereby it is difficult to design a moment joint that would control longitudinal splitting and cross-grain tension effectively while controlling the joint strength and stiffness under sustained load. The use of plywood, an isotropic wood derivative, removes such challenges, making it possible to rigidly connect members at their ends using robust rigid joints and metal connectors. Figure 1.7 presents such a system. The left image in the figure depicts the moment beam-to-column connection in

FIGURE 1.7   Plywood moment frame and beam moment splice.

a plywood member frame with a large span. The right image shows the moment splice connections of two beam members within the frame. The primary standard for the design of wood moment frames is ANSI/AF&PA NDS-2005. Although moment frames featuring reinforced masonry are theoretically possible, they are not practically feasible and not commonly done.

Bearing-wall systems, illustrated schematically in Fig. 1.1*d*, are the most commonly encountered structural system in low-rise buildings constructed of cold-formed steel, structural masonry, and structural wood. The main characteristic of a bearing-wall system is that the structural walls on which the building relies for lateral strength and stiffness also serve as primary vertical elements resisting most or the entire gravity load of the building. Figure 1.8 shows a reinforced-masonry bearing wall system. In this particular example, roof framing consists of a cold-formed steel deck supported by open-web joists. Open-web joists and reactions are transmitted directly into the masonry walls. Masonry walls serving as vertical elements resisting gravity and uplift reactions from the open-web joists also serve as a lateral force resisting system and drift restraints using their in-plane shear and flexural strength and stiffness. The benefits of such a system are that it relies on the inherent strengths of masonry construction, such as the relatively high in-plane strength and stiffness, and has the ability to accommodate openings and the benefit of concurrently acting as a structural system and as partition walls. Furthermore, exterior bearing walls can serve three load-resisting purposes concurrently—they act as lateral force resisting elements, vertical load–resisting elements, and cladding elements resisting wind cladding and component wind pressures. The primary design standards for structural masonry bearing walls are TMS 402–08/ACI 530–08/ASCE 5–08 and TMS 602–08/ACI 530.1–08/ASCE 6–08 with amendments and exceptions thereto provided by Chap. 21 of 2009 IBC.

Bearing-wall systems of cold-formed steel and structural timber come primarily in the form of light-frame construction. For seismic design purposes, Chap. 11 of ASCE/SEI 7–05 defines light-frame construction as "a method of construction where the structural assemblies (e.g., walls, floors, ceilings, and roofs) are primarily formed by a system of repetitive wood or cold-formed steel framing members or subassemblies of those members (e.g., trusses)." Section 2302 of 2009 IBC uses a similarly defined term *conventional light-frame construction* for wood bearing-wall construction. Section 2210 of 2009 IBC stipulates a steel member thickness range of 0.0179 and 0.1180 in. when

Figure 1.8   Reinforced-masonry bearing walls. (*Uzun & Case Engineers, LLC.*)

defining the limits of applicability of its light-frame provisions. Light-frame construction is commonly referred to as *stud-wall construction*. In cold-formed steel applications, stud-wall construction typically is accomplished using C-shaped structural studs typically spaced at 12, 16, or 24 in. depending on the applied load, story height, and number of supported stories. The Steel Stud Manufacturers Association (SSMA) publishes the *Product Technical Information*[40] catalog, which lists and provides properties for the commonly available studs and stud tracks. Structural stud depths range from 2.5 to 16 in., with flange widths ranging from 1.37 to 3.5 in., although typical bearing-wall applications in low-rise buildings involve stud depths from 6 to 10 in. depending on wall height, applied load, number of stories, or other considerations. Individual studs can be stitched into a larger stud member using screws. This is particularly useful for framing large openings or accommodating unique force demands within the structure using the same steel stud size stock. The floor framing typically contains floor joists in the form of cold-formed steel C-shaped members matching the locations of wall stud members. Floor joists typically are sheathed with wood structural panels or steel deck with concrete topping, in each case forming a floor diaphragm. Roof framing depends on the specific architectural design and typically is in the form of preengineered cold-formed steel trusses or roof joists. Steel deck or plywood typically is used to construct the diaphragm at the roof level. Walls are sheathed with wood structural panels, steel, gypsum board, or structural fiberboard sheathing, which provides resistance under lateral load. When sheathing is fastened to both sides of the stud wall, it also can greatly enhance the compressive strength of the wall because it creates a composite assembly of the stud and two-faced sheathing. Each shear-wall segment features chord studs at each end. Chord studs collect tension and compression resulting from the overturning effects of lateral forces. The chord forces are transferred into the foundation via hold-down devices attached to each chord stud. Figure 1.9 presents an example of light-frame bearing-wall construction that relies on sheathing for lateral resistance. When lateral strength cannot be achieved through the use of structural sheathing, tension-only diagonal strap bracing typically is used. This type of bracing is illustrated in Fig. 1.10. This usually occurs when plywood sheathing cannot be used owing to fire-rating requirements or when a specific arrangement of openings precludes an effective shear-wall usage but permits the use of strap bracing. It should be noted that a light-frame system featuring strap bracing is still considered a bearing-wall structural system rather than a braced frame. Specifically, strap bracing is not a distinct and independent bracing system but rather functions as an integral part of a bearing-wall assembly in which the individual studs, including stud chord elements resisting the vertical component of the strap brace force, continue to serve as the sole gravity load–resisting elements of the structure. The benefits of cold-formed steel framing include minimal shop preassembly, ease of construction and member erection, and easy member connectivity, whereby all framing connections typically can be accomplished through the use of self-tapping screws. Also, highly standardized framing and connection methods and the prescriptive nature of much of the design provisions typically result in a relatively expedient and simple structural design process. The primary design standards for light-frame cold-formed steel construction are ANSI/AISI S100–07 and the AISI framing standards referenced in Chap. 4.

With respect to the framing concept employed, conventionally light-framed wood buildings are similar to those featuring cold-formed steel light framing. They typically feature studs nominally 2 × 4, 3 × 4, or 2 × 6 in. in a cross-sectional size spaced 12, 16, or 24 in. on center depending on load demand, story height, or the number of supported

Figure 1.9   Cold-formed steel light-frame bearing wall construction with structural sheathing. (*Don Allen, P.E., Steel Framing Alliance.*)

FIGURE 1.10 Cold-formed steel light-frame bearing-wall construction with strap bracing. (*Don Allen, P.E., Steel Framing Alliance.*)

stories. When required for strength, individual 3- or 2-in. wall stud members can be nailed together to form a larger stud. This allows for field construction of larger members from lumber stock of the same size and is beneficial for framing large openings and chord members. Wall sheathing in the form of wood structural panels provides lateral resistance. Similar to cold-formed steel light-frame construction, the ends of each individual shear-wall element contain stud chords and hold-down devices anchoring the lateral force resisting system against overturning. Floor diaphragms typically are accomplished through structural wood sheathing supported by 2-in. wide floor joists or APA-rated I-joists. Roof diaphragms typically are formed through the use of structural wood sheathing and occasionally steel deck supported by wood joists or trusses. Figure 1.11 illustrates a wood stud frame construction with wood structural panel wall sheathing, whereas Fig. 1.12 shows an example of metal plate connected wood-truss roof framing. The primary benefits of this type of framing include the ease of construction and member fabrication, readily available construction materials, and relatively simple design given its prescriptive nature. The primary design standards for conventional light-frame wood construction include ANSI/AF&PA NDS-2005 and ANSI/AF&PA SDPWS-08. Alternatively, prescriptive design per Sec. 2308 of 2009 IBC can be used.

A non-bearing wall system, or a building frame structural system, is illustrated in Fig. 1.1b. Their main characteristic is that the walls provided for lateral resistance provide little or no support for the building's gravity loads. In essence, the structure featuring this type of system will possess independent mechanisms for resisting lateral and gravity loads. A practical application of such a system is shown in Fig. 1.13. As can be seen, the open-web joist roof framing in this building is supported entirely by a

FIGURE 1.11   Conventional light-frame wood framing. (*Uzun & Case Engineers, LLC.*)

FIGURE 1.12   Wood-truss roof framing. (*Uzun & Case Engineers, LLC.*)

structural steel frame. The reinforced-masonry walls, however, are structurally detached from the vertical force resisting elements and posses only the structural tie to the diaphragm for transfer of lateral forces. The benefit of a non-bearing-wall system is that it allows for the lateral and gravity system to be constructed concurrently, significantly reducing the construction time. In the case illustrated in Fig. 1.13, the structural steel framing can be erected prior to completion of masonry walls, permitting expedited construction. With respect to cold-formed steel and wood applications, non-bearing wall systems are not common but can occur in conjunction with other materials, such as masonry, reinforced concrete, and structural steel. When post-and-beam wood structures are braced by structural wood panels but retain a complete beam and column system for resisting gravity loads, they can be classified as non-bearing-wall systems. Similarly, post-and-beam cold-formed steel and wall frames in the metal building industry, constructed with bolted C shapes and back-to-back double C shapes, represent a non-bearing-wall system when instead of cross-bracing (Fig. 1.14), and they rely on exterior metal sheathing for lateral resistance.

Braced-frame systems, depicted schematically in Fig. 1.1c, contain a vertical truss whose role is to absorb lateral forces and transmit them into the foundation, generally not relying on the flexural strength and stiffness of beam and column elements for lateral strength and drift restraint. Such systems are not used in low-rise buildings

Figure 1.13   Masonry non-bearing wall system with structural steel framing. (*Uzun & Case Engineers, LLC.*)

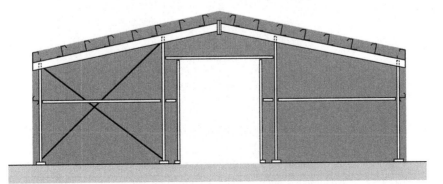

Figure 1.14   Metal post-and-beam cold formed steel braced frame building.

featuring structural masonry and are rarely applied in cold-formed steel or structural timber systems. In cold-formed steel construction, braced frames occur in post-and-beam end-wall frames in the metal building industry (Fig. 1.14). Braced frames are also used in the transverse direction in storage rack structures (Fig. 1.3). With respect to structural timber systems, braced frames can be encountered in post-and-beam construction.

Cantilever column systems, as illustrated in Fig. 1.1e, reflect a system whose entire lateral strength depends on the base fixity of one or more of its columns. In general, such systems are possible for any type of construction that also can accommodate a moment frame, including cold-formed steel and wood. Although such systems have seen application in low-rise light-frame construction, they exhibit fairly poor seismic performance and are not used commonly.

Combinations of various materials and different structural systems are often possible, enhancing design versatility, economy, and sometimes extending practical height and capacity limits of light-frame applications. Figure 1.15 illustrates light-frame construction featuring cold-formed steel stud-framed walls and floors and a wood-framed roof, whereas Fig. 1.16 shows an application of a wood-truss floor system in combination with cold-formed steel stud walls. An increasingly popular type of system is depicted in Fig. 1.17. There, a single- or multistory structural steel or, as is the case in this example, structural concrete structure supports a multistory cold-formed steel light-framed building. This is a particularly useful concept in residential construction, permitting lower parking levels to be constructed with structural steel and concrete, thus achieving the required strength and durability

Figure 1.15 Cold-formed steel light-frame construction with wood-framed roof structure. (*Don Allen, P.E., Steel Framing Alliance.*)

FIGURE 1.16   Cold-formed steel light-frame construction with wood floor joists. (*Don Allen, P.E., Steel Framing Alliance.*)

FIGURE 1.17   Cold-formed steel light-frame and structural concrete vertically combined construction. (*Don Allen, P.E., Steel Framing Alliance.*)

while taking advantage of the economy and lower weight of light-frame construction at the residential level. Combinations of reinforced-masonry walls with moment frames of steel and concrete are also possible and under certain conditions are recognized as valid seismic force resisting systems, as set forth by Chap. 12 of ASCE/SEI 7–05.

1.4.2 Low-Rise Building Properties

In typical low-rise buildings constructed of reinforced masonry, structural wood, and cold-formed steel, relatively low building weights and applied loads result in significantly lower foundation costs than in mid-rise and high-rise construction. Light-frame construction typically can be supported by turned-down slabs on grade, whereas other types of construction in these materials within the low-rise building context very rarely result in foundation systems other than shallow footings. Furthermore, compared with other types of buildings and material applications, such structures do not warrant significant expense associated with formwork in concrete construction, and in most typical applications, they avoid extensive shop fabrication processes, as is the case with structural steel. Load paths and the design process in such buildings generally are simple, and various prescriptive design methods lend themselves to expedient design, further increasing project efficiency. As noted earlier, in terms of cost per square foot of completed space, low-rise buildings typically are significantly more economical than mid-rise or high-rise buildings. Finally, seismic lateral force resisting systems provided by ASCE/SEI 7–05 contain significant building height–related restrictions, especially as they relate to cold-formed steel, structural masonry, and structural wood. Therefore, a greater versatility in available structural systems can be realized in low-rise buildings compared with buildings of greater height.

As can be expected, specific structural properties of low-rise buildings depend on the selected structural system and material of construction.

To better understand the properties critical to the proper seismic behavior of low-rise structures, it is useful to review the fundamental aspects of the current seismic design philosophy. As part of the National Earthquake Hazards Reduction Program (NEHRP), the Building Seismic Safety Council (BSSC) of the National Institute of Building Sciences (NIBS) authors and periodically updates a document recommending provisions for seismic design of buildings and other structures in an effort sponsored by the Federal Emergency Management Agency (FEMA). The 2003 edition of this document entitled *NEHRP-Recommended Provisions for Seismic Regulations for New Buildings and Other Structures* (FEMA 450)[13] represents the underlying basis for the seismic provisions stipulated by ASCE/SEI 7–05 and, consequently, the 2009 IBC. Readers are encouraged to obtain this document from www.nehrp.gov and thoroughly familiarize themselves with its provisions.

The basis of the FEMA 450 approach can best be explained by referring to Fig. 1.18. The primary reference point for determination of seismic effects to be considered in the design at a particular location is the so-called maximum considered earthquake (MCE), which in most of the United States is defined as the earthquake effect for which there is a 2 percent probability of being exceeded in 50 years at the particular location of interest. In other words, it is an earthquake effect that is statistically likely to return at a given location approximately every 2,500 years. Exceptions to this definition are the regions of the country with unique seismologic conditions, such as the state of California. The seismologic conditions in California are such that there is a frequent occurrence of small-and medium-strength earthquakes. Consequently, defining the MCE in California as a seismic event with a 2,500-year return period would result in cost-prohibitive designs. Instead, based on a review of the existing practice and the performance of buildings in actual earthquake events designed under the existing code, it was decided to relate MCE in California to the earthquake level prescribed by the existing code at that time—the *1997 Uniform Building Code (1997 UBC)*.[21] As a result, the MCE in California is defined as an event that has a 10 percent probability of returning in 50 years, or it is statistically likely to return once in approximately 500 years, resulting in an approximately 50 percent smaller magnitude of the computed force than would have been the case considering MCE with a 2-percent probability of being exceeded in 50 years. The consequence of this approach is that while FEMA 450 issues a single map depicting short and 1-second period seismic accelerations for the entire country, such values for most of country are based on earthquakes with a 2-percent probability of occurring in a 50-year period, whereas in California, such mapped acceleration values are based on earthquakes with a 10-percent probability of occurring within a 50-year interval. It was judged subsequently, based on a review of current design practices, that if a building structure were to be designed for two-thirds of the MCE, it still would have a very low likelihood of collapse. Building base shear corresponding to two-thirds of the MCE is represented as V_E in Fig. 1.18. As will become evident in Chap. 2, the one-third reduction of the MCE is incorporated into the routine calculations of seismic loads per ASCE/SEI 7–05. The reader is cautioned to note that ASCE/SEI 7–10 makes a dramatic departure from ASCE/SEI 7–05 in how the seismic effects are quantified in the design. Specifically, ASCE/SEI 7–10 abandons the concept of MCE, defined based on the probability of exceeding certain level of ground motion. Instead, it adopts a more direct concept of probability of collapse, whereby the design levels of ground motion at each locale are benchmarked to the magnitude of ground motion resulting in 1% probability of collapse. Although the preceding discussion may lead the reader to conclude that the nature of determining seismic load effects leads to highly subjective design requirements, it will be apparent from the following discussion that the key aspects of seismic design are the response and performance of the structure under a seismic event, as well as the ability of the designer to control the behavior of the structure by devising specific mechanisms within the structure that effectively will dissipate the energy the structure will absorb during a seismic event.

Referring to Fig. 1.18, if a building were to be designed to respond elastically to the forces corresponding to base shear V_E, it would translate laterally to a displacement D_E. Under this scenario, the structure would absorb vast amounts of energy E equaling the

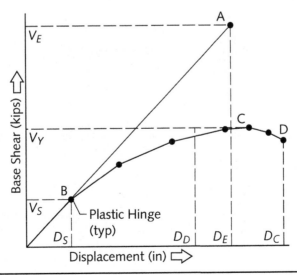

Figure 1.18 Seismic responses of structures.

area under the elastic response curve terminating at point A (i.e., $E = V_E D_E/2$). This energy translates into significant internal forces on the connections and members within the structure, resulting in an extremely inefficient and uneconomical design considering the overall risk of the occurrence of a seismic event that would generate the magnitude of the force V_E. A more useful manner of stating this in a way that would better reflect the overall design objective is as follows: Energy absorbed by the structure during a seismic event is directly related to the forces developed in the structure for which its members and connections must be designed. In other words, any design method that can ensure dissipation of the absorbed energy will result in turn in smaller internal forces for which the members and connections within the structure need to be designed. By definition, elastic structures absorb energy because they retain the ability to return to their undeformed state when the load effect causing deformation ceases. In other words, a load application on an elastic structure causes the structure to absorb potential energy, stored in the structure in the form of member and connection forces. Following application of the load, this potential energy converts into kinetic energy that drives the structure into its original state. Therefore, an obvious mechanism for preventing an accumulation of potential energy within a structure is to permit inelastic deformations within chosen areas and members of the structure. The ability of such a member within the structure and the structure itself to deform inelastically is known as its *ductility*. Ductility can be described in many different ways, such as the ratio of the ultimate strength to the yield strength F_u/F_y of a member or a material or the percentage of elongation expressed as the ratio of the length extension of a member loaded in tension at its ultimate load to its original length. However, in all cases, it is an indicator of the ability of a member, material, or structure to deform significantly following its elastic response. As illustrated in Fig. 1.19, a response, such as rotation, displacement, elongation, or bending, in nonductile, brittle structures is distinguished by a very small deformation following the range of elastic behavior. In general, high-strength materials are more brittle than normal- or low-strength materials. As an example, a plain, unreinforced masonry shear wall will exhibit a brittle, nonductile failure under seismic

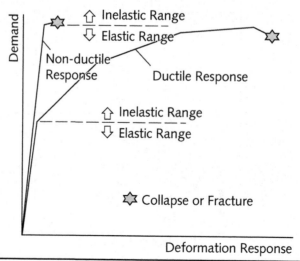

Figure 1.19 Ductile and brittle response.

load, whereas following initial cracking, a well-detailed reinforced-masonry wall will experience a prolonged yielding of ductile steel reinforcement and consequently see a relatively ductile failure.

Actual ductility of a structural system and its ability to dissipate energy through controlled inelastic deformations depend on its design and detailing features and, consequently, its actual performance under seismic loads. Since the magnitude of the internal forces developing in the structural system depends directly on the amount of energy absorbed by the system, the greater the amount of energy dissipated by the structural system, the lower is the magnitude of the base shear V_S in Fig. 1.18, corresponding to the internal forces accumulated in the system. The ratio V_E/V_S is known as the *response-modification coefficient R*. Release of the seismic energy is achieved by identifying and designating a mechanism commonly referred to as a *fuse*. A fuse is a ductile element within the building's lateral force resisting system charged with commencing inelastic deformation as the structure reaches point B on the response curve as depicted in Fig. 1.18. Point B corresponds to the design base shear V_S and the associated displacement D_S. The designation of a fuse element depends on the type of structure. For instance, a strap brace acts as a fuse in a light-frame cold-formed steel building, whereby inelastic deformations are achieved through ductile yielding of the strap. On the other hand, the fuse in a cold-formed steel special bolted moment frame is the beam-to-column connection, which dissipates energy through inelastic deformations and slippage of the bolted joints. Cold-formed steel and wood light-framed-sheathed buildings typically dissipate energy through deformation of sheathing attachments and distortions of the frame, whereas reinforced-masonry wall elements dissipate energy through inelastic yielding of steel reinforcement in a cracked masonry section. Cantilever column systems dissipate energy through plastic hinging at the column base, but they are associated with a very low R factor because such hinging does not offer the same degree of energy dissipation offered by sequential reversal of yielding experienced by moment frames. Unreinforced masonry walls have very low ductility, do not dissipate significant energy prior to failure, and consequently are associated with a very low R factor.

Designation and design of a fuse are said to be a controlled process, and this is a crucial aspect of the design. Specifically, selection of a fuse element does not in itself guarantee the assumed energy-dissipation mechanism. Such a mechanism often must be ensured through a series of well-devised detailing steps. For example, using the example of a cold-formed steel light-framed system with diagonal strap bracing, detailing can involve weakening of the brace at the location of desired inelastic deformation by reducing its cross section at that location. This helps to ensure the assumed inelastic behavior and reduces the possibility of inadvertent overload and inadvertent failure of another, possibly nonductile element.

Figure 1.18 uses the term *hinge* to designate a location of inelastic deformation, which can come in the form of elongation, bending, rotation, slippage, and so on depending on the type of structural system employed. A structural system with low redundancy will feature very few hinges. For instance, a single hinge in a cantilevered-column system will cause its collapse. However, a moment frame consisting of five fixed-top and fixed-base columns will form 10 hinges prior to its collapse. Following the initial hinging at point B, the system depicted in Fig. 1.18 experiences four additional hinges until reaching its maximum strength at point C. This point designates the so-called overstrength of the structure. Overstrength V_Y indicates the maximum load a certain type of structural system can accommodate inherently. The ratio V_Y/V_S is known as the *actual system overstrength factor* Ω.

As important as it is to designate and properly detail a fuse element to serve its purpose of an energy dissipater adequately, it is equally critical, for the same reason, that the remaining components of the lateral force resisting system be designed to remain essentially elastic. This is done to make certain that the assumed, designed, and detailed energy dissipation path remains preserved. Critical elements of the lateral force–distribution system, such as collectors within diaphragms and their splices and connections, must be designed for a capacity exceeding overstrength (i.e., point C in Fig. 1.18). For this reason, FEMA 450 and consequently ASCE/SEI 7–05 prescribe a *system overstrength factor* Ω_0 that exceeds Ω. The resulting design force of $\Omega_0 V_S$ effectively guarantees that the aforementioned elements remain elastic for the duration of energy dissipation through the designated mechanism. The building diaphragm itself must remain elastic to ensure the stability of the vertical elements for which it provides lateral bracing and to preserve the distribution of loads within the lateral force resisting system predicted by the analysis. However, it is worth noting that the current U.S. design practice does not stipulate diaphragm design forces higher than those corresponding with V_S. In designing the lateral force resisting frame members not involved in the energy-dissipating mechanism, the user typically can use either the forces corresponding to $\Omega_0 V_S$ or the maximum expected forces that can be delivered to the member by the energy-dissipating system. Using the original example of a strap-braced light frame, the edge-of-wall chord members can be designed for either the vertical component of the brace force corresponding to $\Omega_0 V_S$ or the higher of the expected brace yield or fracture tensile strength. The term *expected strength* refers to the fact that the actual material strength typically is higher than its nominal value used in structural design. For instance, the expected yield strength of a brace strap is about 10 percent higher than its nominal yield strength. In designing the members and connections attached to the energy dissipation mechanism, it is important to consider the expected strength of such mechanism while calculating the demand on the attached members rather than its nominal strength. The reason for this is an attempt to avoid the realistic scenario where various components in the lateral force resisting system are not strong enough to accommodate the chosen energy dissipation mechanism achieved within other elements to which they are connected.

Figure 1.18 shows a displacement D_D. This displacement corresponds to a translation a building undergoing inelastic deformation owing to a seismic event is likely to experience. The ratio D_D/D_S is referred to as the *deflection amplification factor* C_d. The magnitude of D_D, and consequently C_d depends on the type of structural system. The value of D_D, conveniently estimated as the product of D_S computed from the first-order elastic analysis and C_d, is used in designing against significant damage to nonstructural components and in limiting excessive second-order effects owing to the structure's loss of stiffness during a seismic event. The reader should note from Fig. 1.18 that if $D_D = D_E$, then $C_d = R$. Therefore, in cases where C_d is significantly less than R, the structure possesses significant vibration-damping properties.

The seismic design philosophy presented in the preceding discussion occasionally is referred to as *equal-displacement seismic design*, reflecting the ultimate design objective that the structure be designed and configured such that it has sufficient deformation capacity (i.e., ability to laterally displace without collapsing) to reach displacement D_E, as shown in Fig. 1.18, regardless of what the resulting magnitude of V_S is. Also, following the inelastic deformations (i.e., hinging) that result in overstrength V_Y (i.e., point C in Fig. 1.18), a structure can collapse or form one or more hinges prior to collapsing depending on the specific properties and configuration of the structure. To avoid premature collapse, it is highly desirable that the design of the structure be performed such that the displacement corresponding to V_Y be larger than D_E.

The following list presents the structural systems contained in ASCE/SEI 7–05, Table 12.2–1, that are relevant to the design of low-rise buildings in cold-formed steel, structural masonry, and structural timber. The building frame category contains braced frames and non-bearing-wall systems illustrated in Sec. 1.4.1. Ordinary steel concentrically braced frames apply to cable- and rod-braced cold-formed steel post-and-beam frames in metal building systems. Cantilevered-column systems are used occasionally in the form of large structural steel fixed-base cantilevered columns as a lateral force resisting system in light-framed construction. The steel system not specifically detailed for seismic resistance may be used in light-frame construction or any other system whose lateral force resisting system is made of cold-formed or structural steel with minimal or no seismic detailing. The cold-formed steel special bolted moment frame was not available when ASCE/SEI 7–05 was published but was included in moment frame systems in ASCE/SEI 7–10. The user should refer to Chap. 2 for computational procedures for determining seismic and other forces and to ASCE/ SEI 7–05 for the full listing of available structural systems and corresponding limitations and requirements.

A. Bearing Wall Systems

A7. Special Reinforced Masonry Shear Walls ($R = 5$, $\Omega_0 = 2.5$, $C_d = 3.5$)
A8. Intermediate Reinforced Masonry Shear Walls ($R = 3.5$, $\Omega_0 = 2.5$, $C_d = 2.25$)
A9. Ordinary Reinforced Masonry Shear Walls ($R = 2$, $\Omega_0 = 2.5$, $C_d = 1.75$)
A10. Detailed Plain Masonry Shear Walls ($R = 2$, $\Omega_0 = 2.5$, $C_d = 1.75$)
A11. Ordinary Plain Masonry Shear Walls ($R = 1.5$, $\Omega_0 = 2.5$, $C_d = 1.25$)
A12. Prestressed Masonry Shear Walls ($R = 1.5$, $\Omega_0 = 2.5$, $C_d = 1.75$)
A13. Light-Framed Walls Sheathed with Wood Structural Panels Rated for Shear Resistance or Steel Sheets ($R = 6.5$, $\Omega_0 = 3$, $C_d = 4$)
A14. Light-Framed Walls with Shear Panels of All Other Materials ($R = 2$, $\Omega_0 = 2.5$, $C_d = 4$)
A15. Light-Framed Wall Systems Using Flat Strap Bracing ($R = 4$, $\Omega_0 = 2$, $C_d = 3.5$)

B. Building Frame Systems

B4. Ordinary Steel Concentrically Braced Frames ($R = 3.25$, $\Omega_0 = 2$, $C_d = 3.25$)
B17. Special Reinforced Masonry Shear Walls ($R = 5.5$, $\Omega_0 = 2.5$, $C_d = 4$)
B18. Intermediate Reinforced Masonry Shear Walls ($R = 4$, $\Omega_0 = 2.5$, $C_d = 4$)
B19. Ordinary Reinforced Masonry Shear Walls ($R = 2$, $\Omega_0 = 2.5$, $C_d = 2$)
B20. Detailed Plain Masonry Shear Walls ($R = 2$, $\Omega_0 = 2.5$, $C_d = 2$)
B21. Ordinary Plain Masonry Shear Walls ($R = 1.5$, $\Omega_0 = 2.5$, $C_d = 1.25$)
B22. Prestressed Masonry Shear Walls ($R = 1.5$, $\Omega_0 = 2.5$, $C_d = 1.75$)
B23. Light-Framed Walls Sheathed with Wood Structural Panels Rated for Shear Resistance or Steel Sheets ($R = 7$, $\Omega_0 = 2.5$, $C_d = 4.5$)
B24. Light-Framed Walls with Shear Panels of All Other Materials ($R = 2.5$, $\Omega_0 = 2.5$, $C_d = 2.5$)

D. Dual Systems with Special Moment Frames Capable of Resisting at Least 25 percent of Prescribed Seismic Forces

D10. Special Reinforced Masonry Shear Walls ($R = 5.5$, $\Omega_0 = 3$, $C_d = 5$)
D11. Intermediate Reinforced Masonry Shear Walls ($R = 4$, $\Omega_0 = 3$, $C_d = 3.5$)

E. Dual Systems with Intermediate Moment Frames Capable of Resisting at Least 25 percent of Prescribed Seismic Forces

E3. Ordinary Reinforced Masonry Shear Walls ($R = 3$, $\Omega_0 = 3$, $C_d = 2.5$)
E4. Intermediate Reinforced Masonry Shear Walls ($R = 3.5$, $\Omega_0 = 3$, $C_d = 3$)

G. Cantilevered Column Systems Detailed to Conform to the Requirements for

G1. Special Steel Moment Frames ($R = 2.5$, $\Omega_0 = 1.25$, $C_d = 2.5$)
G2. Intermediate Steel Moment Frames ($R = 1.5$, $\Omega_0 = 1.25$, $C_d = 1.5$)
G3. Ordinary Steel Moment Frames ($R = 1.25$, $\Omega_0 = 1.25$, $C_d = 1.25$)
G7. Timber Frames ($R = 1.5$, $\Omega_0 = 1.5$, $C_d = 1.5$)

H. Steel Systems Not Specifically Detailed for Seismic Resistance, Excluding Cantilever Column Systems ($R = 3$, $\Omega_0 = 3$, $C_d = 3$)

The manifestation of an earthquake is a ground motion that corresponds to a specific ground displacement, velocity, and acceleration. In the design of building structures, acceleration is the most useful of the three parameters because it allows for computation of building design forces by relating building mass and acceleration with force using Newton's second law of motion, which states that the net force exerted on a body is a product of its mass and acceleration. Figure 1.20 presents a generalized tripartite log chart showing a typical displacement, velocity, and acceleration response of buildings of various periods to earthquake-induced ground motion. Such a diagram is known as a *response spectrum*, and it represents a common tool for depicting responses of buildings with various natural periods to specific earthquakes. These diagrams typically are generated from field measurements of building displacement. Since the time intervals of the responses are known, the displacement response can be integrated to generate a velocity and subsequently an acceleration response. Finally, highly irregular response curves are smoothed out for easier use. The reader is referred to Chopra[15] for more in-depth information on the creation of earthquake-specific response spectrums. Also, as will become obvious in Chap. 2, a response spectrum created on the basis of earthquake ground acceleration values from Chap. 5 of ASCE/SEI 7–05 represents a key tool in determining the building-response accelerations used in design. It is intuitively clear from both the approximate period equations presented in Chap. 2 that low-rise buildings have significantly lower natural periods than taller buildings. The same can be discerned from analytical procedures and numerical methods used to compute building periods in systems with multiple degrees of freedom. As can be seen from Fig. 1.20, the response spectrum has three distinct ranges. Range *A* reflects buildings with relatively low natural periods, such as low-rise buildings, and it indicates a constant or nearly constant ground acceleration. Range *B*, characterized by buildings with medium-range periods, such as mid-rise buildings, shows a constant velocity and declining acceleration. Finally, range C, which corresponds to buildings with relatively high periods, such as high-rise buildings, shows a constant ground displacement and rapidly declining ground acceleration. One therefore can conclude that, all else being constant, low-rise buildings, characterized by low natural period, are subjected to significantly higher ground accelerations than mid-rise and high-rise buildings.

As can be inferred from the preceding discussion, the primary role of seismic design is to ensure a certain level of performance rather than merely proportioning the structure to provide a certain level of resistance against prescribed forces. Commonly recognized performance levels include life safety, collapse prevention, immediate occupancy, and operational occupancy. Most buildings are designed with life protection in mind. In summary, it would be irrational and uneconomical to design regular buildings to remain operational and serviceable following an MCE event that is very unlikely to occur during the building's lifespan. To say this another way, it is not reasonable to design a building whose useful lifespan is 30, 40, or 50 years to remain serviceable following an earthquake that is likely to return only once in 2,500 years. Low-rise light-frame buildings in wood and cold-formed steel traditionally have provided excellent performance in accomplishing the life-protection mission given their relatively low mass, very good energy-dissipating properties, and the redundancies inherent in this type of construction. Current research into the seismic behavior of light-frame buildings promises to result in further insights and increased performance in the future. Unreinforced masonry buildings traditionally have performed poorly, but their use was virtually outlawed in California in the first half of the last century, and their use within the current codified seismic design framework is either prescriptively or practically prohibitive in most regions of appreciable seismicity. However, typical buildings featuring properly designed and detailed reinforced masonry walls have performed adequately in seismic events. The higher performance levels are associated with a relatively few buildings representing institutions whose continued functioning following a seismic event is of critical public interest or whose replacement cost would

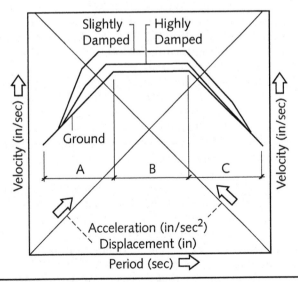

Figure 1.20 Generalized tripartite response spectrum.

outweigh the initial construction cost with a higher performance objective. This is accomplished by increasing the seismic design forces by a coefficient known as the *importance factor* I_E. Further discussion on determination of I_E is provided in Chap. 2.

References

1. American Concrete Institute (ACI). 2008. *Building Code Requirements for Structural Concrete* (ACI 318–08). ACI, Farmington Hills, MI.
2. American Forest and Paper Association (AF&PA). 2005. *National Design Specification for Engineered Wood Construction with 2005 Supplement* (ANSI/AF&PA NDS-2005). AF&PA, Washington, DC.
3. American Forest and Paper Association (AF&PA). 2005. *National Design Specification for Engineered Wood Construction—Commentary.* AF&PA, Washington, DC.
4. American Forest and Paper Association (AF&PA). 2005. *ASD/LRFD Manual for Engineered Wood Construction*, 2005 edition. AF&PA, Washington, DC.
5. American Forest and Paper Association (AF&PA). 2005. *Structural Wood Design Solved Example Problems*, 2005 edition. AF&PA, Washington, DC.
6. American Forest and Paper Association (AF&PA). 2008. *Special Design Provisions for Wind and Seismic* (ANSI/AF&PA SDPWS-08). AF&PA, Washington, DC.
7. American Institute of Steel Construction (AISC). 2006. *Specification for Structural Steel Buildings* (ANSI/AISC 360–05). AISC, Chicago, IL.
8. American Iron and Steel Institute (AISI). 2007. *North American Specification for the Design of Cold-Formed Steel Members* (ANSI/AISI S100–07). AISI, Washington, DC.
9. American Iron and Steel Institute (AISI). 2007. *Commentary to the North American Specification for the Design of Cold-Formed Steel Structural Members.* AISI, Washington, DC.
10. American Iron and Steel Institute (AISI). 2009. *Standard for Seismic Design of Cold-Formed Steel Structural Systems—Special Bolted Moment Frames with Supplement No. 1* (ANSI/AISI S110–07/S1–09). AISI, Washington, DC.
11. American Iron and Steel Institute (AISI). 2009. *Commentary on Standard for Seismic Design of Cold-Formed Steel Structural Systems—Special Bolted Moment Frames with Supplement No. 1* (AISI S110–07-C/S1-C-09). AISI, Washington, DC.
12. American Iron and Steel Institute (AISI). 2008. *Cold-Formed Steel Design Manual* (AISI D100–08). AISI, Washington, DC.
13. Building Seismic Safety Council (BSSC). 2004. *NEHRP Recommended Provisions for Seismic Regulations for New Buildings and Other Structures* (FEMA 450), 2003 edition. Federal Emergency Management Agency (FEMA), Washington, DC.
13a. American Society of Civil Engineers (ASCE). 1991. *Standard for the Structural Design of Composite Slabs* (ANSI/ASCE 3–91). ASCE, New York, NY.
14. California Building Standards Commission (CBSC). 2010. *California Building Code (2010 CBC).* CBSC, Sacramento, CA.
15. Chopra, A. 2006. *Dynamics of Structures*, 3rd ed. Prentice-Hall, Saddle River, NJ.
16. Council for Tall Buildings and Urban Habitat (CTBUH). 2011. *Criteria for the Defining and Measuring of Tall Buildings.* CTBUH, Chicago, IL.
17. Fanella, D. A., and Munshi, J.A. 1998. *Design of Low-Rise Concrete Buildings for Earthquake Forces*, 2nd ed. Portland Cement Association (PCA), Skokie, IL.
18. Georgia Department of Community Affairs (GDCA). 2010. *Georgia State Amendments to the International Building Code*, 2006 edition. GDCA, Atlanta, GA.

19. Gupta, A. K., and Moss, P. J. 1992. *Guidelines for Design of Low-Rise Buildings Subjected to Lateral Forces.* Council on Low-Rise Buildings, North Carolina State University. CRC Press, Boca Raton, FL.
20. International Accreditation Service (IAS). 2010. *Accreditation Criteria for Testing Laboratories* (AC89). IAS, Whittier, CA.
21. International Conference of Building Officials (ICBO). 1997. *Uniform Building Code (1997 UBC).* ICBO, Whittier, CA.
22. International Code Council (ICC). 2006. *International Building Code (2006 IBC).* ICC, Washington, DC.
23. International Code Council (ICC). 2009. *International Building Code (IBC).* ICC, Washington, DC.
23a. International Code Council (ICC). 2012. *International Building Code (IBC).* ICC, Washington, DC.
24. International Code Council (ICC). 2007. *IBC Interpretation No. 37–07.* ICC, Washington, DC.
25. International Code Council Evaluation Service (ICC-ES). 2009. *Acceptance Criteria for Expansion Anchors in Masonry Elements* (AC01). ICC-ES, Whittier, CA.
26. International Code Council Evaluation Service (ICC-ES). 2010. *Acceptance Criteria for Tapping Screw Fasteners* (AC118). ICC-ES, Whittier, CA.
27. International Code Council Evaluation Service (ICC-ES). 2010. *ICC-ES Evaluation Report—Kwik Bolt 3 Masonry Anchors* (ESR-1385). Report holder: Hilti, Inc., ICC-ES, Whittier, CA.
28. McKeever, D. B., Adair, C., and O'Connor, J. 2006. *Wood Products Used in the Construction of Low-Rise Nonresidential Buildings in the United States—2003.* USDA Forest Service, Madison, WI.
29. The Masonry Society (TMS). 2010. *Masonry Designers' Guide,* 6th ed. (MDG-6). TMS, Boulder, CO.
30. Masonry Standards Joint Committee (MSJC). 2008. *Building Code Requirements for Masonry Structures* (TMS 402–08/ACI 530–08/ASCE 5–08). TMS, Boulder, CO; ACI, Farmington Hills, MI; and ASCE, Reston, VA.
31. Masonry Standards Joint Committee (MSJC). 2008. *Specification for Masonry Structures* (TMS 602–08/ACI 530.1–08/ASCE 6–08). TMS, Boulder, CO; ACI, Farmington Hills, MI; and ASCE, Reston, VA.
32. Rack Manufacturers Institute (RMI). 2008. *Specification for Design, Testing and Utilization of Industrial Steel Storage Racks* (RMI ANSI/MH16.1–08). RMI, Charlotte, NC.
33. Steel Deck Institute (SDI). 2006. *Standard for Composite Steel Floor Deck* (ANSI/SDI-C1.0–06). SDI, Fox River Grove, IL.
34. Steel Deck Institute (SDI). 2006. *Standard for Noncomposite Steel Floor Deck* (ANSI/SDI-NC1.0–06). SDI, Fox River Grove, IL.
35. Steel Deck Institute (SDI). 2006. *Standard for Steel Roof Deck* (ANSI/SDI-RD1.0–06). SDI, Fox River Grove, IL.
36. Steel Joist Institute (SJI). 2006. *Standard Specification for Composite Steel Joists, CJ Series* (SJI-CJ-1.0–06). SJI, Forest, VA.
37. Steel Joist Institute (SJI). 2005. *Standard Specification for Joist Girders* (SJI-JG-1.1–05). SJI, Forest, VA.
38. Steel Joist Institute (SJI). 2005. *Standard Specification for Open Web Steel Joists, K Series* (SJI-K-1.1–05). SJI, Forest, VA.

39. Steel Joist Institute (SJI). 2005. *Standard Specification for Longspan Steel Joists, LH Series, and Deep Longspan Steel Joists, DLH Series* (SJI-LH/DLH-1.1–05). SJI, Forest, VA.
40. Steel Stud Manufacturers Association (SSMA). 2010. *Product Technical Information.* SSMA, Chicago, IL.
41. Structural Engineering Institute (SEI) of the American Society of Civil Engineers (ASCE). 2006. *Minimum Design Loads for Buildings and Other Structures, Including Supplements Nos. 1 and 2* (ASCE/SEI 7–05). ASCE, Reston, VA.
42. Structural Engineering Institute (SEI) of the American Society of Civil Engineers (ASCE). 2010. *Minimum Design Loads for Buildings and Other Structures* (ASCE/SEI 7–10). ASCE, Reston, VA.

CHAPTER 2

Loads and Load Paths in Low-Rise Buildings

2.1 Introduction

The information presented in this chapter will assist in the proper determination of structural loads in accordance with the 2009 edition of the *International Building Code* (IBC).[1] Chapter 16 of 2009 IBC contains the minimum magnitudes of some nominal loads and references the 2005 edition of American Society of Civil Engineers (ASCE)/ Structural Engineering Institute (SEI) 7, *Minimum Design Loads for Buildings and Other Structures*[2] for others. Methods on the determination of nominal loads that are applicable to low-rise structures also are presented.

In addition to minimum design load requirements, this chapter also covers design load combinations for strength design (or load- and resistance-factor design) and allowable-stress design. It is shown that the load combinations in 2009 IBC are essentially the same as those in ASCE/SEI 7 with some exceptions.

Load-path concepts applicable to low-rise buildings also are examined. Both vertical and lateral load paths are covered, and the discussion includes fundamental information on diaphragms and collectors.

Throughout this chapter, section numbers from 2009 IBC are referenced as illustrated by the following: Section 1605 of 2009 IBC is denoted as IBC 1605. Similarly, Sec. 2.3 from the 2005 ASCE/SEI 7 is referenced as ASCE/SEI 2.3. Readers who are interested in comprehensive coverage of structural load determination under 2009 IBC and ASCE/ SEI 7–05 are referred to Fanella (2009).[3]

2.2 Load Effects

2.2.1 Introduction

Minimum magnitudes of some nominal loads are contained in Chap. 16 of 2009 IBC, which references ASCE/SEI 7 for others. Table 2.1 contains a list of loads that are addressed in 2009 IBC and ASCE/SEI 7. Except for earthquake load effects, all the loads in this table are nominal loads (commonly referred to as *service loads*), which are multiplied by load factors in the strength-design method. The *earthquake load effect E* is defined to be a strength-level load where the load factor is equal to 1.0.

Notation	Load	Code Section
D	Dead load	IBC 1606
D_i	Weight of ice	Chapter 10 of ASCE/SEI 7
E	Combined effect of horizontal and vertical earthquake-induced forces, as defined in ASCE/SEI 12.4.2	IBC 1613 and ASCE/SEI 12.4.2
E_m	Maximum seismic load effect of horizontal and vertical forces as set forth in ASCE/SEI 12.4.3	IBC 1613 and ASCE/SEI 12.4.3
F	Load due to fluids with well-defined pressures and maximum heights	—
F_a	Flood load	IBC 1612
H	Load due to lateral earth pressures, groundwater pressure, or pressure of bulk materials	IBC 1610 (soil lateral loads)
L	Live load, except roof live load, including any permitted live-load reduction	IBC 1607
L_r	Roof live load, including any permitted live-load reduction	IBC 1607
R	Rain load	IBC 1611
S	Snow load	IBC 1608
T	Self-straining force arising from contraction or expansion resulting from temperature change, shrinkage, moisture change, creep in component materials, movement due to differential settlement, or combinations thereof	—
W	Load due to wind pressure	IBC 1609
W_i	Wind-on-ice load	Chapter 10 of ASCE/SEI 7

TABLE 2.1 Summary of Loads Addressed in 2009 IBC and ASCE/SEI 7

A discussion of loads that are encountered commonly in the design of low-rise buildings follows.

2.2.2 Dead Loads

Nominal dead loads D include the actual weight of construction materials and any fixed service equipment that is attached to or supported by the building or structure. IBC 1602 contains specific examples of such loads under the definition of *dead load*.

Since their magnitude remains essentially constant over time, dead loads are considered to be *permanent* loads. *Variable* loads such as live loads and wind loads are not permanent loads.

Superimposed dead loads are permanent loads other than the weights of the structural members and include floor finishes and/or topping; walls; ceilings; heating, ventilation, and air-conditioning (HVAC) and other service equipment; fixed partitions; and cladding.

It is not uncommon that the weights of materials and service equipment (such as plumbing stacks and risers, HVAC equipment, elevators and elevator machinery, fire protection systems, and similar fixed equipment) are not known during the preliminary design stages. To assist in determining such loads, ASCE/SEI Table C3–1 contains minimum design dead loads for various types of common construction components, including ceilings, roof and wall coverings, floor fill, floors and floor finishes, frame partitions, and frame walls. Minimum densities for common construction materials are given in ASCE/SEI Table C3–2. The weights in these tables are meant to be used as a guide when estimating dead loads. Since actual weights of construction materials and equipment can be greater than tabulated values, it is always prudent to verify weights with manufacturers or other similar resources. Values of dead loads used in design must be approved by the building official in cases where information on dead load is unavailable (IBC 1606.2).

2.2.3   Live Loads

General   Live loads are transient in nature and vary in magnitude over the life of a structure. These loads are produced by the use and occupancy of the building or structure and do not include construction loads, environmental loads (e.g., wind loads, snow loads, rain loads, earthquake loads, and flood loads), or dead loads (IBC 1602).

Nominal design values of uniformly distributed live loads L_o and concentrated live loads as a function of occupancy or use are given in IBC Table 1607.1. The occupancy description listed in the table is not necessarily group-specific (IBC Chap. 3 contains definitions of occupancy groups). For example, an office building with a business group B classification also may have storage areas with a live load of 125 psf, which is greater than the prescribed office live load. Structural members are designed based on the maximum effects due to application of either a uniform load or a concentrated load and need not be designed for the effects of both loads applied at the same time. Live loads that are not specifically listed in the table must be approved by the building official.

Partitions that can be relocated and that are not attached permanently to the structure are considered to be live loads in office and other buildings. A live load equal to at least 15 psf must be included for movable partitions if the nominal uniform floor live load is equal to or less than 80 psf.

A minimum roof live load of 20 psf for typical roof structures is prescribed in IBC Table 1607.1; larger live loads are required for roofs used as gardens or places of assembly.

ASCE/SEI Table 4–1 also contains minimum uniform and concentrated live loads, and some of these values differ from those in IBC Table 1607.1. Live loads for some commonly encountered occupancies are given in ASCE/SEI Tables C4–1 and C4–2.

Reduction in Live Loads

The probability that a structural member will be subjected to the full effects of nominal live loads decreases as the area supported by the member increases because live loads are transient in nature. When certain circumstances are satisfied, minimum uniformly distributed live loads L_o in IBC Table 1607.1 and uniform live loads of special-purpose roofs are permitted to be reduced in accordance with the methods in IBC 1607.9.1 and 1607.9.2. The general method of live-load reduction that is contained in IBC 1607.9.1 is also given in ASCE/SEI 4.8. Reduction of roof loads must conform to IBC 1607.11.2.

It is important to note that because partition live loads are not listed in IBC Table 1607.1, such loads are not permitted to be reduced by any method.

General Method of Live-Load Reduction   The general method of live-load reduction in IBC 1607.9.1 is applicable to uniform live loads other than uniform live loads on roofs and is based on the provisions contained in ASCE/SEI 4.8. IBC Eq. (16–22) can be used to obtain a reduced live load L for members that support an influence area $K_{LL}A_T$ ≥ 400 ft²:

$$L = L_o \left(0.25 + \frac{15}{\sqrt{K_{LL}A_T}} \right) \tag{2.1}$$

where K_{LL} is the live-load element factor given in IBC Table 1607.9.1, and A_T is the tributary area supported by the member in square feet.

The live-load element factor K_{LL} converts the tributary area A_T into an influence area that is considered to be the adjacent floor area from which the member derives its load. In other words,

$$K_{LL} = \text{influence area/tributary area } A_T \tag{2.2}$$

For example, the influence area for an interior column is equal to the area of the four bays that are adjacent to the column, which is equal to four times A_T. Thus, using Eq. (2.2), $K_{LL} = 4$ for an interior column. Similar factors can be derived for other members. ASCE/SEI Fig. C4–1 illustrates influence areas and tributary areas for a structure with regular bay spacing.

Limitations on the use of live-load reduction are given in IBC 1607.9.1.1 through 1607.9.1.4:

- One-way slabs. The live load on one-way slabs is permitted to be reduced, provided that A_T does not exceed an area equal to the slab span times a width normal to the slab of 1.5 times the slab span (IBC 1607.9.1.1).

- Heavy live loads. Live loads greater than 100 psf are not permitted to be reduced except in the following cases (IBC 1607.9.1.2): (1) Live loads for members supporting two or more floors are permitted to be reduced by a maximum of 20 percent, but the reduced live load L shall not be less than that calculated by IBC 1607.9.1 and (2) in occupancies other than storage, additional live-load reduction is permitted if it can be shown by a registered design professional that such a reduction is warranted.

- Passenger vehicle garages. Live load in passenger vehicle garages is not permitted to be reduced, except for members supporting two or more floors (IBC 1607.9.1.3). In such cases, the maximum reduction is 20 percent, and the reduced live load L shall not be less than that calculated by IBC 1607.9.1.

- Group A occupancies. Due to the nature of assembly occupancies (i.e., there is a high probability that the entire floor is subjected to full uniform live load), live loads of 100 psf and live loads at areas where fixed seats are located shall not be reduced (IBC 1607.9.1.4).

- Roof members. Live loads of 100 psf or less are not permitted to be reduced on roof members except as specified in IBC 1607.11.2 for flat, pitched, and curved roofs and for special-purpose roofs (see IBC 1607.9.1.5).

Alternate Method of Live-Load Reduction   IBC 1607.9.2 contains an alternate method of floor live-load reduction that is based on provisions that were contained in the 1997 *Uniform Building Code (UBC)*. A reduction factor R for members that support an area equal to or greater than 150 ft^2 is determined by IBC Eq. (16–23):

$$R = 0.08(A-150)$$

$$\leq \text{ the smallest of} \begin{cases} 40 \text{ percent for horizontal members} \\ 60 \text{ percent for vertical members} \\ 23.1(1+D/L_o) \end{cases} \tag{2.3}$$

where A is equal to the floor area supported by a member in square feet, and D is equal to the dead load per square foot of supported area.

The reduced live load L then is calculated by the following equation:

$$L = L_o\left(1 - \frac{R}{100}\right) \tag{2.4}$$

Limitations on live-load reduction using the alternate method are similar to those specified in the general method of live-load reduction (see IBC 1607.9.2).

From the preceding discussion of live-load reduction, it is evident that regardless of the method, the reduction of live-load effects on vertical elements in a low-rise building generally will be small unless such elements carry a relatively large tributary area.

Distribution of Floor Loads

When analyzing continuous floor members, the effects of partial uniform live load (i.e., alternate-span loading) must be investigated (IBC 1607.10). Such loading produces greatest effects at different locations along the span. When performing this type of analysis, it is permitted to use reduced floor live loads.

Figure 2.1 illustrates four loading patterns that must be investigated for a three-span continuous system that is subjected to uniformly distributed dead and live loads. Similar loading patterns can be derived for other span conditions.

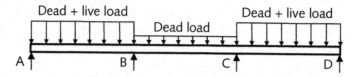

Loading pattern for maximum negative moment at support A or D
and maximum positive moment in span AB or CD

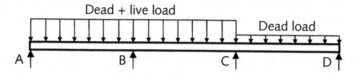

Loading pattern for maximum negative moment at support B

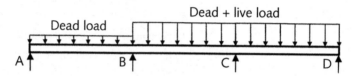

Loading pattern for maximum negative moment at support C

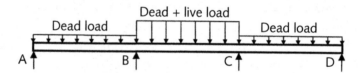

Loading pattern for maximum positive moment in span BC

FIGURE 2.1 Distribution of floor loads for a three-span continuous system (IBC 1607.10).

Roof Loads

Roof members must be designed to resist dead, live, wind, and where applicable, snow and earthquake loads. IBC Table 1607.1 prescribes a minimum roof live load of 20 psf for typical roof structures. Larger minimum live loads are required for roofs used as gardens or places of assembly.

IBC 1607.11.2 permits nominal roof live loads of 20 psf on flat, pitched, and curved roofs to be reduced in accordance with IBC Eq. (16–25):

$$L_r = L_o R_1 R_2 \tag{2.5}$$

where L_r is the reduced roof live load per square foot of horizontal roof projection, and R_1 and R_2 are reduction factors based on the tributary area A_t of the member being considered and the slope of the roof, respectively:

$$R_1 = \begin{cases} 1 \text{ for } A_t \leq 200 \text{ ft}^2 & \text{IBC Eq. (16-26)} \\ 1.2 - 0.001A_t \text{ for } 200 \text{ ft}^2 < A_t < 600 \text{ ft}^2 & \text{IBC Eq. (16-27)} \\ 0.6 \text{ for } A_t \geq 600 \text{ ft}^2 & \text{IBC Eq. (16-28)} \end{cases}$$

$$R_2 = \begin{cases} 1 \text{ for } F \leq 4 & \text{IBC Eq. (16-29)} \\ 1.2 - 0.05F \text{ for } 4 < F < 12 & \text{IBC Eq. (16-30)} \\ 0.6 \text{ for } F \geq 12 & \text{IBC Eq. (16-31)} \end{cases}$$

where the quantity F is the number of inches of rise per foot for a sloped roof and is the rise-to-span ratio multiplied by 32 for an arch or a dome.

It is evident from the preceding discussion that no live-load reduction is permitted for members supporting less than 200 ft^2 as well as for roof slopes equal to or less than 4:12. In no case is the reduced roof live load to be taken as less than 12 psf, and it need not exceed 20 psf.

Elements supporting landscaped roofs must be designed for a minimum roof live load of 20 psf (IBC 1607.11.3). The weight of landscaping material is considered to be dead load, and the saturation level of the soil must be considered when calculating the magnitude of the load.

Awnings and canopies must be designed for a minimum roof live load of 5 psf (IBC 1607.11.4). Such elements also must be designed for the combined effects of snow and wind loads in accordance with the load combinations of IBC 1605.

Distribution of Roof Loads

When analyzing continuous roof members, the effects of partial uniform roof live load (i.e., alternate-span loading) must be investigated (IBC 1607.11.1). Such loading produces its greatest effects at different locations along the span. Partial load analysis is not required when an unreduced roof live load of 20 psf is used (IBC 1607.11.1). Similar to floor live loads, Fig. 2.1 is also illustrative of the roof live-load distribution in a continuous three-span condition.

Crane Loads

IBC 1607.12 contains a general description of the crane loads that must considered where applicable. The support structure must be designed for the maximum wheel load, vertical impact, and horizontal impact as a simultaneous load combination. Provisions on how to determine these loads are given in IBC 1607.12.

Interior Walls and Partitions

Interior walls and partitions (including their finishing materials) greater than 6 ft in height are required to be designed for a horizontal load of 5 psf in accordance with IBC 1607.13. This requirement is intended to provide sufficient strength and durability of the wall framing and its finished construction when subjected to nominal impact loads, such as those from moving furniture or equipment, and from HVAC pressurization. Requirements for fabric partitions are given in IBC 1607.13.1.

2.2.4 Snow Loads

Ground Snow Loads

In accordance with IBC 1608.1, design snow loads on buildings and structures are to be determined by the provisions of Chap. 7 of ASCE/SEI 7–05. The first step is to determine the ground snow load p_g. This quantity is obtained from ASCE/SEI Fig. 7–1 or IBC Fig. 1608.2 for the conterminous United States and from ASCE/SEI Table 7–1 or IBC Table 1608.2 for locations in Alaska. The snow loads on the maps have a 2 percent annual probability of being exceeded (i.e., a 50-year mean recurrence interval). Table C7–1 in the commentary of ASCE/SEI 7 contains ground snow loads at 204 national weather service locations where load measurements are made. Note *a* points out that it is not appropriate to use only the site-specific information in this table to determine design snow loads (see ASCE/SEI C7.2 for more information).

Regions on the maps denoted as "CS" indicate that the ground snow load is too variable to map, and thus a site-specific case study is required. ASCE/SEI C7.2 provides more information on how to conduct a site-specific case study. Also provided on the maps are ground snow loads in mountainous areas based on elevation. Numbers in parentheses represent the upper elevation limits in feet for the ground snow-load values that are given below the elevation. Where a building is located at an elevation greater than that shown on the maps, a site-specific case study must be conducted to establish the ground snow load.

Flat-Roof Snow Loads

Once p_g has been established, a flat-roof snow load p_f is determined by ASCE/SEI Eq. (7–1):

$$p_f = 0.7C_eC_tIp_g \tag{2.6}$$

This equation is applicable to the design of flat roofs (i.e., roofs with a slope equal to or less than 5 degrees, which is approximately 1 in./ft) and is a function of roof exposure (exposure factor C_e; see ASCE/SEI Table 7–2), roof thermal condition (thermal factor C_t; see ASCE/SEI Table 7–3), and occupancy of the structure (importance factor I; see IBC Table 1604.5 or ASCE/SEI Table 7–4).

Minimum values of p_f are given for low-slope roofs in ASCE/SEI 7.3 (a low-slope roof is defined in ASCE/SEI 7.3.4). It is important to note that the minimum roof snow load defined in this section is a separate load case that is not to be combined with drifting, sliding, or other types of snow loading.

Sloped-Roof Snow Loads

According to ASCE/SEI 7.4, snow loads that act on a sloping surface are assumed to act on the horizontal projection of that surface. The sloped-roof snow load p_s is determined by ASCE/SEI Eq. (7–2):

$$p_s = C_s p_f \tag{2.7}$$

where C_s is a factor that depends on the slope and temperature of the roof, the presence or absence of obstructions, and the degree of slipperiness of the roof surface. Roof materials that are considered to be slippery and those which are not given in ASCE/SEI 7.4. ASCE/SEI Fig. 7–2 contains graphs of C_s for various conditions, and equations for

C_s are given in ASCE/SEI C7.4. Note that a *balanced* snow load is defined in the snow-load provisions as the sloped-roof snow load determined by Eq. (2.7).

According to ASCE/SEI 7.4.3, $C_s = 0$ for portions of curved roofs that have a slope exceeding 70 degrees; thus $p_s = 0$ in such cases (see Eq. 2.7). Balanced snow loads for curved roofs are determined from the loading diagrams in ASCE/SEI Fig. 7–3, with C_s determined from the appropriate curve in ASCE/SEI Fig. 7–2. Multiple-folded-plate, sawtooth, and barrel-vault roofs are to be designed using $C_s = 1$, i.e., $p_s = p_f$ (see ASCE/SEI 7.4.4). No reduction in snow load based on roof slope is applied to these types of roofs because they collect additional snow in their valleys by wind drifting and snow sliding.

Partial Snow Loading

The partial loading provisions of ASCE/SEI 7.5 must be considered for roof systems with continuous and/or cantilevered framing members and all other types of roof systems where the removal of snow on one span (e.g., by wind or thermal effects) causes an increase in stress or deflection in an adjacent span.

For simplicity, only the three load cases given in ASCE/SEI Fig. 7–4 need to be investigated; a comprehensive alternate-span (or checkerboard) loading analysis is not required. Partial loading provisions need not be considered for structural members that span perpendicular to the ridgeline of gable roofs with slopes greater than the larger of 2.38 degrees (approximately ½ in./ft) and $(70/W) + 0.5$, where W is the horizontal distance from the eave to the ridge in feet. Also, the minimum roof load requirements of ASCE/SEI 7.3.4 are not applicable in the partial-load provisions.

Unbalanced Roof Snow Loads

Wind and sunlight are the main causes for unbalanced snow loads on sloped roofs. Unbalanced loads are unlike partial loads where snow is removed on one portion of the roof and is not added to another portion. For example, wind tends to reduce the snow load on the windward portion and increase the snow load on the leeward portion (see Fig. 2.2 for unbalanced snow loads on a gable roof due to wind).

Provisions for unbalanced snow loads are given in ASCE/SEI 7.6.1 for hip and gable roofs, in ASCE/SEI 7.6.2 for curved roofs, in ASCE/SEI 7.6.3 for multiple-folded-plate, sawtooth, and barrel-vault roofs, and in ASCE/SEI 7.6.4 for dome roofs. Information on

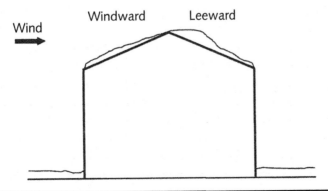

FIGURE 2.2 Unbalanced snow loads on a gable roof due to wind.

how to calculate balanced and unbalanced snow loads based on these provisions is summarized in ASCE/SEI Fig. 7–3 for curved roofs, Fig. 7–5 for hip and gable roofs, and Fig. 7–6 for sawtooth roofs.

Drifts on Lower Roofs

Snow drifts can occur on lower roofs of a building due to

1. Wind depositing snow from higher portions of the same building or an adjacent building or terrain feature (such as a hill) to a lower roof

2. Wind depositing snow from the windward portion of a lower roof to the portion of a lower roof adjacent to a taller part of the building

The first type of drift is called a *leeward drift*, and the second type is called a *windward drift*. Both types of drifts are illustrated in Fig. 2.3, which is adapted from ASCE/SEI Fig. 7–7. Loads from drifting snow p_d are superimposed on balanced snow loads p_s, as shown in ASCE/SEI Fig. 7–8. Note that p_d is determined by multiplying the density of snow ⊠ [which is determined by ASCE/SEI Eq. (7–3)] by the drift height h_d, which is the larger of the two heights defined in ASCE/SEI 7.7.1.

The provisions of ASCE/SEI 7.7.1 are also to be used to determine drift loads caused by a higher structure or terrain feature that is located within 20 ft of the roof (ASCE/SEI 7.7.2). The amount of drift load to be considered on the roof depends on the separation distance s between the roof and the adjacent structure or terrain feature; the applied drift load is equal to that calculated by the provisions of ASCE/SEI 7.7.1 multiplied by the factor $(20 - s)/20$, where s is in feet.

Roof Projections

The provisions of ASCE/SEI 7.8 can be used to determine drift loads on roof projections (e.g., mechanical equipment) and parapet walls. The height of the drift that is used in calculating the drift load is equal to three-quarters of the drift height from ASCE/SEI Fig. 7–9, with the length ℓ_u taken as the length of the roof that is upwind from the projection or parapet wall.

A drift load is not required to be applied to any side of a roof projection that is less than 15 ft long.

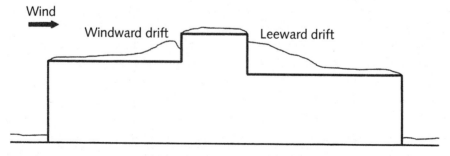

FIGURE 2.3   Windward and leeward snow drifts.

Sliding Snow

The load caused by snow sliding off a sloped roof onto a lower roof is determined by the provisions of ASCE/SEI 7.9. These provisions are applicable to slippery upper roofs with slopes greater than ¼ in./ft and to nonslippery upper roofs with slopes greater than 2 in./ft.

The total sliding load per unit length of eave is equal to $0.4p_fW$, where W is the horizontal distance from the eave to the ridge of the sloped upper roof. This load is to be uniformly distributed on the lower roof a distance of 15 ft from the upper-roof eave. In cases where the lower roof is less than 15 ft, the sliding load can be reduced proportionally. Sliding snow loads are superimposed on the balanced snow load.

Rain-On-Snow Surcharge Load

A rain-on-snow surcharge load of 5 psf is to be added on all roofs that meet the following conditions (see ASCE/SEI 7.10):

- Locations where $0 < p_s \leq 20$ psf
- Roof slope $< W/50$, where W is the horizontal distance from the eave to the ridge

This surcharge load is added directly to the balanced snow load and need not be used in combinations with drift, sliding, unbalanced, or partial loads.

Ponding Instability

Provisions for ponding instability and progressive deflection of roofs with a slope less than ¼ in./ft are given in ASCE/SEI 7.11 and 8.4. A roof must possess adequate stiffness to ensure that deflection instability does not occur (i.e., progressive deflection leading to instability). The larger of the snow load or the rain load is to be used in this analysis (rain loads are covered in Section 2.2.6).

Existing Roofs

Requirements for increased snow loads on existing roofs due to additions or alterations are covered in ASCE/SEI 7.12. This section requires that owners of an existing lower roof must be advised of the potential for increased snow loads in cases where a higher roof is constructed within 20 ft of the existing roof.

2.2.5   Wind Loads

General

According to IBC 1609.1.1, wind loads on buildings and structures are to be determined in accordance with the provisions of Chap. 6 of ASCE/SEI 7 or by the alternate all-heights method of IBC 1609.6. Exceptions listed in IBC 1609.1.1 permit wind forces to be determined using industry standards for certain types of structures. One of the exceptions also permits the use of wind-tunnel tests that conform to the provisions of ASCE/SEI 6.6, provided that the limitations of IBC 1609.1.1.2 are satisfied. Note that the basic wind speed, the exposure category, and the type of opening protection required may be determined by IBC 1609 or ASCE/SEI 7 because the provisions in both documents are essentially the same.

Wind forces are applied to a building in the form of pressures that act normal to the surfaces of the building. *Positive wind pressure* acts toward the surface and is commonly

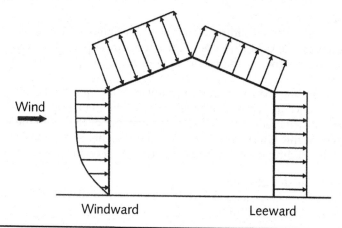

Wind

Windward   Leeward

FIGURE 2.4   The distribution of wind pressures on a building with a gable or a hip roof.

referred to as just *pressure. Negative wind pressure,* which is also called *suction,* acts away from the surface. Positive wind pressure acts on the windward wall of a building, and negative wind pressure acts on the leeward wall, the sidewalls, and the leeward portion of the roof (see Fig. 2.4). Either positive or negative pressure acts on the windward portion of the roof depending on the slope of the roof (flatter roofs will be subjected to negative pressure, whereas more sloped roofs will be subjected to positive pressure). Note that the wind pressure on the windward face varies with respect to height and that the pressures on all other surfaces are assumed to be constant.

In general, pressures must be considered on the main wind force–resisting system (MWFRS) and components and cladding (C&C) of a building or a structure. The MWFRS consists of structural elements that have been assigned to resist the effects from wind loads for the overall structure. Shear walls, moment frames, and braced frames are a few examples of different types of MWFRSs. The elements of the building envelope that do not qualify as part of the MWFRS are the C&C. The C&C receive the wind loads either directly or from the cladding and transfer the loads to the MWFRS. These could be individual exterior walls, roof decking, roof members (e.g., joists or purlins), curtain walls, windows, or doors. A discussion on how wind load propagates in a low-rise building is given in Sec. 2.4 of this book. Certain members must be designed for more than one type of loading; for example, a roof beam that is part of the MWFRS must be designed for the load effects associated with the MWFRS and those associated with the C&C.

Chapter 6 of ASCE/SEI 7–05 contains three methods to determine design wind pressures or loads on the MWFRS and C&C:

- *Method 1:* Simplified procedure
- *Method 2:* Analytical procedure
- *Method 3:* Wind-tunnel procedure

Method 1, which can be found in ASCE/SEI 6.4, is applicable to low-rise buildings that meet certain conditions and is covered in detail below. The provisions of method 2 cover a wide range of buildings and structures; provisions that are applicable to low-rise buildings are also presented in this section (ASCE/SEI 6.5). Wind-tunnel procedures are permitted for any building or structure, but because such tests are not used commonly in the design of low-rise buildings, they are not covered in this book.

The alternate all-heights method in IBC 1609.6 is also applicable in the design of low-rise buildings for wind, and the provisions of this method are covered below as well.

Methods 1 and 2 and the alternate all-heights method are static methods for estimating wind pressures. In such methods, the magnitude of wind pressure on a structure depends on its size, openness, importance, and location, as well as on the height above ground level. Wind gust and local extreme pressures at various locations on a building are also accounted for. Static methods generally yield very accurate results for low-rise buildings.

Basic Wind Speed

Regardless of the wind-load procedure used, the basic wind speed V must be determined. Figure 1609 in *2009* IBC and Fig. 6–1 in ASCE/SEI 7–05 are identical and provide basic wind speeds based on 3-second gusts at 33 ft above ground for exposure C (definitions of all the wind exposures are given in the next section). It is important to note that the design wind speeds on these maps do not include the effects from tornadoes.

Since some referenced standards contain criteria or applications based on fastest-mile wind speed, which was the wind speed used in earlier editions of ASCE/SEI 7 and the legacy codes, IBC 1609.3.1 provides an equation and a table that can be used to convert from one wind speed to the other.

Importance Factor

An importance factor I must be determined for the building or structure using ASCE Table 6–1 based on the occupancy category from IBC Table 1604.5 or ASCE/SEI Table 1–1. This factor is used to adjust the structural reliability of a building or a structure to be consistent with the occupancy classifications given in the tables. In particular, the importance factor adjusts the velocity pressure to different annual probabilities of being exceeded. An importance factor equal to 1 corresponds to a 50-year return period, whereas values of 0.87 and 1.15 in nonhurricane regions correspond to return periods of 25 and 100 years, respectively. ASCE/SEI C6.5.5 contains additional information on importance factors in hurricane regions.

Exposure Categories

An upwind exposure category is to be determined for each direction that wind loads are to be evaluated and shall be based on ground surface roughness from natural topography, vegetation, and facilities constructed in that area. Exposure categories are determined using surface-roughness categories that are defined in ASCE/SEI 6.5.6.2:

- *Surface roughness B:* Urban and suburban areas, wooded areas, or other terrain with numerous closely spaced obstructions having the size of single-family dwellings or larger.

- *Surface roughness C:* Open terrain with scattered obstructions having heights generally less than 30 ft, including flat open country, grasslands, and all water surfaces in hurricane-prone regions (see ASCE/SEI 6.2 for a definition of a hurricane-prone region).
- *Surface roughness D:* Flat, unobstructed areas and water surfaces outside hurricane-prone regions, including smooth mud flats, salt flats, and unbroken ice.

Once the surface-roughness category has been established, an exposure category is determined as follows:

- *Exposure B:* Applies where surface roughness B prevails in the upwind direction for a distance of at least 2,600 ft or 20 times the height of the building, whichever is greater.
- *Exposure C:* Applies for all cases where exposure B or D does not apply.
- *Exposure D:* Applies where surface roughness D prevails in the upwind direction for a distance greater than 5,000 ft or 20 times the height of the building, whichever is greater.

For exposure B, the upwind distance may be reduced from 2,600 to 1,500 ft for buildings with a mean roof height that is equal to or less than 30 ft. Also, exposure D must extend into downwind areas of surface roughness B or C for a distance of 600 ft or 20 times the height of the building, whichever is greater.

Additional information on exposure categories, including photos that illustrate the different types, can be found in ASCE/SEI C6.5.6.

Simplified Procedure (Method 1)

General   The simplified procedure in ASCE/SEI 6.4 is applicable to regularly shaped, enclosed low-rise buildings. This method is based on the low-rise building procedure given in method 2 for a specific type of building, including types that are not sensitive to torsional wind loading. Design methods are presented for both MWFRSs and C&C.

Main Wind Force–Resisting System   For an MWFRS of a building that meets the eight conditions in ASCE/SEI 6.4.1.1, the wind pressures on the walls and roof can be determined using the tabulated values in ASCE/SEI Fig. 6–2. The eight conditions are as follows:

1. *The building is a simple diaphragm building.* A *simple diaphragm building* is defined as a building in which both the windward and leeward wind loads are transmitted through diaphragms to the same MWFRS. No structural expansion joints are permitted because they interrupt the continuity of the diaphragm, resulting in internal pressures that cannot cancel out. Buildings where the wind forces that are applied to walls normal to the direction of wind are transferred to shear walls, moment frames, or braces parallel to the direction of wind through roof and floor diaphragms are examples of simple diaphragm buildings. This is in contrast to a typical metal building, where girts bending horizontally between frames transfer wind loads to those frames, which carry

them to the foundation. Since the wind loads are not transferred through diaphragms, such buildings are not simple diaphragm buildings.

2. *The building is a low-rise building.* A *low-rise building* is defined in ASCE/SEI 6.2 as a building whose mean roof height h is equal to or less than 60 ft, where h does not exceed the least horizontal dimension of the building (see ASCE/SEI 6.2 for the definition of h).

3. *The building is enclosed and conforms to the wind-borne debris provisions of ASCE/ SEI 6.5.9.3.* An *enclosed building* is one that does not comply with the requirements for open or partially enclosed buildings, which are defined in ASCE/SEI 6.2. Based on these definitions, it is evident that most low-rise buildings can be classified as enclosed even if there are large openings in two or more walls.

4. *The building is regularly shaped.* A general definition of a *regularly shaped building* is one having no unusual geometric irregularities in spatial form. Although this condition is somewhat subjective, most buildings would be considered regularly shaped unless they possess clear irregularities.

5. *The building is not classified as a flexible building.* A *flexible building* is defined in ASCE/SEI 6.2 as a building whose fundamental frequency is less than 1 Hz. As a general guideline, rigid buildings, which have a fundamental frequency equal to or greater than 1 Hz, usually have a height to minimum width that is less than 4. ASCE/SEI C6.5.8 provides simple, conservative methods on how to estimate the natural frequency of commonly encountered framing systems.

6. *The building does not have response characteristics making it subject to across-wind loading, vortex shedding, or instability due to galloping or flutter and does not have a site location for which channeling effects or buffeting in the wake of upwind obstructions warrant special conditions.* The phenomena listed in the first part of this condition are usually associated with tall, slender structures and are not encountered commonly in the response of low-rise buildings to wind loads. A special wind investigation should be performed for any size building located where wind forces can be increased by terrain characteristics or other features (e.g., a building located in a gorge between two mountains).

7. *The building has an approximately symmetric cross section in each direction with either a flat roof or a gable or a hip roof, where the roof slope from the horizontal θ is equal to or less than 45 degrees.* Once again, the first part of this condition has to do with the overall regularity of the building: A symmetric cross section is required for a balancing effect in the internal pressure. The second part limits the type of roof and its overall slope: Roofs with relatively large slopes have a reversal in the windward roof pressure from negative (suction) to positive (pressure); this results in loss of the ability to preselect the load case, which is inherent to the simplified method.

8. *The building is exempted from the torsional load cases indicated in note 5 of ASCE/SEI Fig. 6–10, or these torsional load cases do not control the design of any of the MWFRSs of the building.* This condition essentially prevents the use of method 1 for buildings with lateral systems that are sensitive to torsional wind loading. Note 5 of ASCE/SEI Fig. 6–10 contains building types that are exempted from torsional requirements because they are known not to be sensitive to torsion.

Buildings with an MWFRS in each direction that are also not sensitive to torsion can be found in ASCE/SEI C6.4.

Once it has been established that method 1 can be used, the basic wind speed, importance factor, and exposure category are determined using the methods outlined previously. Net design wind pressures p_{s30} are tabulated in ASCE/SEI Fig. 6–2 as a function of the basic wind speed for a specific set of conditions:

- Occupancy category II (I = 1.0), enclosed buildings
- Mean roof height = 30 ft
- Primarily flat ground (topographic factor K_{zt} = 1.0; see ASCE/SEI 6.5.7)
- Exposure B

These parameters were used to generate the pressures in the tables because a vast majority of buildings exist within this universe.

The horizontal pressures in ASCE/SEI Fig. 6–2 are net horizontal wind pressures; that is, they are the summation of the windward and leeward pressures on the MWFRS (the internal pressures on the walls cancel out). The vertical pressures on the horizontal projection of the roof surface include the internal pressures for enclosed buildings (GC_{pi} = ±0.18; see ASCE/SEI Fig. 6–5).

Tabulated pressures p_{s30} are multiplied by the appropriate adjustment factor given in ASCE/SEI Fig. 6–2 based on actual building height, exposure, occupancy, and topography at the site. Simplified design wind pressures p_s are calculated using ASCE/SEI Eq. (6–1):

$$p_s = \lambda K_{zt} I p_{s30} \qquad (2.8)$$

where λ is the adjustment factor for building height and exposure, which is also tabulated in ASCE/SEI Fig. 6–2. These adjusted pressures are applied normal to projected areas (i.e., walls and roof) of the building in accordance with ASCE/SEI Fig. 6–2. The load patterns illustrated in ASCE/SEI Fig. 6–2 for the MWFRS must be applied to each corner of the building in turn as the reference corner for wind in the transverse and longitudinal directions.

ASCE/SEI 6.4.2.1.1 requires that a minimum pressure p_s of 10 psf be applied to the walls (zones A, B, C, and D) while assuming that the pressure on the roof (zones E, F, G, and H) is 0. The application of this minimum wind load, which is a separate load case, is illustrated in ASCE/SEI Fig. C6–1.

Components and Cladding   Method 1 also can be used to determine wind pressures on C&C, provided that the five conditions in ASCE/SEI 6.4.12 are satisfied:

- The mean roof height h is equal to or less than 60 ft.
- The building is enclosed and conforms to the wind-borne debris provisions of ASCE/SEI 6.5.9.3.
- The building is regularly shaped.

- The building does not have response characteristics making it subject to across-wind loading, vortex shedding, instability due to galloping, or flutter and does not have a site location for which channeling effects or buffeting in the wake of upwind obstructions warrants special conditions.

- The building has either a flat roof, a gable roof with $\theta \leq 45$ degrees, or a hip roof with $\theta \leq 27$ degrees.

If a building satisfies only these criteria for C&C, the design wind pressures on the MWFRS must be determined by method 2 or method 3.

Net design wind pressures p_{net30} are tabulated in ASCE/SEI Fig. 6–3 for exposure B at a mean roof height of 30 ft and $I = 1.0$ (the same conditions as those for the MWFRS). Values of p_{net30} are given for interior zones, end zones, and corner zones on a building or a structure.

Similar to the MWFRS, tabulated pressures p_{net30} are multiplied by the appropriate adjustment factor given in ASCE/SEI Fig. 6–3 based on actual building height, exposure, occupancy, and topography at the site. Simplified net design wind pressures p_{net} are calculated using ASCE/SEI Eq. (6–2):

$$p_{net} = \lambda K_{zt} I p_{net30} \tag{2.9}$$

The values of λ in ASCE/SEI Fig. 6–3 for C&C are the same as those in ASCE/SEI Fig. 6–2 for the MWFRS.

ASCE/SEI 6.4.2.2.1 requires that the positive design wind pressures p_{net} determined by Eq. (2.9) not be less than 10 psf and the negative design wind pressures not be less than −10 psf.

Analytical Procedure (Method 2)

General Method 2 provides wind pressures and forces for the design of MWFRSs and C&C of enclosed and partially enclosed rigid and flexible buildings, open buildings, and other structures including freestanding walls, signs, rooftop equipment, and other structures (ASCE/SEI 6.5). Definitions of the different types of buildings can be found in ASCE/SEI 6.2.

In general, this procedure entails the determination of velocity pressures (which are a function of exposure, height, topographic effects, wind directionality, wind velocity, and building occupancy), gust-effect factors, external and internal pressure coefficients, and force coefficients. Pressures and forces can be determined for a wide range of buildings and structures provided that

1. The building or structure is regularly shaped; that is, the building has no unusual geometric irregularity in spatial form (both vertical and horizontal).

2. The building or structure responds to wind primarily in the same direction as that of the wind; that is, it does not have response characteristics that make it subject to across-wind loading, vortex shedding, or any other dynamic load effects that are common in tall, slender buildings and structures and cylindrical buildings and structures.

3. The building or structure is located on a site where there are no channeling effects or buffeting in the wake of upwind obstructions.

The discussion that follows will focus on the low-rise building provisions contained in this method for both MWFRSs and C&C. As noted previously, a low-rise building is one in which the following two conditions are satisfied (ASCE/SEI 6.2): (1) The mean roof height is equal to or less than 60 ft and (2) the mean roof height does not exceed the least horizontal dimension of the building.

Main Wind Force–Resisting System

The design wind pressure p that is used in the determination of wind loads for enclosed or partially enclosed low-rise buildings is determined by ASCE/SEI Eq. (6–18):

$$p = q_h[(GC_{pf}) - (GC_{pi})] \tag{2.10}$$

Design wind pressures are calculated for the windward wall, leeward wall, sidewalls, and roof. Each of the quantities in Eq. (2.10) is covered in detail below:

- *Velocity pressure* q_h. According to this method, the velocity pressure is evaluated at the mean roof height of a building or a structure for a particular exposure and is determined by ASCE/SEI Eq. (6–15):

$$q_h = 0.00256 K_h K_{zt} K_d V^2 I \tag{2.11}$$

  o *Air density.* The constant 0.00256 in Eq. (2.11) is related to the mass density of air for the standard atmosphere (59°F and sea-level pressure of 29.92 in Hg) and is obtained as follows (constant = one-half times the density of air times the velocity squared, where the velocity is in miles per hour and the pressure is in pounds per square foot):

$$\text{Constant} = 0.5\left[\left(0.0765\frac{\text{lb}}{\text{ft}^3}/32.2\frac{\text{ft}}{\text{s}^2}\right)\right] \times \left[\left(1\frac{\text{mi}}{\text{h}}\right) \times 5,280\frac{\text{ft}}{\text{mi}} \times \frac{1\,\text{h}}{3,600\,\text{s}}\right]^2 = 0.00256 \tag{2.12}$$

  Ambient air densities at various altitudes can be found in ASCE/SEI Table C6–13.

  o *Velocity pressure exposure coefficient* K_h. This coefficient modifies wind velocity (or pressure) with respect to exposure and height above ground. Since the coefficient is evaluated at the mean roof height for low-rise buildings, it is assumed that the pressure on the windward face is a constant and does not vary with respect to height. ASCE/SEI Table 6–3 contains values of K_h for various heights and exposures and also gives equations that can be used to calculate this coefficient. Note that the tabulated values under "Case 1" in the table are applicable to MWFRSs of low-rise buildings designed by the low-rise provisions of method 2.

  o *Topographic factor* K_{zt}. The topographic factor modifies the velocity pressure exposure coefficients for buildings located on the upper half of an isolated hill or an escarpment. For buildings located on relatively flat terrain, $K_{zt} = 1.0$. ASCE/SEI Eq. (6–3) is used to calculate K_{zt} for buildings that meet the conditions listed in ASCE/SEI 6.5.7.1:

$$K_{zt} = (1 + K_1 K_2 K_3)^2 \tag{2.13}$$

The multipliers K_1, K_2, and K_3 are given in ASCE/SEI Fig. 6–4:

- o *Directionality factor K_d.* This factor accounts for the following: (1) reduced probability of maximum winds coming from any given direction and (2) reduced probability of the maximum pressure coefficient occurring for any given wind direction. Values of K_d are tabulated in ASCE/SEI Table 6–4 for different structure types; this factor is equal to 0.85 for the MWFRSs and C&C of buildings.

- *External pressure coefficient (GC_{pf}).* This coefficient represents the product of the gust-effect factor and the equivalent external pressure coefficient. ASCE/SEI Fig. 6–10 contains values of this coefficient for the walls and roof (zones 1 through 6 and 1E, 2E, 3E, and 4E) as a function of roof angle. Basic load cases that must be considered are also illustrated in the figure for wind blowing in both the transverse and longitudinal directions. In general, the building must be designed for all wind directions by considering in turn each corner as a reference corner (see the eight load cases depicted in the figure). Unlike method 1, the torsional load cases shown in ASCE/SEI Fig. 6–10 must be considered (where applicable) in addition to the basic load cases when designing low-rise buildings for wind in accordance with method 2.

- *Internal pressure coefficient (GC_{pi}).* This coefficient represents the product of the gust-effect factor and the equivalent internal pressure coefficient. Values of this coefficient, which can be found in ASCE/SEI Table 6–5, were determined from wind-tunnel tests on various building shapes and are given as a function of the enclosure classification (i.e., open, partially enclosed, and enclosed). As expected, internal pressure is not considered in open buildings. Internal pressure effects cancel out when considering the net horizontal wind pressures on the MWFRS, so this term is not needed when calculating the horizontal-design wind pressure by Eq. (2.10).

Net design pressures for the MWFRSs of open buildings with monoslope, pitched, or troughed roofs are determined by ASCE/SEI Eq. (6–25):

$$p = q_h GC_N \tag{2.14}$$

where G is the gust-effect factor determined in accordance with ASCE/SEI 6.5.8. For rigid buildings or structures (fundamental frequency equal to or greater than 1 Hz), ASCE/SEI 6.5.8.1 permits G to be taken as 0.85. More involved calculations are required to determine G for flexible or dynamically sensitive buildings or structures (see ASCE/SEI 6.5.8.2). The coefficient C_N is the net pressure coefficient determined from ASCE/SEI Fig. 6–18*a* through *d* for different roof types.

Minimum wind pressure to be used in the design of the MWFRS is 10 psf (ASCE/SEI 6.1.4.1). As in method 1, this minimum pressure, which is a separate load case, is projected onto a vertical plane normal to the direction of the wind (see ASCE/SEI Fig. C6–1).

Components and Cladding

For low-rise buildings with a mean roof height equal to or less than 60 ft, ASCE/SEI Eq. (6–22) is used to determine the design wind pressures on the C&C:

$$p = q_h[(GC_p) - (GC_{pi})] \tag{2.15}$$

where all the terms in this equation are the same as those used in the determination of design wind pressure for the MWFRS with the exception of the external pressure coefficients GC_p; these coefficients are given in ASCE/SEI Figs. 6–11 through 6–16 for walls and different types of roofs.

Net design wind pressures on the C&C of open buildings with monoslope, pitched, or troughed roofs are determined by ASCE/SEI Eq. (6–26):

$$p = q_h GC_N \tag{2.16}$$

where in this case the coefficient C_N is the net pressure coefficient determined from ASCE/SEI Fig. 6–19A through C for different roof types.

The minimum design wind pressure for C&C is 10 psf acting in either direction normal to the surface of the component or cladding (ASCE/SEI 6.1.4.2).

Alternate All-Heights Method

General   The *alternate all-heights method* in IBC 1609.6 is based on the analytical method (method 2) in ASCE/SEI 6.5 and can be used when the following conditions are satisfied:

1. The height of the building or structure is equal to or less than 75 ft with a height-to-least-width ratio of 4 or less, or the building or structure has a fundamental frequency equal to or greater than 1 Hz.

2. The building or structure is not sensitive to dynamic effects.

3. The building or structure is not located on a site for which channeling effects or buffeting in the wake of upwind obstructions warrants special consideration.

4. The building meets the requirements of a simple diaphragm building according to the provisions of ASCE/SEI 6.2, where wind loads are transmitted to the MWFRS through the diaphragms.

5. For open buildings, multispan gable roofs, stepped roofs, sawtooth roofs, domed roofs, roofs with slopes greater than 45 degrees, solid free-standing walls and solid signs, and rooftop equipment, the applicable provisions of ASCE/SEI 7 must be used.

It is evident from the preceding conditions that this method is applicable to regularly shaped low-rise buildings (contrary to what the title of the method suggests) that are rigid. It is shown in the following sections that the determination of design wind pressures on MWFRSs and C&C is simpler than that using method 2 for buildings that meet the preceding conditions.

Main Wind Force–Resisting System   The net design wind pressure is determined by IBC Eq. (16–34):

$$p_{\text{net}} = q_s K_z C_{\text{net}} I K_{zt} \tag{2.17}$$

where q_s is the wind stagnation pressure in pounds per square foot and is equal to the air density constant 0.00256 multiplied by the basic wind speed in miles per hour squared. IBC Table 1609.6.2(1) contains values of q_s at a standard height of 33 ft above ground level as a function of the basic wind speed.

The velocity exposure coefficient K_z is determined from ASCE/SEI Table 6–3, and the topographic factor K_{zt} is determined in accordance with ASCE/SEI 6.5.7. For the windward face of the building, these terms are determined as a function of the height z above ground level, whereas for the leeward wall, sidewalls, and roof they are determined based on the mean roof height h.

The net pressure coefficient C_{net} is equal to $K_d(GC_p - GC_{pi})$, which are the same terms in the design pressure equation of method 2: (1) The directionality factor K_d is equal to 0.85 for buildings (ASCE/SEI Table 6–4), (2) the gust-effect factor G is equal to 0.85 for rigid buildings (ASCE/SEI 6.5.8.1), (3) the pressure coefficients C_p are given in ASCE/SEI Fig. 6–6, and (4) the internal pressure coefficients GC_{pi} are given in ASCE/SEI Fig. 6–5. Tabulated values of C_{net} are given in IBC Table 1609.6.2(2) for the walls and roofs of enclosed and partially enclosed buildings. For example, on the windward wall of an enclosed building with positive internal pressure, $C_{net} = 0.85[(0.85 \times 0.8) - 0.18] = 0.43$, which matches the value in IBC Table 1609.6.2(2). In cases where more than one value of C_{net} is listed in the table, the more severe wind-load condition shall be used for design.

The net pressures determined by Eq. (2.17) are applied simultaneously on and in a direction normal to all wall and roof surfaces. Since the pressures are net pressures, they are applied to the building similar to those in method 1. However, unlike method 1, IBC 1609.6.4.1 requires that the torsional load cases identified in ASCE/SEI Fig. 6–9 be considered in design.

IBC 1609.6.3 requires a minimum design pressure of 10 psf projected on a plane normal to the assumed wind direction.

Components and Cladding Net design wind pressures on components and cladding (C&C) are also determined using Eq. (2.17) and the tabulated values of C_{net} in IBC Table 1609.6.2(2) for C&C. Proper application of these pressures is summarized in IBC 1609.6.4.4.1. A minimum design net wind pressure of 10 psf is required that acts in either direction normal to the surface.

2.2.6 Rain Loads

Requirements for design rain loads are given in IBC 1611. Roofs equipped with hardware that controls the rate of drainage are required to have a secondary drainage system at a higher elevation that limits accumulation of water on the roof above that elevation. Such roofs must be designed to sustain the load of rainwater that will accumulate to the elevation of the secondary drainage system plus the uniform load caused by water that rises above the inlet of the secondary drainage system at its design flow.

The nominal rain load R is determined by IBC Eq. (16–35):

$$R = 5.2(d_s + d_h) \tag{2.18}$$

where 5.2 is the unit load per inch depth of rainwater (pounds per square foot per inch). The quantities d_s and d_h are the depth of the rainwater on the undeflected roof up to the inlet of the secondary drainage system when the primary drainage system is blocked

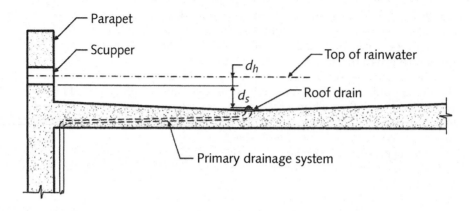

Figure 2.5 Example of rainwater depths for perimeter scuppers.

and the additional depth of rainwater on the undeflected roof above the inlet of the secondary drainage system is at its design flow, respectively. Figure 2.5 illustrates the rainwater depths for the case of perimeter scuppers as the secondary drainage system.

The nominal rain load R represents the weight of accumulated rainwater on the roof assuming that the primary roof drainage system is blocked. The primary roof drainage system can include, for example, roof drains, leaders, conductors, and horizontal storm drains and is designed for the 100-year hourly rainfall rate indicated in IBC Fig. 1611.1, as well as the area of the roof that it drains. Secondary drainage systems can occur at the perimeter of the roof (scuppers) or at the interior (drains).

In the case of relatively flat roofs, rainwater can accumulate and cause excessive deflection that can progress into failure if the supporting members are not stiff enough. IBC 1611.2 requires that such ponding instability be checked for all roofs that have a slope of less than a ¼ in./ft. The larger of the rain load or snow load should be used in the analysis (ASCE/SEI 8.4).

2.2.7 Flood Loads

Introduction
According to IBC 1612.1, all structures and portions of structures located in flood hazard areas must be designed and constructed to resist the effects of flood hazards and flood loads. Flood hazards may include erosion and scour, whereas flood loads include flotation, lateral hydrostatic pressures, hydrodynamic pressures (due to moving water), wave impact, and debris impact.

Where a building or a structure is located in more than one flood zone or is partially located in a flood zone, the entire building or structure must be designed and constructed according to the requirements of the more restrictive flood zone.

Flood Hazard Areas
A *flood hazard area* is defined as the greater of the following two areas:

1. The area within a floodplain subject to a 1 percent or greater chance of flooding in any year

2. The area designated as a flood hazard area on a community's flood hazard map or otherwise legally designated

The first of these two areas typically is acquired from flood insurance rate maps (FIRMs) that are prepared by the Federal Emergency Management Agency (FEMA) through the National Flood Insurance Program (NFIP). FIRMs show flood hazard areas along bodies of water where there is a risk of flooding by a base flood, that is, a flood having a 1 percent chance of being equaled or exceeded in any given year. In addition to showing the extent of flood hazards, the maps also show *base flood* elevations (BFEs) (the heights to which flood waters will rise during passage or occurrence of the base flood) and floodways.

Some local jurisdictions develop and subsequently adopt flood hazard maps that are more extensive than FEMA maps; in such cases, flood design and construction requirements must be satisfied in the areas delineated by the more extensive maps. Thus a *design flood* is a flood associated with the greater of the area of a base flood or the area legally designated as a flood hazard area by a community.

The NFIP divides flood hazard areas into flood hazard zones beginning with the letter A or V. *A zones* are those areas within inland or coastal floodplains where high-velocity wave action is not expected during the base flood. In contrast, *V zones* are those areas within a coastal floodplain where high-velocity wave action can occur during the base-flood event. Table 2.2 contains general descriptions of these zones. Such zone designations are contained in FIRMs and essentially indicate the magnitude and severity of flood hazards. Comprehensive definitions for each zone in this table can be found on the FEMA Map Service Center Web site (http://msc.fema.gov/).

Zone	Description
Moderate- to low-risk areas	
X	These zones identify areas outside the flood hazard area. Shaded zone X identifies areas that have a 0.2 percent probability of being equaled or exceeded during any given year. Unshaded zone X identifies areas where the annual exceedance probability of flooding is less than 0.2 percent.
High-risk areas	
A, AE, A1–30, A99, AR, AO, and AH	These zones identify areas of flood hazard that are not within the coastal high-hazard area.
High risk—coastal areas	
V, VE, and V1 to V30	These zones identify the coastal high-hazard area, which extends from offshore to the inland limit of a primary frontal dune along an open coast and any other portion of the flood hazard zone that is subject to high-velocity wave action from storms or seismic sources and to the effects of severe erosion and scour. Such zones generally are based on wave heights (3 ft or greater) or wave run-up depths (3 ft or greater).

TABLE 2.2 FEMA Flood Hazard Zones (Flood Insurance Zones)

Design and Construction

General   The design and construction of buildings and structures located in flood hazard areas shall be in accordance with Chap. 5 of ASCE/SEI 7–05 and ASCE/SEI 24–05, *Flood Resistant Design and Construction*[4] (IBC 1612.4). Section 1.6 of ASCE/SEI 24 requires that design flood loads and their combination with other loads be determined by ASCE/SEI 7.

The provisions of ASCE/SEI 24 are intended to meet or exceed the requirements of the NFIP. Figures 1–1 and 1–2 in ASCE/SEI 24 illustrate the application of the standard and application of the sections in the standard, respectively. The provisions contained in IBC Appendix G, Flood-Resistant Construction, are intended to fulfill the floodplain management and administrative requirements of NFIP that are not included in the IBC. Other provisions related to construction in flood hazard areas worth noting are found in IBC Chap. 18.

Flood Loads   Flood waters can create the following loads (see ASCE/SEI 5.4):

- Hydrostatic loads (ASCE/SEI 5.4.2)
- Hydrodynamic loads (ASCE/SEI 5.4.3)
- Wave loads (ASCE/SEI 5.4.4)
- Impact loads (ASCE/SEI 5.4.5)

Determination of these loads is based on the design flood, which was defined earlier. The *design flood elevation* (DFE) is the elevation of the design flood including wave height. For communities that have adopted minimum NFIP requirements, the DFE is identical to the BFE. The DFE exceeds the BFE in communities that have adopted requirements that exceed minimum NFIP requirements; communities that have chosen to exceed minimum NFIP requirements typically require a specified freeboard above the BFE. In-depth information on the various types of flood loads follows.

Hydrostatic Loads   Hydrostatic loads occur when stagnant or slowly moving (velocity less than 5 fps; see ASCE/SEI C5.4.2) water comes into contact with a building or a building component. The water can be above or below the ground surface.

Hydrostatic loads are commonly subdivided into lateral loads, vertical downward loads, and vertical upward loads (uplift or buoyancy). The hydrostatic pressure at any point on the surface of a structure or a component is equal in all directions and acts perpendicular to the surface.

Lateral hydrostatic force F_{sta} can be calculated from the following equation:

$$F_{sta} = \gamma_w d_s^2 w / 2 \qquad (2.19)$$

where γ_w is the unit weight of water (62.4 pcf for fresh water and 64.0 pcf for salt water), d_s is the design stillwater flood depth (see Fig. 2.6), and w is the width of a vertical element. This force acts at the point that is two-thirds below the stillwater surface of the water.

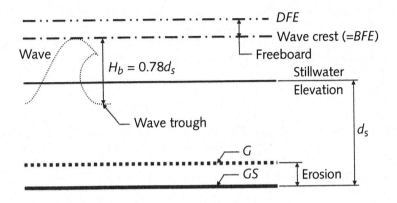

BFE = Base Flood Elevation

DFE = Design Flood Elevation

d_s = Design stillwater flood depth

G = Ground elevation

GS = Lowest eroded ground elevation adjacent to structure

H_b = Breaking wave height

FIGURE 2.6   Flood parameters.

The *vertical hydrostatic force,* commonly referred to as the *buoyant force* F_{buoy}, is equal to γ_w times the volume of flood water displaced by a submerged object. Such forces on a building can be of concern where the actual stillwater flood depth exceeds the design stillwater flood depth.

Hydrodynamic Loads   Hydrodynamic loads are caused by water moving at a moderate to high velocity above the ground level. Similar to wind loads, the loads produced by moving water include an impact load on the upstream face of a building, drag forces along the sides, and suction on the downstream face.

ASCE 5.4.3 permits the dynamic effects of moving water to be converted into an equivalent hydrostatic load for a water velocity equal to or less than 10 fps. This is accomplished by increasing the DFE by an equivalent surcharge depth d_h calculated by ASCE/SEI Eq. (5–1):

$$F_{dyn} = \gamma_w d_s d_h w \qquad (2.20)$$

where $d_h = (aV^2/2g)$, where V is the average velocity of water [which can be estimated by ASCE/SEI Eqs. (C5–1) and (C5–2)], g is the acceleration due to gravity (32.2 ft/s²), and a is the drag coefficient (or shape factor) that must be equal to or greater than 1.25 (guidelines on how to determine a are given in ASCE/SEI C5.4.3). It is assumed in this case that the total force F_{dyn} on the width w of a vertical element acts at the point that is two-thirds below the stillwater surface of the water.

For a water velocity greater than 10 fps, basic concepts of fluid mechanics must be used to determine loads imposed by moving water. The total load F_{dyn} in this case can be determined as follows:

$$F_{dyn} = a\rho V^2 d_s w / 2 \tag{2.21}$$

where ρ is equal to the mass density of water ($= \gamma_w / g$), and A is equal to the surface area normal to the water flow ($= wd_s$). The recommended value of the drag coefficient a is 2.0 for square or rectangular piles and 1.2 for round piles. In this case, F_{dyn} is assumed to act at the stillwater mid-depth (halfway between the ground surface and the stillwater elevation).

Wave Loads Wave loads result from water waves propagating over the surface of the water and striking a building or other object. The following loads must be accounted for when designing buildings and other structures for wave loads:

- Waves breaking on any portion of a building or a structure
- Uplift forces caused by shoaling waves beneath a building or a structure (or any portion thereof)
- Wave run-up striking any portion of a building or a structure
- Wave-induced drag and inertia forces
- Wave-induced scour at the base of a building or a structure or at its foundation

The effects of nonbreaking waves and broken waves can be determined using the procedures in ASCE/SEI 5.4.2 and 5.4.3 for hydrostatic and hydrodynamic loads, respectively.

Of the wave loads noted earlier, the loads from breaking waves are the highest. Thus this load is used as the design wave load where applicable.

Wave loads must be included in the design of buildings and other structures located in both zone V (wave heights equal to or greater than 3 ft) and zone A (wave heights less than 3 ft). ASCE/SEI 5.4.4 permits three methods to determine wave loads: (1) analytical procedures (ASCE/SEI 5.4.4), (2) advanced numerical modeling procedures, and (3) laboratory test procedures. The analytical procedures of ASCE/SEI 5.4.4 for breaking wave loads are discussed next.

- *Breaking wave loads on vertical pilings and columns.* The net force F_D resulting from a breaking wave acting on a rigid vertical pile or a column is determined by ASCE/SEI Eq. (5–4):

$$F_D = \gamma_w C_D D H_b^2 / 2 \tag{2.22}$$

where C_D = drag coefficient for breaking waves

= 1.75 for round piles or columns

= 2.25 for square or rectangular piles or columns

D = pile or column diameter for circular sections

= 1.4 times the width of the pile or column for rectangular or square sections

= breaking wave height (see Fig. 2.6)

= $0.78d_s = 0.51(BFE - G)$

This load is assumed to act at the stillwater elevation.

- *Breaking wave loads on vertical walls.* Two cases are considered in ASCE/SEI 5.4.4.2. In the first case, a wave breaks against a vertical wall of an enclosed dry space (i.e., the space behind the vertical wall is dry). ASCE/SEI Eqs. (5–5) and (5–6) give the maximum pressure P_{max} and net force F_t, respectively, resulting from waves that are normally incident to the wall (the direction of the wave approach is perpendicular to the face of the wall):

$$P_{max} = C_p \gamma_w d_s + 1.2 \gamma_w d_s \tag{2.23}$$

$$F_t = 1.1 C_p \gamma_w d_s^2 + 2.4 \gamma_w d_s^2 \tag{2.24}$$

The hydrostatic and dynamic pressure distributions are illustrated in ASCE/SEI Fig. 5–1.

In the second case, a wave breaks against a vertical wall where the stillwater levels on both sides of the wall are equal (this can occur, for example, where a wave breaks against a wall equipped with openings (such as flood vents) that allow flood waters to equalize on both sides). The maximum combined wave pressure is computed by ASCE/SEI Eq. (5–5) and the net breaking wave force F_t is computed by ASCE/SEI Eq. (5–7):

$$F_t = 1.1 C_p \gamma_w d_s^2 + 1.9 \gamma_w d_s^2 \tag{2.25}$$

Values of the dynamic pressure coefficient C_p are given in ASCE/SEI Table 5–1 based on the building category. ASCE/SEI C5.4.4.2 contains information on the probabilities of exceedance that correspond to the different building categories listed in the table.

- *Breaking wave loads on nonvertical walls.* ASCE/SEI Eq. (5–8) can be used to determine the horizontal component of a breaking wave load F_{nv} on a wall that is not vertical:

$$F_{nv} = F_t \sin^2 \alpha \tag{2.26}$$

The angle α is the vertical angle between the nonvertical surface of a wall and the horizontal.

- *Breaking wave loads from obliquely incident waves.* Maximum breaking wave loads occur where a wave strikes perpendicular to a surface. ASCE/SEI Eq. (5–9) can be used to determine the horizontal component of an obliquely incident breaking wave force F_{oi}:

$$F_{oi} = F_t \sin^2 \alpha \qquad (2.27)$$

The angle α is the horizontal angle between the direction of the wave approach and a vertical surface. In coastal areas, it is usually assumed that the direction of wave approach is perpendicular to the shoreline. Therefore, Eq. (2.27) provides a method for reducing breaking wave loads on vertical surfaces that are not parallel to the shoreline.

Impact Loads Impact loads occur where objects carried by moving water strike a building or a structure. Normal impact loads result from isolated impacts of normally encountered objects, whereas special impact loads result from large objects such as broken up ice floats and accumulated debris. These two types of impact loads are commonly considered in the design of buildings and similar structures.

The recommended method for calculating normal impact loads on buildings is given in ASCE/SEI C5.4.5. ASCE/SEI Eq. (C5–3) can be used to determine the impact force F:

$$F = \frac{\pi W V_b C_I C_O C_D C_B R_{max}}{2g(\Delta t)} \qquad (2.28)$$

where the parameters and coefficients in this equation are discussed in ASCE/SEI C5.4.5.

ASCE/SEI C5.4.5 also contains ASCE/SEI Eq. (C5–4), which can be used to determine the drag force due to debris accumulation (i.e., special impact load):

$$F = C_D \rho A V^2 / 2 \qquad (2.29)$$

It is assumed that objects are at or near the water surface level when they strike a building. Thus the loads determined by ASCE/SEI Eqs. (C5–3) and (C5–4) usually are assumed to act at the stillwater flood level; in general, these loads should be applied horizontally at the most critical location at or below the DFE.

2.2.8 Earthquake Loads

Introduction

According to IBC 1613.1, the effects of earthquake motion on structures and their components are to be determined in accordance with ASCE/SEI 7, excluding Chap. 14 and App. 11A. The design spectral accelerations, which are proportional to the magnitude of the earthquake forces, can be determined by either IBC 1613 or ASCE/SEI Chap. 11 because the provisions are the same.

The forces that a building must resist during a seismic event are caused by ground motion: As the base of a building moves with the ground, the inertia of the building mass resists this movement, which causes the building to distort. This distortion wave travels along the height of the building. With continued shaking, the building undergoes a series of complex oscillations.

The manner in which a building responds to an earthquake depends on its mass, stiffness, and strength (the strength of the structure generally plays a role beyond the stage of elastic response). There are as many natural modes of vibration as there are

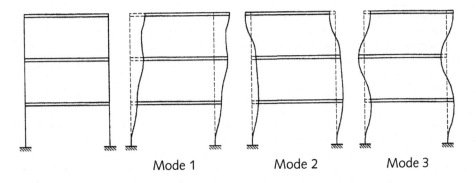

Mode 1   Mode 2   Mode 3

FIGURE 2.7 Modes of vibration for idealized structures.

degrees of freedom. The seismic response of short, stiff buildings is dominated by the first (fundamental) mode of vibration, where all the masses move in the same direction in response to the ground motion (see Fig. 2.7). Higher modes of vibration contribute significantly to the response of tall, flexible buildings.

The *equivalent lateral force procedure* of ASCE/SEI 12.8 can be used to determine earthquake forces for structures that meet the conditions in ASCE/SEI Table 12.6–1. It applies to essentially regular buildings that respond to earthquake forces primarily in the first mode. The simplified alternative method of ASCE/SEI 12.14 can be used to determine design earthquake forces for bearing-wall and building-frame systems that meet the conditions of ASCE/SEI 12.14.1.1. Both methods are covered after basic information on seismic design criteria is presented.

Seismic Design Criteria

Seismic Ground-Motion Values   IBC Figs. 1613.5(1) and 1613.5(2) and ASCE/SEI Figs. 22–1 and 22–2 contain contour maps of the conterminous United States giving S_S and S_1, which are the mapped *maximum considered earthquake* (MCE) spectral response accelerations at periods of 0.2 and 1.0 second, respectively, for a site class B soil profile and 5 percent damping. These MCE spectral response accelerations, which are directly related to base shear, reflect the maximum level of earthquake ground shaking that is considered reasonable for the design of new structures. IBC Figs. 1613.5(3) through 1613.5(14) and ASCE/SEI Figs. 22–3 through 22–14 contain similar contour maps for specific regions of the conterminous United States, Alaska, Hawaii, Puerto Rico, and U.S. commonwealths and territories.

In lieu of the maps, MCE spectral response accelerations can be obtained from the *ground motion parameter calculator* that can be accessed on the U.S. Geological Survey (USGS) Web site (http://earthquake.usgs.gov/research/hazmaps/design/).

Where $S_S \leq 0.15$ and $S_1 \leq 0.04$, the structure is permitted to be assigned to *seismic design category (SDC) A* (IBC 1613.5.1). These areas are considered to have very low seismic risk based solely on the mapped ground motions. Additional information on how to determine the SDC for a building is covered below.

Site Class Six site classes are defined in IBC Table 1613.5.2 and ASCE/SEI Table 20.3–1. A site is to be classified as one of these six site classes based on one of three soil properties: (1) soil shear-wave velocity, (2) standard penetration resistance or blow count, and (3) soil undrained shear strength. These properties must be measured over the top 100 ft of the site. Steps for classifying a site are given in IBC 1613.5.5.1 and ASCE/SEI 20.3. Methods of determining the site class where the soil is not homogeneous over the top 100 ft are provided in IBC 1613.5.5 and ASCE/SEI 20.4.

Site class A is hard rock, which typically is found in the eastern United States, whereas site class B is a softer rock that is found commonly in western parts of the country. Site classes C, D, and E indicate progressively softer soils, whereas site class F indicates soil so poor that a site-specific geotechnical investigation and dynamic site-response analysis are required to determine site coefficients (site-specific ground-motion procedures for seismic design are given in ASCE/SEI Chap. 21).

When soil properties are not known in sufficient detail to determine the site class in accordance with code provisions, site class D must be used, unless the building official requires that site class E or F must be used at the site.

Site Coefficients and Adjusted MCE Spectral Response Acceleration Parameters The MCE spectral response acceleration for short periods S_{MS} and at 1-second periods S_{M1} adjusted for site-class effects are determined by IBC Eqs. (16–36) and (16–37), respectively, or ASCE/SEI Eqs. (11.4–1) and (11.4–2), respectively:

$$S_{MS} = F_a S_S \tag{2.30}$$

$$S_{M1} = F_v S_1 \tag{2.31}$$

where F_a is the short-period site coefficient determined from IBC Table 1613.5.3(1) or ASCE/SEI Table 11.4–1, and F_v is the long-period site coefficient determined from IBC Table 1613.5.3(2) or ASCE/SEI Table 11.4–2. These factors essentially adjust the mapped spectral response accelerations for site classes other than B.

Design Spectral Response Acceleration Parameters Five percent damped design spectral response accelerations at short periods S_{DS} and at 1-second periods S_{D1} are determined by IBC Eqs. (16–38) and (16–39), respectively, or ASCE/SEI Eqs. (11.4–3) and (11.4–4), respectively:

$$S_{DS} = \frac{2}{3} S_{MS} \tag{2.32}$$

$$S_{D1} = \frac{2}{3} S_{M1} \tag{2.33}$$

The design ground motion is $2/3 = 1/1.5$ times the soil-modified MCE ground motion; the basis of this factor is that it is highly unlikely that a structure designed by the code provisions will collapse when subjected to ground motion that is 1.5 times as strong as the design ground motion.

Occupancy Category and Importance Factor

Occupancy categories are defined in IBC Table 1604.5 and ASCE/SEI Table 1–1. An importance factor I is assigned to a building or a structure in accordance with ASCE/SEI Table 11.5–1 based on its occupancy category.

Larger values of I are assigned to high-occupancy and essential facilities to increase the likelihood that such structures would suffer less damage and continue to function during and following a design earthquake.

Seismic Design Category

In accordance with IBC 1613.5.6 or ASCE/SEI 11.6, all buildings and structures must be assigned to an SDC. In general, an SDC is a function of occupancy or use and the design spectral accelerations at the site.

The SDC is determined twice: first as a function of S_{DS} by IBC Table 1613.5.6(1) or ASCE/SEI Table 11.6–1 and second as a function of S_{D1} by IBC Table 1613.5.6(2) or ASCE/SEI Table 11.6–2. The more severe SDC of the two governs.

In cases where S_1 is less than 0.75, it is permitted to determine the SDC by IBC Table 1613.5.6(1) or ASCE/SEI Table 11.6–1 based solely on the $S_{DS'}$ provided that all four conditions listed under IBC 1613.5.6.1 or ASCE/SEI 11.6 are satisfied:

1. The approximate fundamental period of the building or structure T_a in each of the two orthogonal directions determined in accordance with ASCE/SEI 12.8.2.1 is less than $0.8T_{S'}$ where $T_S = S_{D1}/S_{DS}$.

2. The fundamental period of the building or structure in each of the two orthogonal directions used to calculate the story drift is less than T_S.

3. ASCE/SEI Eq. (12.8–2) is used to determine the seismic response coefficient C_s.

4. The diaphragms are rigid (as defined in ASCE/SEI 12.3.1), or for diaphragms that are flexible, the distance between vertical elements of the seismic force–resisting system does not exceed 40 ft.

This exception always should be considered, especially when determining the SDC of low-rise buildings. The SDC is a trigger for many seismic requirements, including (1) permissible seismic force–resisting systems, (2) limitations on building height, (3) consideration of structural irregularities, and (4) the need for additional special inspections, structural testing, and structural observation for seismic resistance. Thus this exception can be extremely beneficial when establishing the SDC of low-rise buildings that meet these four conditions because it is likely that the building may be assigned to a lower SDC.

Design Requirements for SDC A

Buildings and structures that are assigned to SDC A must comply with the requirements of ASCE/SEI 11.7 only. Static seismic forces F_x are applied at each floor level in accordance with ASCE/SEI Eq. (11.7–1):

$$F_x = 0.01w_x \qquad (2.34)$$

where w_x is the portion of the total dead load of the structure that is tributary to level x (the total dead load of the structure for use in seismic calculations is defined in ASCE/

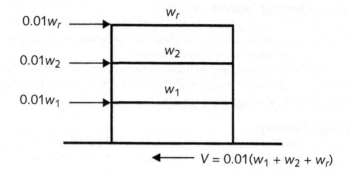

$$V = 0.01(w_1 + w_2 + w_r)$$

Figure 2.8   Design seismic force distribution for a low-rise building assigned to SDC A.

SEI 12.7.2 and is covered in the next section). Figure 2.8 illustrates the seismic force distribution for a low-rise building assigned to SDC A.

The reactions in the structural members of the seismic force–resisting system due to the seismic forces determined by Eq. (2.34) are combined with other load effects using applicable load combinations (see Sec. 2.3).

Provisions for load-path connections, connections to supports, and anchorage of concrete or masonry walls are also given in ASCE/SEI 11.7.

IBC 1604.8.2 requires that all walls be sufficiently anchored to provide positive direct connections capable of resisting the horizontal forces specified in ASCE/SEI 11.7.3. Where anchor spacing exceeds 4 ft, concrete and masonry walls must be designed to resist bending between anchors.

Seismic Force–Resisting Systems

ASCE/SEI Table 12.2–1 contains the basic structural systems that may be used to resist earthquake forces. Included in this table is the following information:

- Response modification coefficient R
- System overstrength factor Ω_o
- Deflection amplification factor C_d
- System limitations

The specific type of seismic force–resisting system that is permitted to be used in a particular building is based on the SDC. It is evident from ASCE/SEI Table 12.2–1 that some systems can be used in any situation regardless of the SDC, whereas others are either limited in height or not permitted at all depending on the severity of the SDC.

Bearing-wall systems, building frame systems, and moment-resisting frame systems are used commonly as the seismic force–resisting systems in low-rise buildings. A *bearing-wall system* is a structural system without an essentially complete space frame that provides support for gravity loads. Bearing walls provide support for most or all of the gravity loads and provide resistance to lateral forces.

A *building frame system* is a structural system with an essentially complete space frame that supports gravity loads. Resistance to lateral forces is provided solely by shear walls; it is assumed that there is no interaction between the frames and walls. The deformation compatibility requirements in ASCE/SEI 12.12.4 must be satisfied for buildings assigned to SDC D, E, or F.

Like a building frame system, a *moment-resisting frame system* is a structural system with an essentially complete space frame that supports gravity loads. Lateral forces are resisted primarily by flexural action of the frame members. The entire space frame or selected portions of the frame may be designated as the seismic force–resisting system. The requirements of ASCE/SEI 12.12.4 must be satisfied in the latter case, where applicable.

ASCE/SEI 12.2 contains additional requirements on combinations of framing systems in different directions and in the same direction. In the latter case, specific provisions are provided on how to account for vertical and horizontal combinations of seismic force–resisting systems in the same direction of analysis.

Equivalent Lateral Force Procedure

General   The *equivalent lateral force procedure* (ELFP) is a general-purpose method that can be used to determine the seismic forces for buildings assigned to SDC B and C, as well as some buildings assigned to SDC D, E, and F. ASCE/SEI Table 12.6–1 contains the permitted analytical procedures that must be used to determine seismic forces as a function of the SDC. In this table, reference is made to different types of structural irregularities; detailed information on horizontal and vertical structural irregularities can be found in ASCE/SEI Tables 12.3–1 and 12.3–2, respectively.

This method is applicable to buildings that respond to earthquake motions primarily in the first mode of vibration. It essentially reduces a complicated overall dynamic response to a set of static forces: A base shear is calculated based on the seismicity of the site, the type and structural characteristics of the seismic force–resisting system, and the importance of the building, and it is distributed over the height of the building. In the case of low-rise buildings, the vertical force distribution is assumed to be linear. The vertical forces at each level then are distributed horizontally to the elements of the seismic force–resisting system.

Seismic Base Shear   The *seismic base shear V* in a given direction is determined by ASCE/SEI Eq. (12.8–1):

$$V = C_s W \qquad (2.35)$$

where W is the effective seismic weight (which is discussed in detail below), and C_s is the seismic response coefficient, which is determined by ASCE/SEI Eq. (12.8–3):

$$C_s = \frac{S_{D1}}{T(R/I)} \qquad (2.36)$$

This equation is applicable in cases where the fundamental period of the building T is equal to or less than the long-period transition period T_L, which is given in ASCE/SEI Figs. 22–15 through 22–20 for the conterminous United States, Alaska, Hawaii, and other islands in the Caribbean and Pacific. It is evident that ASCE/SEI Eq. (12.8–4) usually is applicable in the design of tall buildings because the magnitude of T_L in these figures is relatively large.

The response modification coefficient R is given in ASCE/SEI Table 12.2–1 for various seismic force–resisting systems and in general terms is intended to account for differences in the inelastic deformability or energy-dissipation capacity of the various systems. The importance factor I is determined from ASCE/SEI Table 11.5–1 as a function of the occupancy category. Larger values of I are assigned to buildings that represent a substantial hazard to human life in the event of failure and to essential buildings that must remain operational after the design seismic event.

The value of C_s need not exceed that determined by ASCE/SEI Eq. (12.8–2):

$$C_s = \frac{S_{DS}}{R/I} \tag{2.37}$$

Also, C_s must not be less than that determined by ASCE/SEI Eq. (12.8–5):

$$C_s = 0.044 S_{DS} I \geq 0.01 \tag{2.38}$$

For buildings that are located where S_1 is equal to or greater than 0.6, C_s must not be less than that determined by ASCE/SEI Eq. (12.8–6):

$$C_s = \frac{0.5 S_1}{R/I} \tag{2.39}$$

The design spectrum defined in the preceding equations is depicted in Fig. 2.9. Equation (2.37) represents the constant-acceleration portion of the spectrum, whereas Eq. (2.36) represents the constant-velocity portion. It is evident that most low-rise buildings will fall within the constant-acceleration portion of the spectrum. The minimum seismic base shear represented by Eqs. (2.38) and (2.39) is usually applicable to taller buildings with longer periods.

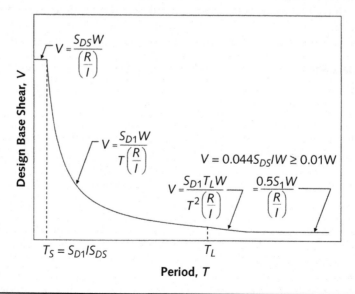

FIGURE 2.9 Design response spectrum according to the ELFP.

The fundamental period T can be established for a building using fundamental principles of dynamics based on the structural properties of its members and the deformational characteristics of the seismic force–resisting system. Since member sizes may not be readily available during the preliminary design stage, ASCE/SEI Eq. (12.8–7) can be used to determine an approximate fundamental period T_a:

$$T_a = C_t h_n^x \tag{2.40}$$

Values of the approximate period parameters C_t and x are given in ASCE/SEI Table 12.8–2 as a function of the structure type. The quantity h_n is the height in feet above the base to the highest level of the structure. Fundamental periods determined by Eq. (2.40) will result in conservative values of base shear.

The effective seismic weight W includes the total dead load of the building or structure and the following other loads (ASCE/SEI 12.7.2):

1. In areas used for storage, 25 percent of the floor live load (floor live load in public garages and public spaces need not be included in this live load)

2. Where partitions must be included in the floor load design per ASCE/SEI 4.2.2, the actual partition weight or a minimum of 10 psf, whichever is greater

3. Total operating weight of permanent equipment

4. Where the flat-roof snow load p_f exceeds 30 psf, 20 percent of the uniform design snow load, regardless of the roof slope

Vertical Distribution of Seismic Forces   Once the base shear V has been determined, it is distributed over the height of the building in accordance with ASCE/SEI 12.8.3. The lateral seismic force F_x at any level x above the ground level is determined by ASCE/SEI Eq. (12.8–11):

$$F_x = C_{vx} V \tag{2.41}$$

where the vertical distribution factor C_{vx} is determined by ASCE/SEI Eq. (12.8–12):

$$C_{vx} = \frac{w_x h_x^k}{\sum_{i=1}^{n} w_i h_i^k} \tag{2.42}$$

where w_i and w_x are the portions of the total effective seismic weight W located or assigned to levels i or x, respectively. The exponent related to the structure period k is equal to 1 for structures with a fundamental period equal to or less than 0.5 second and is equal to 2 where the fundamental period is equal to or greater than 2.5 seconds. Thus V is distributed linearly over the height of the building in cases where $T \le 0.5$ second, varying from zero at the base to a maximum value at the top (see Fig. 2.10); this type of distribution is common for low-rise buildings. A parabolic distribution is to be used where $T \ge 2.5$ seconds, and for a period between 0.5 and 2.5 seconds, a linear interpolation between a linear and parabolic distribution is permitted or a parabolic distribution may be used.

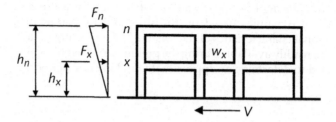

FIGURE **2.10**   Vertical distribution of seismic forces where $T \le 0.5$ second.

Horizontal Distribution of Seismic Forces   The seismic design story shear V_x in story x is the sum of the lateral forces acting at the floor or roof level supported by that story and all the floor levels above, including the roof (see ASCE/SEI Eq. 12.8–13). The story shear is distributed to the vertical elements of the seismic force–resisting system in the story based on the lateral stiffness of the diaphragm.

For flexible diaphragms, V_x is distributed to the vertical elements of the seismic force–resisting system based on the area of the diaphragm tributary to each line of resistance (ASCE/SEI 12.3 provides information on how to determine the degree of diaphragm flexibility). Untopped steel decking or wood structural panels are permitted to be idealized as flexible diaphragms in structures where the vertical elements are steel or composite steel and concrete braced frames or concrete, masonry, steel, or composite shear walls.

For diaphragms that are not flexible, V_x is distributed based on the relative stiffness of the vertical resisting elements and the diaphragm. Inherent and accidental torsion must be considered in the overall distribution (see ASCE/SEI 12.8.4.1 and 12.8.4.2). Where type 1a or 1b irregularity is present in structures assigned to SDC C, D, E, or F, the accidental torsional moment is to be amplified in accordance with ASCE/SEI 12.8.4.3 (see ASCE/SEI Fig. 12.8–1).

Other Considerations   In addition to the seismic force requirements outlined earlier, overturning, story drift, and P-delta effects also must be considered in the seismic design of any building or structure. These topics are covered in ASCE/SEI 12.8.5, 12.8.6, and 12.8.7, respectively, and applicable provisions must be satisfied in all cases where applicable, including low-rise buildings.

Diaphragms, Chords, and Collectors

General   According to IBC 1602.1, a *diaphragm* is a horizontal or sloped system that transfers lateral forces to the vertical-resisting elements of the lateral system. In general, a floor or a roof diaphragm acts as a beam that spans between vertical members of the lateral system and transfers lateral forces in the plane of the diaphragm to such members (see Fig. 2.11 for the case of collector beams and shear walls).

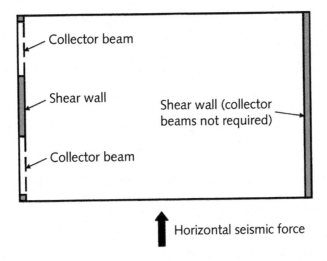

Horizontal seismic force

Figure 2.11 Collector beams and shear walls.

Diaphragms also transfer gravity loads that are perpendicular to the diaphragm surface to floor or roof members (i.e., beams, joists, and columns) and, when properly attached to the top surface of horizontal framing members, increase the flexural lateral stability of such members. Further discussion on horizontal load transfer is given in Sec. 2.4.

Diaphragm Design Forces Floor and roof diaphragms are to be designed to resist the design seismic forces from structural analysis of the building or the force F_{px} determined by ASCE/SEI Eq. (12.10–1):

$$0.4S_{DS}Iw_{px} \geq F_{px} = \frac{\sum\limits_{i=x}^{n} F_i}{\sum\limits_{i=1}^{n} w_i} w_{px} \geq 0.2S_{DS}Iw_{px} \tag{2.43}$$

where F_i is the design seismic force applied to level i, which is obtained from Eq. (2.41), w_i is the weight tributary to level i, and w_{px} is the weight tributary to the diaphragm at level x.

In cases where there are offsets in the vertical seismic force–resisting system or where there are changes in lateral stiffness of the vertical elements, the diaphragm must be designed for the force from Eq. (2.43) plus the force due to the offset or change in lateral stiffness. Figure 2.12 illustrates the case of an offset in the vertical components of the seismic force–resisting system.

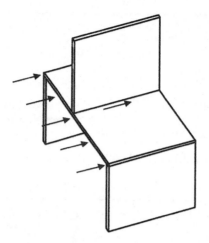

FIGURE 2.12   Offset of vertical components of the seismic force–resisting system.

Collector Elements   *Collectors,* which are also commonly referred to as *drag struts,* collect and transfer lateral forces from the diaphragm to the vertical elements of the seismic force–resisting system in cases where the seismic force–resisting system does not extend the full length of the building in the direction of analysis (see Fig. 2.11). Such elements also distribute forces within a diaphragm.

In structures assigned to SDC C and above, collectors, splices, and their connections to the resisting elements must be designed to resist the load combinations with an overstrength factor (see Sec. 2.3). The purpose of this provision is to ensure that inelastic energy dissipation occurs in the properly detailed elements of the seismic force–resisting system instead of in these crucial elements.

Structural Walls

Out-of-Plane Forces   In addition to forces in the plane of the wall, structural walls and their anchorage in structures assigned to SDC B and higher must be designed for an out-of-plane force equal to $0.4S_{DS}I$ times the weight of the wall or 10 percent of the weight of the wall, whichever is greater (ASCE/SEI 12.11.1). This force can control the design of walls, especially in low-rise buildings where the horizontal length of the wall is many times that of the height.

Anchorage of Concrete or Masonry Structural Walls   Concrete or masonry structural walls in structures assigned to SDC B and higher must be anchored to the roof and floor members that provide lateral support for the walls. The anchorage must be designed to resist the greater of the following forces (ASCE/SEI 12.11.2):

- The force prescribed in ASCE/SEI 12.11.1
- $400S_{DS}I$ pounds per linear foot of wall
- 280 pounds per linear foot of wall

Where the anchor spacing is greater than 4 ft, the walls must be designed to resist bending between the anchors.

The requirements of ASCE/SEI 12.11.2.1 must be satisfied for concrete or masonry walls anchored to flexible diaphragms. In particular, in structures assigned to SDC C and above, the minimum out-of-plane anchorage force is determined by ASCE/SEI (12.11–1):

$$F_p = 0.8 S_{DS} I W_p \qquad (2.44)$$

where W_p is the weight of the wall tributary to an anchor.

Additional requirements for diaphragms in structures assigned to SDC C through F are given in ASCE/SEI 12.11.2.2.

Simplified Alternative Structural Design Criteria for Simple Bearing-Wall or Building Frame Systems

General   The simplified design procedure in ASCE/SEI 12.14 can be used to determine seismic forces for relatively small, regular structures with bearing-wall or building frame systems that meet the conditions in ASCE/SEI 12.14.1.1. In essence, these conditions define a set of simple, redundant structures that can be analyzed using a lesser number of prescriptive requirements than the more comprehensive analytical procedures, including the ELFP, of ASCE/SEI 7.

ASCE/SEI 12.14 is a stand-alone section and contains requirements for seismic load effects and combinations, seismic force–resisting systems, diaphragm flexibility, application of loading, design and detailing, and application of a simplified lateral force analysis.

Limitations   ASCE/SEI 12.14.1.1 gives 12 limitations that must be satisfied for use of the simplified procedure. A summary of these limitations is given in Table 2.3. Even though there are 12 conditions to be met, it is important to note that the procedure is applicable to a wide range of relatively stiff, low-rise structures that fall under occupancy categories I and II and possess seismic force–resisting systems that are arranged in a torsionally resistant, regular layout.

Number	Limitation
1	Occupancy category I or II
2	Site classes A through D
3	Structure height equal to or less than three stories
4	Seismic force–resisting system must be bearing wall or building frame
5	Two lines of walls or frames are required in each of two major axis directions
6	At least one line of walls or frames is required on each side of the center of mass in each direction
7	ASCE/SEI Eq. (12.14–1) must be satisfied for structures with flexible diaphragms

8	ASCE/SEI Eqs. (12.14–2A) and (12.14–2B) must be satisfied for structures with diaphragms that are not flexible
9	Lines of walls or frames must be oriented at angles of no more than 15 degrees from the major orthogonal axes of the building
10	The simplified procedure must be used for each major orthogonal horizontal axis of the building
11	In-plane or out-of-plane structural irregularities are not permitted
12	Lateral load resistance of any story must be at least 80 percent of the story above

TABLE 2.3   Limitations of the Simplified Seismic Procedure of ASCE/SEI 12.14

Seismic Force–Resisting System   ASCE/SEI Table 12.14–1 reproduces portions of ASCE/SEI Table 12.2–1 for the systems permitted to be designed using the simplified procedure and provides references to sections of the standard pertaining to detailing requirements, the response modification coefficient R, and system limitations for SDCs B through E. The system overstrength factor Ω_o is taken as 2.5, so it is not included in the table (see ASCE/SEI 12.14.3). Also, because drift calculations are not required (see ASCE/SEI 12.14.8.5), the deflection amplification factor C_d is not included in the table.

Diaphragms   Untopped metal deck, wood structural panels, and similar panelized construction are permitted to be considered flexible diaphragms, and lateral load is distributed to the vertical elements of the seismic force–resisting system using the tributary area (ASCE/SEI 12.14.8.3.1). For diaphragms that are not flexible, a simple rigidity analysis is required, which includes torsional moments resulting from eccentricity between the locations of center mass and center of rigidity (ASCE/SEI 12.14.8.3.2). Analysis of accidental torsion and dynamic amplification of torsion are not required, representing another significant simplification of calculations.

Simplified Lateral Force Analysis Procedure

Seismic Base Shear   The seismic base shear V is determined by ASCE/SEI Eq. (12.14–11), which represents the horizontal short-period segment of the design response spectrum that is independent of the period of the structure (see Fig. 2.9):

$$V = \frac{FS_{DS}}{R}W \qquad (2.45)$$

where $S_{DS} = 2F_aS_S/3$. In lieu of determining the acceleration-based site coefficient F_a in accordance with ASCE/SEI 11.4.3, which requires knowledge of the soil profile to a depth of 100 ft below the surface of the site, F_a is permitted to be taken as 1.0 for rock sites and as 1.4 for soil sites. A site is considered to be a rock site where there is no more than 10 ft of soil between the rock surface and the bottom of the spread footing or mat foundation. Limiting applicability of the simplified design procedure to these sites results in the need for only a basic geotechnical investigation; 100-ft-deep borings and seismic shear velocity tests are not necessary.

The factor F is related to the number of stories in the building and is equal to 1.0 for one-story buildings, 1.1 for two-story buildings, and 1.2 for three-story buildings. It is evident that the value of the seismic base shear is increased by 10 and 20 percent for two- and three-story buildings, respectively. These increases primarily account for the method that is used for vertical distribution of the base shear, which is based on tributary weight, and have been shown by parametric studies to be adequate without being overly conservative (see below).

Vertical Distribution   The seismic force that is to be applied at each floor level x is determined by ASCE/SEI Eq. (12.14–12):

$$F_x = \frac{w_x}{W} V \qquad\qquad (2.46)$$

where w_x is equal to the portion of the effective seismic weight W (which is defined in ASCE/SEI 12.14.8.1 and is the same definition as in ASCE/SEI 12.7.2) at level x.

Drift   According to ASCE/SEI 12.14.8.5, the simplified design procedure does not require a drift check because it is assumed that bearing-wall and building frame systems within the prescribed height range will not require one (unlike moment-frame systems, where drift is a major concern in design). For requirements such as those for structural separations between buildings or the design of cladding, the allowable drift is to be taken as 1 percent of the building height.

Nonstructural Components

General   ASCE/SEI Chap. 13 contains minimum design criteria for nonstructural components that are permanently attached to structures and for their supports and attachments. Provisions are included for

- Architectural components
- Mechanical and electrical components
- Anchorage

Nonstructural components that weigh greater than 25 percent of the effective seismic weight of the structure W must be designed as nonbuilding structures in accordance with ASCE/SEI 15.3.2. Nonstructural components are assigned to the same SDC as they occupy or to which they are attached.

ASCE/SEI 13.1.4 contains a list of nonstructural components that are exempt from the requirements of this section, and ASCE/SEI Table 13.2–1 provides a summary of applicable requirements for architectural, mechanical, and electrical components.

The component importance factor I_p is equal to 1.0 except when the following conditions apply, where it is equal to 1.5:

- The component is required to function for life-safety purposes after an earthquake (including fire-protection sprinkler systems).
- The component contains hazardous materials.

- The component is in or attached to an occupancy category IV structure and is needed for continued operation of the facility, or its failure could impair the continued operation of the facility.

Seismic Demands on Nonstructural Components   ASCE/SEI Eqs. (13.3–1), (13.3–2), and (13.3–3) are used to calculate the horizontal seismic design force F_p that is to be applied at the center of gravity of a component:

$$1.6S_{DS}I_pW_p \geq F_p = \frac{0.4a_pS_{DS}W_p}{R_p/I_p}\left(1+\frac{2z}{h}\right) \geq 0.3S_{DS}I_pW_p \qquad (2.47)$$

where a_p is the component amplification factor, which varies between 1.00 and 2.50 (see ASCE/SEI Table 13.5–1 for architectural components and ASCE/SEI Table 13.6–1 for mechanical and electrical components), W_p is the component operating weight, R_p is the component response modification factor that varies from 1.00 to 12.0 (see ASCE/SEI Table 13.5–1 for architectural components and ASCE/SEI Table 13.6–1 for mechanical and electrical components), z is the height above the base of the structure where the component is attached to the structure, and h is the average roof height of the structure with respect to the base.

Component seismic forces F_p are to be applied independently in at least two orthogonal horizontal directions in combination with the service loads that are associated with the component. Requirements are also given in this section for vertically cantilevered systems.

ASCE/SEI Eq. (13.3–4) can be used to calculate F_p based on accelerations determined by a modal analysis. Requirements for seismic relative displacements are given in ASCE/SEI 13.3.2.

Nonstructural Component Anchorage   Components and their supports are to be attached or anchored to the structure in accordance with the requirements of ASCE/SEI 13.4. Forces in the attachments are to be determined using the prescribed forces and displacements given in ASCE/SEI 13.3.1 and 13.3.2. Anchors in concrete or masonry elements must satisfy the provisions of ASCE/SEI 13.4.2.

Architectural Components   General design and detailing requirements for architectural components are given in ASCE/SEI 13.5. All architectural components and attachments must be designed for the seismic forces defined in ASCE/SEI 13.3.1. Specific requirements are specified for

- Exterior nonstructural wall panels
- Glass
- Suspended ceilings
- Access floors
- Partitions
- Glass in glazed curtain walls, glazed storefronts, and glazed partitions

Mechanical and Electrical Components   The requirements of ASCE/SEI 13.6 are to be satisfied for mechanical and electrical components and their supports. ASCE/SEI Eq. (13.6–1) can be used to determine the fundamental period T_p of the mechanical or electrical component:

$$T_p = 2\pi\sqrt{\frac{W_p}{K_p g}}$$

(2.48)

where K_p is the stiffness of the resilient support system of the component and attachment, which is determined in terms of the load per unit deflection at the center of gravity of the component, and g is gravitational acceleration.

Requirements are provided for the following systems:

- Utility and service lines
- Heating, ventilation, and air-conditioning (HVAC) ductwork
- Piping systems
- Boilers and pressure vessels
- Elevators and escalators

Nonbuilding Structures

General   Chapter 15 of ASCE/SEI 7 contains the minimum seismic forces that must be resisted by nonbuilding structures supported by the ground or supported by other structures. The selection of a structural analysis procedure for a nonbuilding structure is based on its similarity to buildings. Nonbuilding structures that are similar to buildings exhibit behaviors similar to those of building structures; however, their function and performance are different. According to ASCE/SEI 15.1.3, structural analysis procedures for such buildings are to be selected in accordance with ASCE/SEI 12.6 and ASCE/SEI Table 12.6–1, which are applicable to building structures. Guidelines and recommendations on the use of these methods are given in ASCE/SEI C15.1.3. The provisions for building structures need to be examined carefully before they are applied to nonbuilding structures.

Nonbuilding structures that are not similar to buildings exhibit behaviors that are markedly different from those of building structures. Most of these types of structures have reference documents that address their unique structural performance and behaviors. Such reference documents are permitted to be used to analyze the structure (see ASCE/SEI 15.1.3). In addition, the following procedures may be used:

- Equivalent lateral force procedure (ASCE/SEI 12.8)
- Modal analysis procedure (ASCE/SEI 12.9)
- Linear response history analysis procedure (ASCE/SEI 16.1)
- Nonlinear response history analysis procedure (ASCE/SEI 16.2)

As in the case of nonbuilding structures similar to buildings, guidelines and recommendations on the proper analysis method to use for nonbuilding structures that are not similar to buildings are given in ASCE/SEI C15.1.3.

Reference Documents   Reference documents may be used to design nonbuilding structures for earthquake load effects. Documents that have seismic requirements based on the same force and displacement levels used in ASCE/SEI 7–05 are listed in ASCE/SEI Chap. 23 (ASCE/SEI 15.2). The provisions in the reference documents are subject to the amendments given in ASCE/SEI 15.4.1. Additional references that cannot be referenced directly by ASCE/SEI 7–05 are given in ASCE/SEI C15.2.

Provisions of an industry standard or a document must not be used unless the seismic ground accelerations and seismic coefficients are in conformance with the requirements of ASCE/SEI 11.4. The values for total lateral force and total base overturning moment from the reference documents must be taken to be greater than or equal to 80 percent of the corresponding values determined by the seismic provisions of ASCE/SEI 7–05.

Seismic-Load Effects

General   A building or a structure is analyzed for earthquake effects once the base shear has been determined and has been distributed vertically and horizontally throughout the structure. In general, both horizontal and vertical effects must be considered in design. The seismic-load effect E that is used in the load combinations discussed in the next section consists of the summation of the effect of horizontal seismic forces E_h and vertical seismic forces E_v. Both types of seismic effects are covered next.

Horizontal Seismic-Load Effect   The *horizontal seismic-load effect E_h* is determined by ASCE/SEI Eq. (12.4–3):

$$E_h = \rho Q_E \tag{2.49}$$

where Q_E is the effect of the horizontal seismic forces on the members due to the code-prescribed base shear, and ρ is the redundancy factor defined in ASCE/SEI 12.3.4, which, as the name implies, provides a relative measure of the redundancy inherent in the seismic force–resisting system. For structures assigned to SDC B or C, ρ can be taken equal to 1.0, and for structures assigned to SDC D, E, or F, it shall be set equal to 1.3 unless one of the conditions in ASCE/SEI 12.3.4.2 is met; if such conditions are met, ρ is permitted to be taken as 1.0. The redundancy factor is also permitted to be taken as 1.0 for structures designed by the simplified method (ASCE/SEI 12.14.3).

Vertical Seismic-Load Effect   The *vertical seismic-load effect E_v* is determined by ASCE/SEI Eq. (12.4–4):

$$E_v = 0.2 S_{DS} D \tag{2.50}$$

ASCE/SEI 12.4.2.2 contains two exceptions where it is permitted to take E_v to be equal to zero.

2.3   Load Combinations

2.3.1   Introduction

Structural members of buildings and other structures must be designed to resist the load combinations of IBC 1605.2, 1605.3.1, or 1605.3.2 (IBC 1605.1). Load combinations that are specified in Chaps. 18 through 23 of *2009* IBC, which contain provisions for soils and foundations, concrete, aluminum, masonry, steel, and wood, also must be considered. The structural elements identified in ASCE/SEI 12.2.5.2, 12.3.3.3, and 12.10.2.1 must be designed for the load combinations with overstrength factor of ASCE/ SEI 12.4.3.2.

The load combinations that are to be used when strength design or load and resistance factor design is used are contained in IBC 1605.2, whereas those using allowable stress design are given in IBC 1605.3. In addition to design for strength, the combinations of IBC 1605.2 or 1605.3 can be used to check the overall structural stability, including stability against overturning, sliding, and buoyancy (IBC 1605.1.1).

Load combinations must be investigated with one or more of the variable loads set equal to zero (IBC 1605.1). It is possible that the most critical load effects on a member occur when variable loads are not present.

ASCE/SEI 2.3 and 2.4 contain load combinations using strength design and allowable stress design, respectively. The load combinations are essentially the same as those in IBC 1605.2 and 1605.3 with some exceptions. Differences in the IBC and ASCE/ SEI 7 load combinations are covered in the following sections.

All the aforementioned load combinations and their applicability are examined below.

2.3.2   Load Combinations Using Strength Design or Load and Resistance Factor Design

Table 2.4 contains a summary of the basic load combinations where strength design or load and resistance factor design is used (IBC 1605.2). These combinations of factored loads establish the required strength that needs to be provided in the structural members of a building or a structure.

The constants f_1 and f_2 in the table are defined as follows:

> $f_1 = 1$ for floors in places of public assembly, for live loads in excess of 100 psf, and for parking garage live load
>
> $= 0.5$ for other live loads
>
> $f_2 = 0.7$ for roof configurations (such as sawtooth) that do not shed snow off the structure
>
> $= 0.2$ for other roof configurations

Factored load combinations that are specified in other provisions of the IBC take precedence to those listed in IBC 1605.2 (see the exception in this section).

The load combinations given in IBC 1605.2.1 are the same as those in ASCE/SEI 2.3.2 with the following exceptions:

- The variable f_1 that is present in IBC Eqs. (16–3), (16–4), and (16–5) is not found in ASCE/SEI combinations 3, 4, and 5. Instead, the load factor on the live load

Equation No.	Load Combination
16–1	$1.4(D+F)$
16–2	$1.2(D+F+T)+1.6(L+H)+0.5(L_r \text{ or } S \text{ or } R)$
16–3	$1.2D+1.6(L_r \text{ or } S \text{ or } R)+(f_1 L \text{ or } 0.8W)$
16–4	$1.2D+1.6W+f_1 L+0.5(L_r \text{ or } S \text{ or } R)$
16–5	$1.2D+1.0E+f_1 L+f_2 S$
16–6	$0.9D+1.6W+1.6H$
16–7	$0.9D+1.0E+1.6H$

TABLE 2.4 Summary of Load Combinations Using Strength Design or Load and Resistance Factor Design (IBC 1605.2.1)

L in the ASCE/SEI combinations is equal to 1.0, with the exception that the load factor on L is permitted to equal 0.5 for all occupancies where the live load is less than or equal to 100 psf, except for parking garages or areas occupied as places of public assembly (see exception 1 in ASCE/SEI 2.3.2). This exception makes these load combinations the same in ASCE/SEI 7 and the IBC.

- The variable f_2 that is present in IBC Eq. (16–5) is not found in ASCE/SEI combination 5. Instead, a load factor of 0.2 is applied to S in the ASCE/SEI combination. The third exception in ASCE/SEI 2.3.2 states that in combinations 2, 4, and 5, S shall be taken as either the flat-roof snow load p_f or the sloped-roof snow load p_s. This essentially means that the balanced snow load defined in ASCE/SEI 7.3 for flat roofs and ASCE/SEI 7.4 for sloped roofs can be used in combinations 2, 4, and 5. Drift loads and unbalanced snow loads are covered by combination 3.

The load combinations of ASCE/SEI 2.3.3 are to be used where flood loads F_a must be considered in design (IBC 1605.2.2). In particular, $1.6W$ in IBC Eqs. (16–4) and (16–6) shall be replaced by $1.6W + 2.0F_a$ in zone V or coastal A zones. In noncoastal A zones, $1.6W$ in IBC Eqs. (16–4) and (16–6) shall be replaced by $0.8W + 1.0F_a$.

2.3.3 Load Combinations Using Allowable-Stress Design

IBC 1605.3 contains the basic load combinations where allowable-stress design (i.e., working-stress design) is used. A set of basic load combinations is given in IBC 1605.3.1, and a set of alternative basic load combinations is given in IBC 1605.3.2. Both sets are examined below.

Basic Load Combinations

The basic load combinations of IBC 1605.3.1 are summarized in Table 2.5. Since the probability is low that two or more of the variable loads will reach their maximum

Equation No.	Load Combination
16–8	$D + F$
16–9	$D + H + F + L + T$
16–10	$D + H + F + (L_r \text{ or } S \text{ or } R)$
16–11	$D + H + F + 0.75(L + T) + 0.75(L_r \text{ or } S \text{ or } R)1$
16–12	$D + H + F + (W \text{ or } 0.7E)$
16–13	$D + H + F + 0.75(W \text{ or } 0.7E) + 0.75L + 0.75(L_r \text{ or } S \text{ or } R)$
16–14	$0.6D + W + H$
16–15	$0.6D + 0.7E + H$

TABLE 2.5 Summary of Basic Load Combinations Using Allowable-Stress Design (IBC 1605.3.1)

values at the same time, a factor of 0.75 is applied where these combinations include more than one variable load.

The combined effect of horizontal and vertical earthquake-induced forces E is a strength-level load. A factor of 0.7 (which is approximately equal to $1/1.4$) is applied to E in IBC Eqs. (16–12), (16–13), and (16–15) to convert the strength-level load to a service-level load.

Two exceptions are given in IBC 1605.3.1:

1. Crane hook loads need not be combined with roof live load or with more than three-fourths of the snow load or one-half of the wind load. In particular, the following load combinations must be investigated where crane live loads L_c are present:

   - IBC Eq. (16–11): $D + H + F + 0.75L + 0.75(L_r \text{ or } S \text{ or } R)$ and
     $D + H + F + 0.75(L + L_c) + 0.75(0.75S \text{ or } R)$
   - IBC Eq. (16–13) with E:
     $D + H + F + 0.75(0.7E) + 0.75L + 0.75(L_r \text{ or } S \text{ or } R)$ and
     $D + H + F + 0.75(0.7E) + 0.75(L + L_c) + 0.75(0.75S \text{ or } R)$
   - IBC Eq. (16–13) with W: $D + H + F + 0.75W + 0.75L + 0.75(L_r \text{ or } S \text{ or } R)$ and
     $D + H + F + 0.75(0.5W) + 0.75(L + L_c) + 0.75(0.75S \text{ or } R)$

2. Flat-roof snow loads p_f that are equal to or less than 30 psf and roof live loads that are equal to or less than 30 psf need not be combined with seismic loads. Also, where p_f is greater than 30 psf, 20 percent of the snow load must be combined with seismic loads.

According to IBC 1605.3.1.1, increases in allowable stresses that are given in the materials chapters of the IBC or in referenced standards are not permitted when the load combinations of IBC 1605.3.1 are used. However, it is permitted to use the duration of load factor when designing wood structures in accordance with Chap. 23 of 2009 IBC.

The load combinations of ASCE/SEI 2.4.2 are to be used where flood loads F_a must be considered in design (IBC 1605.3.1.2). In particular, $1.5F_a$ must be added to the other loads in IBC Eqs. (16–12), (16–13), and (16–14), and E is set equal to zero in IBC Eqs. (16–12) and (16–13) in V zones or coastal A zones. In noncoastal A zones, $0.75F_a$ must be added to the other loads in IBC Eqs. (16–12), (16–13), and (16–14), and E is set equal to zero in IBC Eqs. (16–12) and (16–13).

The load combinations of *IBC* 1605.3.1 and ASCE/SEI 2.4.1 are the same except for the following:

- There is no specific exception for crane loads in ASCE/SEI 2.4.1.

- The exception in ASCE/SEI 2.4.1 states that in combinations 4 and 6, S shall be taken as either the flat-roof snow load p_f or the sloped-roof snow load p_s. The balanced snow load defined in ASCE/SEI 7.3 for flat roofs and in ASCE/SEI 7.4 for sloped roofs can be used in combinations 4 and 6, and drift loads and unbalanced snow loads are covered by combination 3.

Alternative Basic Load Combinations

An alternative set of load combinations is provided in *IBC* 1605.3.2 for allowable-stress design, and the set is summarized in Table 2.6. These load combinations are based on the allowable-stress load combinations that appeared in the *Uniform Building Code* for many years. ASCE/SEI 7–05 does not contain provisions for the alternative basic load combinations of the IBC.

IBC 1605.3.2 states that for load combinations that include counteracting effects of dead and wind loads, only two-thirds of the minimum dead load that is likely to be in place during a design wind event is to be used in the load combination [the alternative allowable-stress design load combinations do not include a load combination

Equation No.	Load Combination
16–16	$D + L + (L_r \text{ or } S \text{ or } R)$
16–17	$D + L + \omega W$
16–18	$D + L + \omega W + S/2$
16–19	$D + L + S + \omega W/2$
16–20	$D + L + S + E/1.4$
16–21	$0.9D + E/1.4$

TABLE 2.6 Summary of Alternative Basic Load Combinations Using Allowable Stress Design (IBC 1605.3.2)

comparable with IBC Eq. (16–14) for dead load counteracting wind-load effects]. This is equivalent to a load combination of $0.67D + W$.

As noted previously, the combined effect of horizontal and vertical earthquake-induced forces E is a strength-level load. This strength-level load is divided by 1.4 in IBC Eqs. (16–20) and (16–21) to convert it to a service-level load.

ASCE/SEI 12.13.4 permits a reduction in foundation overturning due to earthquake forces, provided that the criteria of that section are satisfied. Such a reduction is not permitted when the alternative basic load combinations are used to evaluate sliding, overturning, and soil bearing at the soil-structure interface. Also, the vertical seismic-load effect E_v in ASCE/SEI Eq. (12.4–4) may be taken as zero when proportioning foundations using these load combinations.

The coefficient ω in IBC Eqs. (16–17), (16–18), and (16–19) is equal to 1.3 where wind loads are calculated in accordance with ASCE/SEI Chap. 6. As noted previously, the wind directionality factor, which is equal to 0.85 for building structures, is explicitly included in the velocity-pressure equation for wind. In earlier editions of ASCE/SEI 7 and in the legacy codes, the directionality factor was part of the load factor, which was equal to 1.3 for wind. Thus, for allowable-stress design, $\omega = 1.3 \times 0.85 \approx 1.0$, and for strength design, $\omega = 1.6 \times 0.85 \approx 1.3$.

Allowable stresses are permitted to be increased or load combinations are permitted to be reduced where permitted by the material chapters of the IBC (Chaps. 18 through 23) or by referenced standards when the alternative basic load combinations of IBC 1605.3.2 are used (as discussed previously, this type of reduction is not permitted when using the basic load combinations of IBC 1605.3.1). This applies to load combinations that include wind or earthquake loads.

The two exceptions in IBC 1605.3.2 for crane hook loads and for combinations of snow loads, roof live loads, and earthquake loads are the same as those in IBC 1605.3.1, which were discussed previously.

IBC 1605.3.2.1 requires that where F, H, or T must be considered in design, each applicable load is to be added to the load combinations in IBC Eqs. (16–16) through (16–21).

2.3.4   Load Combinations with Overstrength Factor

General
According to IBC 1605.1, item 3, the load combinations that are given in ASCE/SEI 12.4.3.2, must be used where required by ASCE 12.2.5.2, 12.3.3.3, or 12.10.2.1 instead of the corresponding load combinations in IBC 1605.2 and 1605.3. It is important to note that these load combinations are applicable to the design of the members noted in these sections (e.g., cantilevered column systems, structural members that support discontinuous frames or shear-wall systems, and collector elements, including splices and connections to resisting elements); other members need not be designed and proportioned based on these requirements.

Basic Combinations for Strength Design with Overstrength Factor
Where applicable, the following strength design load combinations are to be used instead of the corresponding ones in IBC 1605.2:

- IBC Eq. (16–5): $(1.2 + 0.2S_{DS})D + \Omega_o Q_E + L + 0.2S$
- IBC Eq. (16–7): $(0.9 - 0.2S_{DS})D + \Omega_o Q_E + 1.6H$

where Ω_o is the system overstrength factor, which is given in ASCE/SEI Table 12.2–1 based on the seismic force–resisting system. This factor essentially increases the design-level horizontal effects Q_E to represent the actual forces that may be experienced in a structural member as a result of the design ground motion.

Basic Combinations for Allowable-Stress Design with Overstrength Factor

Where applicable, the following allowable-stress design load combinations are to be used instead of the corresponding ones in IBC 1605.3.1:

- IBC Eq. (16–12): $(1.0 + 0.14S_{DS})D + H + F + 0.7\Omega_o Q_E$
- IBC Eq. (16–13): $(1.0 + 0.105S_{DS})D + H + F + 0.525\Omega_o Q_E + 0.75L + 0.75(L_r \text{ or } S \text{ or } R)$
- IBC Eq. (16–15): $(0.6 - 0.14S_{DS})D + 0.7\Omega_o Q_E + H$

ASCE/SEI 12.4.3.3 permits allowable stresses to be increased by a factor of 1.2 where allowable-stress design is used with seismic-load effect including overstrength factor. This increase is not to be combined with increases in allowable stresses or reductions in load combinations that are otherwise permitted in ASCE/SEI 7 or in other referenced materials standards. However, the duration-of-load factor is permitted to be used when designing wood members in accordance with the referenced standard.

Alternative Basic Load Combinations for Allowable-Stress Design with Overstrength Factor

Where applicable, the following allowable-stress design load combinations are to be used instead of the corresponding ones in IBC 1605.3.2:

- IBC Eq. (16–20): $\left(1.0 + \dfrac{0.2S_{DS}}{1.4}\right)D + \dfrac{\Omega_o Q_E}{1.4} + L + S$

- IBC Eq. (16–21): $\left(0.9 - \dfrac{0.2S_{DS}}{1.4}\right)D + \dfrac{\Omega_o Q_E}{1.4}$

2.3.5 Load Combinations for Extraordinary Events

The strength and stability of a structure must be checked to ensure that it can withstand the effects from extraordinary events (i.e., low-probability events) such as fires, explosions, and vehicular impact (ASCE/SEI 2.5). Information on these types of events and recommended load combinations can be found in ASCE/SEI C2.5. Reference is made in that commentary section to ASCE/SEI 1.4 and C1.4, which address general structural integrity requirements. Included is a discussion on resistance to progressive collapse.

Structural-integrity provisions are contained in IBC 1614 for buildings classified as high-rise buildings per IBC 403 and assigned to occupancy category III or IV with bearing-wall structures or frame structures (a high-rise building is defined in IBC 202 as a building with an occupied floor located more than 75 ft above the lowest level of fire department vehicle access, and occupancy categories III and IV are defined in IBC Table 1604.5). The purpose of these prescriptive requirements is to improve the redundancy and ductility of these types of framing systems in the event of damage due to an abnormal loading event; specific load combinations are not included.

2.4 Load Paths

2.4.1 Introduction

In order to ensure the strength and stability of a building as a whole and all its members, it is important to understand the paths that loads take through a structure. A continuous load path is required from the top of a building or a structure to the foundation and into the ground. Structural systems must be capable of providing a clearly defined and uninterrupted load path for the effects due to both gravity and lateral loads. The general requirements for providing a complete load path capable of transferring all loads from their point of origin to the resisting elements are found in IBC 1604.4 and ASCE/SEI 12.1.3.

In general, the path that loads take through a structure includes the structural members, the connections between the members, and the interface between the foundation and the soil. Each element in the path must be designed and detailed to resist the applicable combinations of gravity and lateral-load effects (see, for example, National Council of Structural Engineers Associations[5] and Mays[6]). The detailing required to ensure a complete load path depends to some extent on the material and type of structural system involved. For example, the load path in wood-frame structures consists of many individual elements; as such, considerable detailing is often required to ensure that lateral forces are transferred from the out-of-plane walls, through the diaphragm and collectors, and into the supporting shear walls and foundation. Additionally, the structure as a whole must be capable of resisting the effects of overturning and sliding.

The load paths for gravity loads are usually very easy to identify, especially in low-rise buildings. Vertical elements, such as walls and columns, or sometimes bracing elements (that are not vertical) carry the gravity loads down to the foundation, which transfer them to the soil. The following sections cover typical load paths for both wind and earthquake loads.

2.4.2 Wind Loads

General

Figure 2.13 illustrates the continuous load path that carries wind loads through a low-rise building. The windward wall, which is supported laterally at the roof by a diaphragm, receives the wind pressures and transfers the resulting force to the diaphragm. The wind loads then are transferred from the diaphragm to the elements of the MWFRS that are parallel to the direction of wind, which in this case are walls. The walls, in turn, transfer the wind loads to the spread-footing foundation, which is supported by the soil beneath the footing.

The same sequence of events would occur if moment frames or any other type of lateral force–resisting system were used instead of or in conjunction with the walls or if floor diaphragms also were present.

The windward wall (and/or windows and doors) that initially receives the wind loads generally is perpendicular to the direction of wind. This wall is classified as C&C for wind in this direction, and the wind pressure on it is determined differently from that on the MWFRS, as shown in Sec. 2.2. For wind in the perpendicular direction, this wall would be part of the MWFRS of the building.

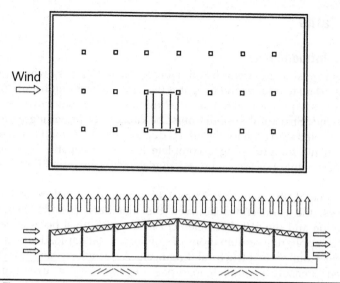

Figure 2.13   The propagation of wind loads in a low-rise building.

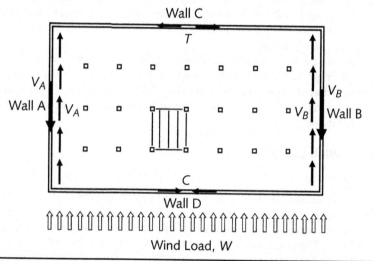

Figure 2.14   Wind-load propagation from a diaphragm to the MWFRS.

Diaphragms

Wind loads from the windward wall are transferred to diaphragms, which essentially act as beams that span between the walls that are oriented parallel to the direction of the wind. Typical diaphragms include wood sheathing, corrugated-metal deck, concrete fill over corrugated-metal deck, concrete topping slab over precast planks, and cast-in-place concrete slab. Figure 2.14 illustrates wind-load propagation at the roof level of the example building in Fig. 2.13 for wind loads in the perpendicular direction.

Wall D is the windward wall, which receives the wind load W. The roof diaphragm, which behaves essentially as a deep flexural member that spans between walls A and B that are oriented parallel to the direction of the wind, laterally supports wall D (and wall C) and receives the wind load from this wall along its windward edge. The web element of the beam transfers the wind loads (shear forces v_A and v_B) to the supports at walls A and B. The sum of the forces V_A and V_B at walls A and B, respectively, is equal to the total wind load W at this level. Chords that are perpendicular to the wind loads at the edges of the diaphragm behave as flange elements that must be designed to resist the axial tension and compression forces T and C, respectively, resulting from flexural behavior of the diaphragm.

The manner in which the wind loads from the diaphragm are transferred to the elements of the MWFRS depends on the flexibility of the diaphragm. Diaphragms usually can be idealized as either flexible or rigid. According to ASCE/SEI 12.3.1.1, diaphragms that are constructed of untopped steel decking or wood structural panels are permitted to be idealized as flexible diaphragms if the vertical elements of the MWFRS consist of (1) steel or composite steel and concrete-braced frames or (2) concrete, masonry, steel, or composite shear walls (i.e., relatively stiff elements). Diaphragms constructed of concrete slabs or concrete-filled metal deck with span-to-depth ratios of 3 or less in structures that have no horizontal irregularities are permitted to be idealized as rigid (ASCE/SEI 12.3.1.2). Note that the provisions given in ASCE/SEI 12.3 are applicable to diaphragms subjected to wind loads, even though they are written explicitly for diaphragms subjected to seismic loads.

ASCE/SEI 12.3.1.3 can be used to determine the flexibility of a diaphragm in cases where the idealized conditions are not met or are not evident. Requirements are based on the maximum diaphragm deflection and the average drift of the vertical elements of the MWFRS. If the MWFRS consists of a relatively flexible system, such as a moment frame, the diaphragm may behave as a rigid element, whereas if the MWFRS consists of stiffer elements, such as concrete or masonry walls, the diaphragm may behave as a flexible element. In particular, a diaphragm may be considered flexible where the maximum in-plane deflection of the diaphragm under lateral load is more than two times the average drift of the adjoining vertical elements of the MWFRS.

It is commonly assumed that flexible diaphragms transfer the wind loads to the elements of the MWFRS on a tributary-area basis without any participation of the walls that are perpendicular to the direction of analysis. Thus, for the building in Fig. 2.14, walls A and B would each carry one-half the total wind load W at the roof level. In the case of rigid diaphragms, wind loads are transferred based on the relative stiffness of the members in the MWFRS considering all such elements at that level in both directions: The stiffer the element, the greater the load it must resist. Additionally, the effects of torsion must be considered in rigid diaphragms; torsion occurs where the center of rigidity of the elements of the MWFRS and the centroid of the wind pressure are at distinct locations. In any situation, a finite-element model of the system can be used to determine the forces in the diaphragm and in the elements of the MWFRS.

Openings in a diaphragm can have a dramatic effect on the flow of forces through the diaphragm to the MWFRS. Typical elevator and stair openings in a diaphragm usually will not have a significant impact on the overall behavior and load distribution. Larger openings, especially those which can be defined as a horizontal diaphragm discontinuity irregularity in accordance with ASCE/SEI 12.3.2.1, can have a significant influence on overall behavior, and a more refined analysis generally is warranted to ensure that a complete load path is achieved around the opening. Regardless of the opening size, secondary chord forces will occur due to local bending of the diaphragm segments on either side of the opening; these chord forces need to be accounted for in the design.

To ensure a continuous load path, the connection between the diaphragm and the elements of the MWFRS must be designed and detailed properly. Figure 2.15 illustrates the typical connection of the diaphragm to the masonry wall of the building depicted in Fig. 2.13. The load path is as follows:

1. The shear force from the diaphragm is transferred to a continuous angle via a welded connection between these elements.

2. The angle transfers the load to the top chord of the joists, which is welded to the base plate that transfers the horizontal and vertical loads to the bond beam through the headed shear studs.

3. Vertical reinforcement in the wall transfers the loads through the load-bearing wall, which is the MWFRS in this direction.

4. At the base of the wall, the vertical reinforcement is spliced to reinforcement in the reinforced-concrete footing that transfers the loads to the ground.

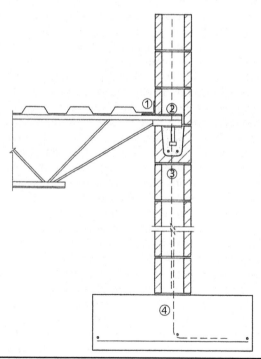

Figure 2.15 Load path from roof to foundation.

Collectors

In situations where the MWFRS does not extend the full length of the building in the direction of analysis, collectors must be used (see Fig. 2.11). The main purpose of a collector is to collect the shear force from the diaphragm over the length where there are no elements of the MWFRS and to transfer it to the MWFRS.

Collector forces are obtained from equilibrium. Figure 2.16 illustrates the forces along the line of the wall and collector beams in Fig. 2.11. Assuming that the calculated shear force in the wall is equal to V_w, the unit shear force per length in the diaphragm is equal to $V_w/(\ell_1 + \ell_2 + \ell_3)$, and the unit shear force per length in the wall is equal to V_w/ℓ_2. The net shear forces per unit length are determined by subtracting the unit shear force in the diaphragm from that in the wall. The force at any point along the length of a collector is equal to the net shear force in that segment times the length to that point. For example, at point B, the force in the collector is equal to $v_{AB,net}\ell_1$, and at point C, it is $v_{CD,net}\ell_3$ or, equivalently, $v_{BC,net}\ell_2 - v_{AB,net}\ell_1$.

The collector beams must be designed for the appropriate gravity and wind-load combinations where the axial force in the collector due to wind-load effects is tension or compression. Connections between the collector beams and elements of the MWFRS also must be designed and detailed properly to ensure a continuous load path between these members.

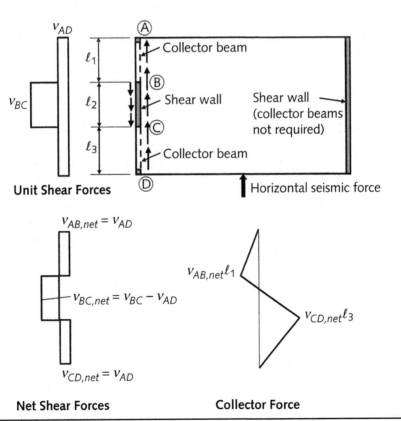

Figure 2.16 Unit shear forces, net shear forces, and collector force diagrams.

2.4.3   Seismic Loads

Unlike wind loads, the magnitude of which are proportional to the surface area of a building, seismic loads are generated by the inertia of the building mass that counteracts ground motion. Although the loads are generated differently, the propagation of the loads is assumed to be essentially the same under particular conditions.

It was shown in Sec. 2.2 that for certain types of buildings, including regular, low-rise buildings, the effects of complex seismic ground motion can be represented adequately by applying a set of static forces over the height of the building. These seismic loads are distributed to the elements of the seismic force–resisting system in essentially the same way wind loads are distributed by considering the relative stiffness of the diaphragm.

Section 2.2 also contains information on how to calculate diaphragm design forces at roof and floor levels (see ASCE/SEI 12.10). Special design and detailing requirements must be satisfied for collector elements and their connections in buildings assigned to SDC C and above; similar load combinations are not applicable in the case of wind loads.

Similar to when wind loads are applied to buildings with rigid diaphragms, torsional effects must be considered when distributing seismic loads. In particular, torsion occurs where the center of mass and the center of rigidity of the elements of the seismic force–resisting system are at different locations in plan at a particular floor level (see ASCE/SEI 12.8.4.1). Additionally, accidental torsion must be considered: The center of mass must be displaced each way from its actual location a distance equal to 5 percent of the dimension of the structure perpendicular to the direction of the applied load.

2.5   Examples

Example 2.1: Live-Load Reduction

Determine the reduced live load on a typical interior column for the building depicted in Fig. 2.17. Assume that the roof is an ordinary flat roof that is not used as a place of public assembly or for any special purposes. Neglect rain and snow loads.

Solution

Since the roof is an ordinary flat roof that is not used as a place of public assembly or for any special purposes, the nominal roof live load is 20 psf in accordance with IBC Table 1607.1.

The reduced live load is determined by IBC Eq. (16–25):

$$L_r = L_o R_1 R_2$$

The tributary area A_t of an interior column = 35 × 32 = 1,120 ft².

Since $A_t > 600$ ft², R_1 is determined by IBC Eq. (16–28): $R_1 = 0.6$. Since the quantity $F = 1/2 < 4$, $R_2 = 1$ (IBC Eq. 16–29). Thus

$$L_r = 20 \times 0.6 \times 1 = 12.0 \text{ psf}$$

Axial live load on the column $= 12.0 \times 1120 / 1000 = 13.4$ kips

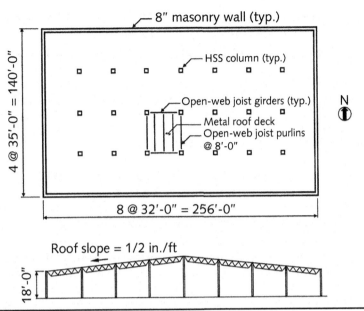

FIGURE 2.17   Plan and elevation of example building.

Example 2.2: Snow Loads

Determine the design snow loads for the one-story building illustrated in Fig. 2.17. Assume the following design data:

- Ground snow load p_g: 40 psf.
- Terrain category C (open terrain with scattered obstructions less than 30 ft in height).
- Occupancy: Warehouse use. Fewer than 300 people congregate in one area, and the building is not used to store hazardous or toxic material.
- Thermal condition: Structure is kept just above freezing.
- Roof exposure condition: Partially exposed.
- Roof surface: Rubber membrane.
- Roof framing: All members are simply supported.

Solution

- Determine the flat-roof snow load p_f by ASCE/SEI Eq. (7–1).

1. Determine exposure factor C_e from ASCE/SEI Table 7–2. Since the terrain category is C and the roof exposure is partially exposed, $C_e = 1.0$.

2. Determine the thermal factor C_t from ASCE/SEI Table 7–3. Since the structure is kept just above freezing, $C_t = 1.1$.

3. Determine the importance factor I from ASCE/SEI Table 7–4. From IBC Table 1604.5, the occupancy category is II. Thus $I = 1.0$.

4. Therefore, $p_f = 0.7C_eC_tIp_g = 0.7 \times 1.0 \times 1.1 \times 1.0 \times 40 = 30.8$ psf.

Check whether the minimum snow-load requirements are applicable: Minimum values of p_f apply in accordance with ASCE/SEI 7.3 to hip and gable roofs with slopes of less than the larger of 2.38 degrees (1/2 on 12 governs in this case) and $(70/W) + 0.5 = (70/128) + 0.5 = 1.05$ degrees. Since the roof slope in this example is equal to 2.38 degrees, minimum roof snow loads do not apply.

- Determine the sloped-roof snow load p_s by ASCE/SEI Eq. (7–1).

1. Determine whether the roof is warm or cold. Since $C_t = 1.1$, the roof is defined as a cold roof in accordance with ASCE/SEI 7.4.2.

2. Determine whether the roof is unobstructed or not and whether the roof is slippery or not. The roof is unobstructed because there are no obstructions (such as large vent pipes, large mechanical equipment, or snow guards) on the roof that prohibit the snow from sliding off the eaves. The roof is slippery because the surface is a rubber membrane (see ASCE/SEI 7.4).

Use the dashed line in ASCE/SEI Fig. 7–2b to determine C_s. For a roof slope of 2.38 degrees, $C_s = 1.0$. Therefore, $p_s = C_s p_f = 30.8$ psf.

- Consider partial loading. Since all the roof members are simply supported, partial loading is not considered (ASCE/SEI 7.5).

- Consider unbalanced snow loads. Unbalanced snow loads must be considered for this roof because the slope is equal to or greater than the larger of $(70/W) + 0.5 = (70/128) + 0.5 = 1.05$ degrees and 2.38 degrees (governs in this case). Since $W = 128$ ft > 20 ft, the unbalanced load consists of the following (see ASCE/SEI Fig. 7–5):

1. Windward side: $0.3p_s = 9.2$ psf.

2. Leeward side: $p_s = 30.8$ psf along the entire leeward length plus a uniform pressure of $h_d\gamma/\sqrt{S} = (4.3 \times 19.2)/\sqrt{24} = 16.9$ psf that extends from the ridge a distance of $8h_d\sqrt{S}/3 = 56.2$ ft.

   a. Where h_d = drift length from ASCE/SEI Fig. 7–9 with $W = 128$ ft substituted for $\ell_u = 0.43(W)^{1/3}(p_g + 10)^{1/4} - 1.5 = 4.3$ ft.

   b. γ = snow density (see ASCE/SEI Eq. 7–3) $= 0.13p_g + 14 = 19.2$ pcf < 30 pcf.

   c. S = roof slope for a rise of 1 = 24.

- Consider rain-on-snow loads. A rain-on-snow surcharge of 5 psf is required for locations where p_g is 20 psf or less (but not zero) with roof slopes of less than $W/50$ (ASCE/SEI 7.10). In this example, since $p_g = 40$ psf, a rain-on-snow surcharge is not required.

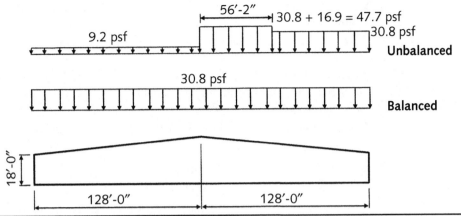

Figure 2.18   Balanced and unbalanced snow loads for example warehouse building.

- Consider ponding instability. Since the roof slope in this example is greater than 1.4 in./ft, progressive roof deflection and ponding instability need not be investigated (see ASCE/SEI 7.11 and 8.4).

The balanced and unbalanced snow loads are depicted in Fig. 2.18.

Example 2.3: Wind Loads, Simplified Procedure

For the one-story building illustrated in Fig. 2.17, determine design wind pressures on (1) the MWFRS, (2) a masonry wall, and (3) an open-web joist using the simplified procedure (method 1).

Assume the following design data:

- Basic wind speed: $V = 90$ mph.
- Surface roughness: C (open terrain with scattered obstructions less than 30 ft in height).
- Topography: Not situated on a hill or an escarpment.
- Occupancy: Fewer than 300 people congregate in one area, and the building is not used to store hazardous or toxic material.
- Enclosure: Enclosed.

Solution

Part 1: Design Wind Pressures on the MWFRS

- Check whether the building meets all the conditions of ASCE/SEI 6.4.1.1 so that method 1 can be used to determine the wind pressures on the MWFRS:

1. The building is a simple diaphragm building as defined in ASCE/SEI 6.2 because the windward and leeward wind loads are transmitted through the

metal deck roof (diaphragm) to the masonry walls (MWFRS), and there are no structural separations in the MWFRS.

2. Three conditions must be checked to determine whether a building is a low-rise building:

    a. Mean roof height = 18 ft < 60 ft (for buildings with roof angles of less than 10 degrees, the mean roof height is equal to the roof eave height per the definition given in ASCE/SEI 6.2).

    b. Mean roof height = 18 ft < least horizontal dimension = 140 ft.

    c. The enclosure classification of the building is assumed to be enclosed.

    Thus the building is a low-rise building.

3. It is assumed that the building is enclosed.

4. The building is regularly shaped; that is, it does not have any unusual geometric irregularities in spatial form.

5. A flexible building is defined in ASCE/SEI 6.2 as one in which the fundamental natural frequency of the building n_1 is less than 1 Hz. Although it is evident by inspection that the building is not flexible, the natural frequency will be determined and compared with 1 Hz.

    In lieu of obtaining the natural frequency of the building from a dynamic analysis, ASCE/SEI Eq. (C6–18) is used to determine an approximate lower-bound value of n_1:

$$n_1 = \frac{75}{H} = \frac{75}{18} = 4.2 \text{ Hz} > 1 \text{ Hz}$$

    Thus the building is not flexible.

6. The building does not have response characteristics that make it subject to across-wind loading or other similar effects, and it is not sited at a location where channeling effects or buffeting in the wake of upwind obstructions needs to be considered.

7. The building has a symmetric cross section in each direction and has a relatively flat roof.

8. The building is exempted from torsional load cases, as indicated in note 5 of ASCE/SEI Fig. 6–10 (the building is one story high with a height h of less than 30 ft, and it has a flexible roof diaphragm).

Since all the conditions of ASCE/SEI 6.4.1.1 are satisfied, method 1 may be used to determine the design wind pressures on the MWFRS.

- Determine the net design wind pressures p_s on the MWFRS:

1. Determine importance factor I from ASCE/SEI Table 6–1 based on the occupancy category from IBC Table 1604.5. From IBC Table 1604.5, the occupancy category is II based on the occupancy given in the design data. From ASCE/SEI Table 6–1, $I = 1.0$.

2. Determine the exposure category. In the design data, the surface roughness is given as C. It is assumed that exposures B and D are not applicable, so exposure C applies (see ASCE/SEI 6.5.6.3).

3. Determine the adjustment factor for height and exposure λ from ASCE/SEI Fig. 6–2. For a mean roof height of 18 ft and exposure C, λ = 1.26 from linear interpolation.

4. Determine the topographic factor K_{zt}. As noted in the design data, the building is not situated on a hill, ridge, or escarpment. Thus topographic factor $K_{zt} = 1.0$ (ASCE/SEI 6.5.7.2).

5. Determine simplified design wind pressures p_{s30} from ASCE/SEI Fig. 6–2 for zones A through H on the building. Wind pressures p_{s30} can be read directly from ASCE/SEI Fig. 6–2 for V = 90 mph and a roof angle of between 0 and 5 degrees. Since the roof is essentially flat, only load case 1 is considered (see note 4 in ASCE/SEI Fig. 6–2). These pressures, which are based on exposure B, h = 30 ft, K_{zt} = 1.0, and I = 1.0, are given in Table 2.7.

6. Determine net design wind pressures $p_s = \lambda K_{zt} I p_{s30}$ by ASCE/SEI Eq. (6–1) for zones A through H:

$$p_s = 1.26 \times 1.0 \times 1.0 \times p_{s30} = 1.26 p_{s30}$$

The horizontal pressures in Table 2.8 represent the combination of the windward and leeward net (sum of internal and external) pressures. Similarly, the vertical pressures represent the net (sum of internal and external) pressures.

7. The net design pressures p_s in Table 2.8 are to be applied to the surfaces of the building in accordance with ASCE/SEI Fig. 6–2. According to note 7 in ASCE/SEI Fig. 6–2, the total horizontal load must not be less than that determined by assuming p_s = 0 in zones B and D. Since the net pressures in zones B and D in this example act in the direction opposite to those in A and C, they decrease the horizontal load. Thus the pressures in zones B and D are set equal to 0 when analyzing the structure for wind in the transverse direction.

According to note 2 in ASCE/SEI Fig. 6–2, the load patterns for the transverse and longitudinal directions are to be applied to each corner of the building; that is, each

Horizontal Pressures (psf)				Vertical Pressures (psf)			
A	B	C	D	E	F	G	H
12.8	−6.7	8.5	−4.0	−15.4	−8.8	−10.7	−6.8

TABLE 2.7 Wind Pressures p_{s30} on MWFRS

Horizontal Pressures (psf)				Vertical Pressures (psf)			
A	B	C	D	E	F	G	H
16.1	−8.4	10.7	−5.0	−19.4	−11.1	−13.5	−8.6

TABLE 2.8 Wind Pressures p_s on MWFRS

corner of the building must be considered a reference corner. Eight different load cases need to be examined (four in the transverse direction and four in the longitudinal direction). One load pattern in the transverse direction and one in the longitudinal direction are illustrated in Fig. 2.19.

The width of the end zone $2a$ in this example is equal to 14.4 ft, where $a =$ least of 0.1(least horizontal dimension) $= 0.1 \times 140 = 14.0$ ft or $0.4h = 0.4 \times 18 = 7.2$ ft (governs). This value of a is greater than 0.04 (least horizontal dimension) $= 0.04 \times 140 = 5.6$ ft or 3 ft (see note 10a in ASCE/SEI Fig. 6–2).

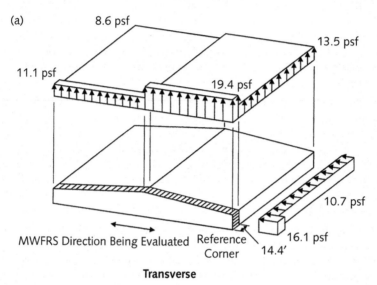

Transverse

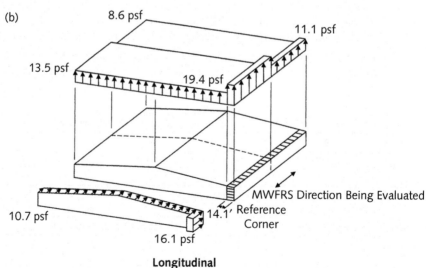

Longitudinal

Figure 2.19 Design wind pressures on MWFRS: (a) transverse direction; (b) longitudinal direction.

The minimum design wind-load case of ASCE/SEI 6.4.2.1.1 also must be considered: The load effects from the design wind pressures calculated earlier must not be less than the load effects assuming that $p_s = +10$ psf in zones A through D and $p_s = 0$ psf in zones E through H (see ASCE/SEI Fig. C6–1 for application of load).

Part 2: Design Wind Pressures on a Masonry Wall

- Determine the net design wind pressures p_{net} on the C&C.

1. Determine net design wind pressures p_{net30} from ASCE/SEI Fig. 6–3 for zones 4 and 5, which are the interior and end zones of walls, respectively. Wind pressures p_{net30} can be read directly from ASCE/SEI Fig. 6–3 for $V = 90$ mph and an effective wind area.

   The effective wind area is defined as the span length multiplied by an effective width that need not be less than one-third the span length: $18 \times (18/3) = 108.0$ ft². Note that the smallest span length corresponding to the east and west walls is used because this results in larger pressures.

   According to note 4 in ASCE/SEI Fig. 6–3, tabulated pressures may be interpolated for effective wind areas between those given, or the value associated with the lower effective wind area may be used. The latter of these two options is used in this example. The pressures p_{net30} in Table 2.9 are obtained from ASCE/SEI Fig. 6–3 for $V = 90$ mph and an effective wind area of 100 ft² and are based on exposure B, $h = 30$ feet, $K_{zt} = 1.0$, and $I = 1.0$.

2. Determine net design wind pressures $p_{net} = \lambda K_{zt} I p_{net30}$ by ASCE/SEI Eq. (6–2) for zones 4 and 5:

$$p_{net} = 1.26 \times 1.0 \times 1.0 \times p_{net30} = 1.26 p_{net30}$$

The pressures in Table 2.10 represent the net (sum of internal and external) pressures that are applied normal to the masonry walls.

In zones 4 and 5, the computed positive and negative (absolute) pressures are greater than the minimum values prescribed in ASCE/SEI 6.4.2.2.1 of +10 psf and −10 psf, respectively.

Zone	p_{net30} (psf)	
4	12.4	−13.6
5	12.4	−15.1

TABLE 2.9 Wind Pressures p_{net30} on Masonry Walls

Zone	p_{net} (psf)	
4	15.6	−17.1
5	15.6	−19.0

TABLE 2.10 Wind Pressures p_{net} on Masonry Walls

Part 3: Design Wind Pressures on an Open-Web Joist Purlin

- Determine the net design wind pressures p_{net} on the C&C.

1. Determine net design wind pressures p_{net30} from ASCE/SEI Fig. 6–3 for zones 1, 2, and 3, which are the interior, end, and corner zones of the roof, respectively. The effective wind area is equal to the larger of the purlin tributary area = 35 × 8 = 280.0 ft² or the span length multiplied by an effective width that need not be less than one-third the span length = 35 × (35/3) = 408.3 ft² (governs). The pressures p_{net30} in Table 2.11 are obtained from ASCE/SEI Fig. 6–3 for V = 90 mph, a roof angle of between 0 and 7 degrees, and an effective wind area of 100 ft² (where actual effective wind areas are greater than 100 ft², the tabulated pressure values associated with an effective wind area of 100 ft² are applicable). These pressures are based on exposure B, h = 30 ft, K_{zt} = 1.0, and I = 1.0.

2. Determine net design wind pressures $p_{net} = \lambda K_{zt} I p_{net30}$ by ASCE/SEI Eq. (6–2) for zones 1, 2, and 3:

$$p_{net} = 1.26 \times 1.0 \times 1.0 \times p_{net30} = 1.26 p_{net30}$$

The pressures in Table 2.12 represent the net (sum of internal and external) pressures that are applied normal to the open-web joist purlins and that act over the tributary area of each purlin, which is equal to 35 × 8 = 280.0 ft².

The positive net design pressures in zones 1, 2, and 3 must be increased to the minimum value of 10 psf in accordance with ASCE/SEI 6.4.2.2.1. Figure 2.20 contains the loading diagrams for typical open-web joist purlins located within the various zones of the roof.

Zone	p_{net30} (psf)	
1	4.7	−13.3
2	4.7	−15.8
3	4.7	−15.8

TABLE 2.11   Wind Pressures p_{net30} on Open-Web Joist Purlins

Zone	p_{net} (psf)	
1	5.9	−16.8
2	5.9	−19.9
3	5.9	−19.9

TABLE 2.12   Wind Pressures p_{net} on Open-Web Joist Purlins

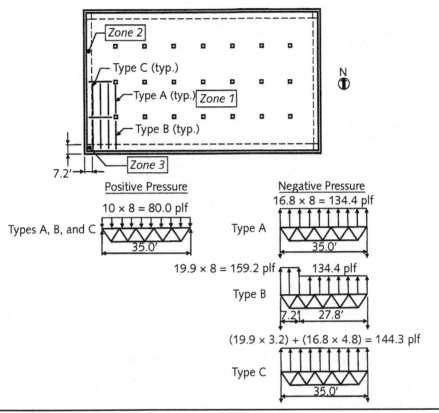

FIGURE 2.20   Open-web joist purlin loading diagrams.

Example 2.4: Wind Loads, Low-Rise Building Provisions of Method 2

For the one-story building illustrated in Fig. 2.17, determine design wind pressures on (1) the MWFRS, (2) a masonry wall, and (3) an open-web joist using the low-rise building provisions of method 2. Use the same design data as Example 2.3.

Solution

Part 1: Design Wind Pressures on the MWFRS

- *Check whether the low-rise building provisions of ASCE/SEI 6.5.12.2.2 can be used to determine the design wind pressures on the MWFRS.*

The provisions of ASCE/SEI 6.5.12.2.2 may be used to determine design wind pressures provided that the building is a regularly shaped low-rise building as defined in ASCE/SEI 6.2. It was shown in Example 2.3 that this warehouse building is a low-rise building, and the building is regularly shaped. Also, the building does not have response characteristics that make it subject to across-wind loading or other similar effects, and it is not sited at a location where channeling effects or buffeting in the wake of upwind obstructions need to be considered. Therefore, the low-rise building provisions of ASCE/SEI 6.5.12.2.2 can be used to determine the design wind pressures on the MWFRS.

- Determine the velocity pressure q_h.

1. Determine the wind directionality factor K_d. From ASCE/SEI Table 6–4, $K_d =$ 0.85 for the MWFRS of a building structure.
2. Determine the velocity exposure coefficient K_h. For exposure C and a mean roof height of 18 ft, $K_h = 0.88$ from linear interpolation of the values given in ASCE/ SEI Table 6–3.
3. Determine the velocity pressure q_h at the mean roof height by ASCE/SEI Eq. (6–15):

$$q_h = 0.00256 K_h K_{zt} K_d V^2 I$$
$$= 0.00256 \times 0.88 \times 1.0 \times 0.85 \times 90^2 \times 1.0 = 15.5 \text{ psf}$$

- Determine the external pressure coefficients GC_{pf} for zones 1 through 6, 1E, 2E, 3E, and 4E.

External pressure coefficients GC_{pf} can be read directly from ASCE/SEI Fig. 6–10 using a roof angle of between 0 and 5 degrees for wind in the transverse direction. For wind in the longitudinal direction, the pressure coefficients corresponding to a roof angle of 0 degrees are to be used (see note 7 in ASCE/SEI Fig. 6–10). The pressure coefficients summarized in Table 2.13 are applicable in both the transverse and longitudinal directions in this example.

- Determine the internal pressure coefficients GC_{pi}.

Zone	GC_{pf}
1	0.40
2	−0.69
3	−0.37
4	−0.29
5	−0.45
6	−0.45
1E	0.61
2E	−1.07
3E	−0.53
4E	−0.43

TABLE 2.13   External Pressure Coefficients GC_{pf} for MWFRS

For an enclosed building, $GC_{pi} = +0.18, -0.18$ from ASCE/SEI Fig. 6–5.

- Determine the design wind pressure p by ASCE/SEI Eq. (6–18) on zones 1 through 6, 1E, 2E, 3E, and 4E.

$$p = q_h(GC_{pf} - GC_{pi}) = 15.5[GC_{pf} - (\pm 0.18)]$$

Calculation of design wind pressures is illustrated for zone 1: For positive internal pressure, $p = 15.5(0.40 - 0.18) = 3.4$ psf. For negative internal pressure, $p = 15.5[0.40 - (-0.18)] = 9.0$ psf.

A summary of the design wind pressures is given in Table 2.14. Pressures are applicable to wind in the transverse and longitudinal directions and are provided for both positive and negative internal pressures. These pressures act normal to the surface.

According to note 8 in ASCE/SEI Fig. 6–10, when the roof pressure coefficients GC_{pf} are negative in zones 2 or 2E, they shall be applied in zone 2/2E for a distance from the edge of the roof equal to 50 percent of the horizontal dimension of the building that is parallel to the direction of the MWFRS being designed or 2.5 times the eave height h_e at the windward wall, whichever is less. The remainder of zone 2/2E extending to the ridge line must use the pressure coefficients GC_{pf} for zone 3/3E.

For this building:

Transverse direction: $0.5 \times 256 = 128$ ft.

Longitudinal direction: $0.5 \times 140 = 70$ ft.

$2.5h_e = 2.5 \times 18 = 45$ ft (governs in both directions).

Therefore, in the transverse direction, zone 2/2E applies over a distance of 45 ft from the edge of the windward roof, and zone 3/3E applies over a distance of $128 - 45 = 83$ ft

Zone	GC_{pf}	Design Pressure p (psf)	
		$GC_{pi} = +0.18$	$GC_{pi} = -0.18$
1	0.40	3.4	9.0
2	−0.69	−13.5	−7.9
3	−0.37	−8.5	−3.0
4	−0.29	−7.3	−1.7
5	−0.45	−9.8	−4.2
6	−0.45	−9.8	−4.2
1E	0.61	6.7	12.3
2E	−1.07	−19.4	−13.8
3E	−0.53	−11.0	−5.4
4E	−0.43	−9.5	−3.9

TABLE 2.14   Design Wind Pressures p on the MWFRS

in what is normally considered to be zone 2/2E. In the longitudinal direction, zone 3/3E is applied over a distance of $70 - 45 = 25$ ft.

The design pressures are to be applied on the building in accordance with the eight load cases illustrated in ASCE/SEI Fig. 6–10. As shown in the figure, each corner of the building is considered to be a reference corner for wind loading in both the transverse and longitudinal directions.

According to note 4 in ASCE/SEI Fig. 6–10, combinations of external and internal pressures are to be evaluated to obtain the most severe loading. Thus, when both positive and negative pressures are considered, a total of 16 separate loading conditions must be evaluated for this building (in general, the number of load cases can be reduced for symmetric buildings).

Figures 2.21 and 2.22 illustrate the design wind pressures for one load case in the transverse direction and one load case in the longitudinal direction, respectively, including positive and negative internal pressure.

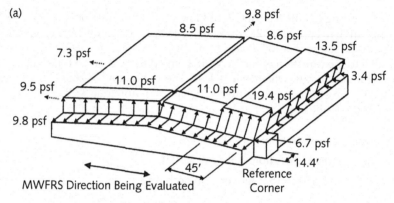

Note: Dashed arrows represent uniformly distributed loads over surfaces 4E, 4, and 6

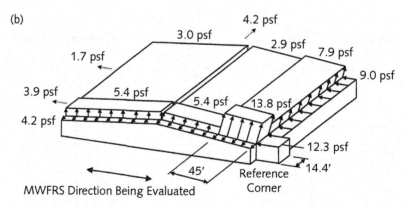

Note: Dashed arrows represent uniformly distributed loads over surfaces 4E, 4, and 6

FIGURE 2.21 Design wind pressures on MWFRS in transverse direction: (a) positive internal pressure; (b) negative internal pressure.

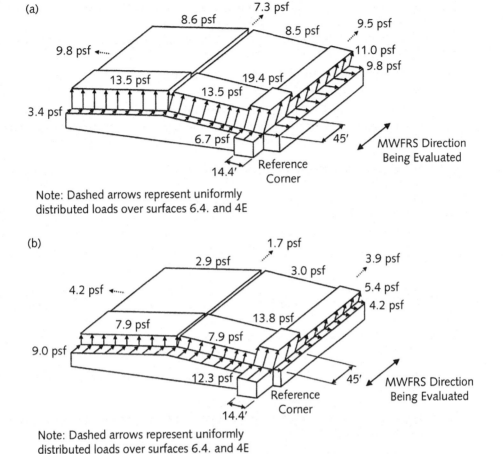

FIGURE 2.22 Design wind pressures on MWFRS in longitudinal direction: (a) positive internal pressure; (b) negative internal pressure.

Torsional load cases, which are given in ASCE/SEI Fig. 6–10, must be considered in addition to the basic load cases noted earlier, unless one or more of the conditions under the exception in note 5 of the figure are satisfied. The first condition is satisfied because this building is one story high with a mean roof height h of less than 30 ft, and torsional load cases need not be considered. The building also satisfies the third condition because it is two stories high or less in height and has a flexible diaphragm.

The minimum design loading of ASCE/SEI 6.1.4.1 also must be investigated (see ASCE/SEI Fig. C6–1).

Part 2: Design Wind Pressures on a Masonry Wall

- Determine external pressure coefficients GC_p for zones 4 and 5.

1. Pressure coefficients for zones 4 and 5 can be determined from ASCE/SEI Fig. 6–11A based on the effective wind area.

2. The effective wind area is defined as the span length multiplied by an effective width that need not be less than one-third the span length: $18 \times (18/3) = 108.0$ ft^2. Note that the smallest span length corresponding to the east and west walls is used because this results in larger pressures.

3. The pressure coefficients from the figure are summarized in Table 2.15.

Note 5 in ASCE/SEI Fig. 6–11A states that values of GC_p for walls are to be reduced by 10 percent when the roof angle is less than or equal to 10 degrees. Modified values of GC_p based on note 5 are provided in Table 2.16.

• Determine the internal pressure coefficients GC_{pi}.

For an enclosed building, $GC_{pi} = +0.18, -0.18$ from ASCE/SEI Fig. 6–5.

• Determine the design wind pressure p by ASCE/SEI Eq. (6–22) on zones 4 and 5.

$$p = q_h(GC_p - GC_{pi}) = 15.5[GC_p - (\pm 0.18)]$$

Calculation of design wind pressures is illustrated for zone 4: For positive GC_p, $p = 15.5[0.74 - (-0.18)] = 14.3$ psf

For negative GC_p,

$$p = 15.5[-0.83 - (+0.18)] = -15.7 \text{ psf}$$

These pressures act perpendicular to the walls.

The maximum design wind pressures for positive and negative internal pressures are summarized in Table 2.17.

Zone	GC_p	
	Positive	Negative
4	0.82	−0.92
5	0.82	−1.04

TABLE 2.15   External Pressure Coefficients GC_p for Masonry Walls

Zone	GC_p	
	Positive	Negative
4	0.74	−0.83
5	0.74	−0.94

TABLE 2.16   Modified External Pressure Coefficients GC_p for Masonry Walls

Zone	GC_p	Design pressure p (psf)
4	0.74	14.3
	−0.83	−15.7
5	0.74	14.3
	−0.94	−17.4

TABLE 2.17 Design Wind Pressures p on Masonry Walls

Zone	GC_p	
	Positive	Negative
1	0.20	−0.90
2	0.20	−1.10
3	0.20	−1.10

TABLE 2.18 External Pressure Coefficients GC_p for Open-Web Joist Purlins

Zone	GC_p	Design pressure p (psf)
1	0.20	5.9
	−0.90	−16.7
2 and 3	0.20	5.9
	−1.10	−19.8

TABLE 2.19 Design Wind Pressures p on Open-Web Joist Purlins

In zones 4 and 5, the computed positive and negative pressures are greater than the minimum values prescribed in ASCE/SEI 6.1.4.2 of +10 psf and −10 psf, respectively.

Part 3: Design Wind Pressures on an Open-Web Joist Purlin

- Determine external pressure coefficients GC_p for zones 1, 2, and 3 for gable roofs with a roof slope of less than or equal to 7 degrees.

1. Pressure coefficients for zones 1, 2, and 3 can be determined from ASCE/SEI Fig. 6–11B based on the effective wind area.
2. Effective wind area = larger of $35 \times 8 = 280.0$ ft^2 or $35 \times (35/3) = 408.3$ ft^2 (governs).
3. The pressure coefficients from the figure are summarized in Table 2.18.

- Determine design wind pressure p by ASCE/SEI Eq. (6–22) on zones 1, 2, and 3.

$$p = q_h(GC_p - GC_{pi}) = 15.5[GC_p - (\pm 0.18)]$$

The maximum design wind pressures for positive and negative internal pressures are summarized in Table 2.19.

The pressures in Table 2.19 are applied normal to the open-web joist purlins and act over the tributary area of each purlin, which is equal to 280 ft². If the tributary area were greater than 700 ft², the purlins could have been designed using the provisions for MWFRSs (see ASCE/SEI 6.5.12.1.3).

The positive pressures on zones 1, 2, and 3 must be increased to the minimum value of 10 psf in accordance with the provisions of ASCE/SEI 6.1.4.2.

The pressures determined by this method are for all intents and purposes the same as those determined by the simplified method in Example 2.3. As such, the loading diagrams in Figure 2.20 are applicable in this example.

Example 2.5: Wind Loads, Alternate All-Heights Method

For the one-story building illustrated in Fig. 2.17, determine design wind pressures on (1) the MWFRS, (2) a masonry wall, and (3) an open-web joist using the alternate all-heights method of IBC 1609.6. Use the same design data as in Example 2.3.

Solution

Part 1: Design Wind Pressures on the MWFRS

- Check whether the provisions of IBC 1609.6 can be used to determine the design wind pressures on this building.

The provisions of IBC 1609.6 may be used to determine design wind pressures on this regularly shaped building, provided that the conditions of IBC 1609.6.1 are satisfied:

1. The height of the building is 18 ft, which is less than 75 ft, and the height-to-least-width ratio $= 18/140 = 0.13 < 4$. Also, it was shown in Example 2.3 that the fundamental frequency $n_1 > 1$ Hz in both directions.

2. As was discussed in Example 2.3, this building is not sensitive to dynamic effects.

3. This building is not located on a site where channeling effects or buffeting in the wake of upwind obstructions need to be considered.

4. As was shown in Example 2.3, the building meets the requirements of a simple diaphragm building as defined in ASCE/SEI 6.2.

5. The fifth condition is not applicable in this example.

Thus the provisions of the alternate all-heights method of IBC 1609.6 can be used to determine the design wind pressures on the MWFRS as well as the C&C.

- Determine the design wind pressures on the MWFRS.

1. Determine the wind stagnation pressure q_s. From IBC Table 1609.6.2(1), $q_s = 20.7$ psf for $V = 90$ mph.

2. Determine the velocity pressure exposure K_z from ASCE/SEI Table 6–3. Values of K_z for exposure C are summarized in Table 2.20.

Height above ground level z (ft)	K_z
18	0.88
15	0.85

TABLE 2.20 Velocity Pressure Exposure Coefficient K_z

3. Determine the net pressure coefficients C_{net} from the walls and roof from IBC Table 1609.6.2(2) assuming that the building is enclosed.

For wind in the E-W (transverse) direction:

Windward wall: $C_{net} = 0.43$ for positive internal pressure
$C_{net} = 0.73$ for negative internal pressure

Leeward wall: $C_{net} = -0.51$ for positive internal pressure
$C_{net} = -0.21$ for negative internal pressure

Sidewalls: $C_{net} = -0.66$ for positive internal pressure
$C_{net} = -0.35$ for negative internal pressure

Leeward roof (wind perpendicular to ridge):
$C_{net} = -0.66$ for positive internal pressure
$C_{net} = -0.35$ for negative internal pressure

Windward roof (wind perpendicular to ridge with roof slope <2:12):
$C_{net} = -1.09, -0.28$ for positive internal pressure
$C_{net} = -0.79, 0.02$ for negative internal pressure

For wind in the N-S (longitudinal) direction:

Windward wall: $C_{net} = 0.43$ for positive internal pressure
$C_{net} = 0.73$ for negative internal pressure

Leeward wall: $C_{net} = -0.51$ for positive internal pressure
$C_{net} = -0.21$ for negative internal pressure

Sidewalls: $C_{net} = -0.66$ for positive internal pressure
$C_{net} = -0.35$ for negative internal pressure

Roof (wind parallel to ridge):
$C_{net} = -1.09$ for positive internal pressure
$C_{net} = -0.79$ for negative internal pressure

4. Determine the net design wind pressures p_{net} by IBC Eq. (16–34):

$$p_{net} = q_s K_z C_{net} I K_{zt} = 20.7 K_z C_{net}$$

Location		Height Above Ground Level z (ft)	K_z	C_{net}		Net design pressure p_{net} (psf)	
				+ Internal Pressure	− Internal Pressure	+ Internal Pressure	− Internal Pressure
Windward wall		18	0.88	0.43	0.73	7.8	13.3
		15	0.85	0.43	0.73	7.6	12.8
Leeward wall		All	0.88	−0.51	−0.21	−9.3	−3.8
Sidewalls		All	0.88	−0.65	−0.35	−11.8	−6.4
Roof	Windward	18	0.88	−1.09	−0.79	−19.9	−14.4
		18	0.88	−0.28	0.02	−5.1	0.4
	Leeward	18	0.88	−0.66	−0.35	−12.0	−6.4

TABLE 2.21   Net Design Wind Pressures p_{net} in E-W (Transverse) Direction

Location	Height Above Ground Level z (ft)	K_z	C_{net}		Net design pressure p_{net} (psf)	
			+ Internal Pressure	− Internal Pressure	+ Internal Pressure	− Internal Pressure
Windward wall	18	0.88	0.43	0.73	7.8	13.3
	15	0.85	0.43	0.73	7.6	12.8
Leeward wall	All	0.88	−0.51	−0.21	−9.3	−3.8
Side walls	All	0.88	−0.65	−0.35	−11.8	−6.4
Roof	18	0.88	−1.09	−0.79	−19.9	−14.4

TABLE 2.22   Net Design Wind Pressures p_{net} in N-S (Longitudinal) Direction

A summary of the net design wind pressures in the E-W and N-S directions is given in Tables 2.21 and 2.22, respectively.

Illustrated in Figs. 2.23 and 2.24 are the net design wind pressures in the E-W (transverse) and N-S (longitudinal) directions, respectively, for positive and negative internal pressures. Note that for wind in the E-W direction, only condition 1 wind pressures on the roof are illustrated in these figures. Although the pressures from the other conditions are not shown in the figures, they must be considered in the overall design.

The wind-load cases defined in ASCE/SEI Fig. 6–9 must be considered in the design of the MWFRS of buildings whose wind loads have been determined by IBC 1609.6 (IBC 1609.6.4.1). The wind pressures on the windward and leeward walls shown in Figs. 2.23 and 2.24 fall under case 1. *IBC* 1609.6.4.1 requires consideration of torsional effects, as indicated in ASCE/SEI Fig. 6–9. One-story buildings with a mean roof height equal to or less than 30 ft need to be designed only for cases 1 and 3 (see the exception in ASCE/SEI 6.5.12.3). In case 3, 75 percent of the wind pressures on the windward and leeward walls of case 1, which are shown in Figs. 2.23 and 2.24, act simultaneously on

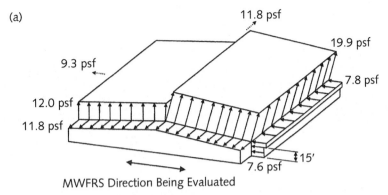

(a)

11.8 psf

19.9 psf

9.3 psf

7.8 psf

12.0 psf

11.8 psf

15'

7.6 psf

MWFRS Direction Being Evaluated

Note: Dashed arrows represent uniformly
distributed loads over leeward and side surfaces

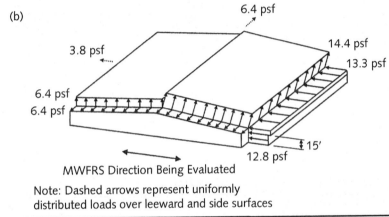

(b)

6.4 psf

3.8 psf

14.4 psf

13.3 psf

6.4 psf

6.4 psf

15'

12.8 psf

MWFRS Direction Being Evaluated

Note: Dashed arrows represent uniformly
distributed loads over leeward and side surfaces

FIGURE 2.23   Design wind pressures on the MWFRS in the E-W (transverse) direction: (a) positive
internal pressure; (b) negative internal pressure.

the building (see ASCE/SEI Fig. 6–9). This load case, which needs to be considered in addition to the load cases in Figs. 2.23 and 2.24, accounts for the effects due to wind along the diagonal of the building.

The minimum design wind loading of IBC 1609.6.3 must be considered as a separate load case in addition to the load cases described earlier. The 10-psf wind pressure acts on the area of the building projected on a plane normal to the direction of wind, as illustrated in ASCE/SEI Fig. C6–1.

Part 2: Design Wind Pressures on a Masonry Wall

- Determine net pressure coefficients C_{net} for zones 4 and 5 in ASCE/SEI Fig. 6–11A from IBC Table 1609.6.2(2).

(a)

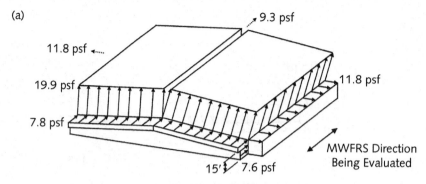

Note: Dashed arrows represent uniformly distributed loads over side and leeward surfaces.

(b)

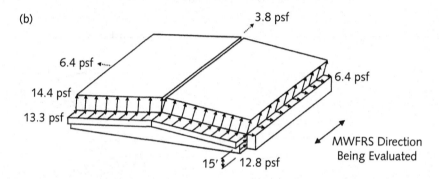

Note: Dashed arrows represent uniformly distributed loads over side and leeward surfaces.

Figure 2.24 Design wind pressures on the MWFRS in the N-S (longitudinal) direction: (a) positive internal pressure; (b) negative internal pressure.

Zone	C_{net}	
	Positive	Negative
4	0.95	−1.04
5	0.95	−1.24

Table 2.23 Net Pressure Coefficients C_{net} for Masonry Walls

1. The effective wind area is defined as the span length multiplied by an effective width that need not be less than one-third the span length: $18 \times (18/3) = 108.0$ ft^2. Note that the smallest span length corresponding to the east and west walls is used because this results in larger pressures.

2. The net pressure coefficients from IBC Table 1609.6.2(2) for C&C (walls) not in areas of discontinuity (item 4, $h \leq 60$ ft, zone 4) and in areas of discontinuity (item 5, $h \leq 60$ ft, zone 5) are summarized in Table 2.23. Linear interpolation was used to determine these values [see note a in IBC Table 1609.6.2(2)].

- Determine the net design wind pressures p_{net} by IBC Eq. (16–34).

$$p_{net} = q_s K_z C_{net} I K_{zt} = (20.7 \times 0.88) C_{net} = 18.2 C_{net}$$

A summary of the design wind pressures on the masonry walls is given in Table 2.24. These pressures act perpendicular to the face of the walls.
In zones 4 and 5, the calculated positive and negative pressures are greater than the minimum values prescribed in IBC 1609.6.3 of +10 psf and –10 psf, respectively.

Part 3: Design Wind Pressures on an Open-Web Joist Purlin

- Determine the net pressure coefficients C_{net} for zones 1, 2, and 3 in ASCE/SEI Fig. 6–11B from IBC Table 1609.6.2(2).

1. Effective wind area = larger of $35 \times 8 = 280.0$ ft² or $35 \times (35/3) = 408.3$ ft² (governs).

2. The net pressure coefficients from IBC Table 1609.6.2(2) for C&C (roofs) not in areas of discontinuity (item 2, gable roof with flat < slope < 6:12, zone 1) and in areas of discontinuity (item 3, gable roof with flat < slope < 6:12, zones 2 and 3) are summarized in Table 2.25.

- Determine net design wind pressures p_{net} by IBC Eq. (16–34).

$$p_{net} = q_s K_z C_{net} I K_{zt} = (20.7 \times 0.88) C_{net} = 18.2 C_{net}$$

A summary of the design wind pressures on the open-web joist purlins is given in Table 2.26.

Zone	C_{net}	Design pressure p_{net} (psf)
4	0.95	17.3
	–1.04	–18.9
5	0.95	17.3
	–1.24	–22.6

TABLE 2.24   Design Wind Pressures on Masonry Walls

Zone	C_{net}	
	Positive	Negative
1	0.41	–0.92
2	0.41	–1.17
3	0.41	–1.85

TABLE 2.25   Net Pressure Coefficients C_{net} for Open-Web Joist Purlins

Zone	C_{net}	Design pressure p_{net} (psf)
1	0.41	7.5
	−0.92	−16.7
2	0.41	7.5
	−1.17	−21.3
3	0.41	7.5
	−1.85	−33.7

TABLE 2.26 Design Wind Pressures on Open-Web Joist Purlins

The pressures in Table 2.26 are applied normal to the open-web joist purlins and act over the tributary area of each purlin, which is equal to $35 \times 8 = 280$ ft². If the tributary area were greater than 700 ft², the purlins could have been designed using the provisions for MWFRSs (ASCE/SEI 6.5.12.1.3).

The positive pressures in zones 1, 2, and 3 must be increased to the minimum value of 10 psf in accordance with IBC 1609.6.3.

Example 2.6: Rain Loads

Determine the rain load on a roof located in Chicago, Illinois, similar to the one depicted in Fig. 2.5 given the following design data:

- Tributary area of primary roof drain: 5,000 ft²
- Closed scupper size: 6 in. wide (b) by 4 in. high (h)
- Vertical distance from primary roof drain to inlet of scupper (d_s): 6 in.
- Rainfall rate: 3.1 in./h (see IBC Fig. 1611.1)

Solution

- Determine the hydraulic head d_h based on the flow rate Q.

$Q =$ tributary area of roof drain × rainfall rate = $5,000 \times 3.1/12 = 1,292$ ft³/h = 21.5 ft³/min = 161.0 gal/min, where 1 gal = 0.1337 ft³.

The hydraulic head d_h is obtained from the following equation, which is applicable to closed scuppers where the free surface of the water is above the top of the scupper:

$$Q = 2.9b\left(d_h^{1.5} - h_1^{1.5}\right) = 2.9b[(h + h_1)^{1.5} - h_1^{1.5}]$$

In this equation, $h_1 = d_h - h$, which is the distance from the free surface of the water to the top of the scupper.

For a flow rate of 161.0 gal/min, this equation can be solved for h_1: $h_1 = 0.53$ in. Thus $d_h = 0.53 + 4 = 4.53$ in. Note that by interpolating the values in ASCE/SEI Table C8–1 for a flow rate of 161 gal/min, the hydraulic head d_h is found to be 4.57 in.

- Determine the rain load R.

Use IBC Eq. (16–35):

$$R = 5.2(d_s + d_h) = 5.2(6.0 + 4.53) = 54.8 \text{ psf}$$

Example 2.7 Earthquake Loads, Equivalent Lateral Force Procedure

For the warehouse building depicted in Fig. 2.17, determine the following: (1) the seismic base shear, (2) the design seismic force on the diaphragm, and (3) the out-of-plane design seismic forces on a masonry wall given the following design data:

- Seismic ground motion values: $S_S = 1.51$ and $S_1 = 0.76$.
- Soil classification: Site class D.
- Occupancy: Fewer than 300 people congregate in one area, and the building is not used to store hazardous or toxic material.
- Structural system: Building frame system with special reinforced masonry shear walls.

Assume that the weight of the roof is 15 psf and the weight of the masonry walls is 83 psf.

Solution

Part 1: Seismic Base Shear

- Determine the design spectral response accelerations.

1. Using ASCE/SEI Tables 11.4–1 and 11.4–2, determine the soil-modified accelerations:

   From ASCE/SEI Table 11.4–1, $F_a = 1.0$, so $S_{MS} = 1.0 \times 1.51 = 1.51$.
   From ASCE/SEI Table 11.4–2, $F_v = 1.5$, so $S_{M1} = 1.5 \times 0.76 = 1.14$.

2. Therefore,

$$S_{DS} = \frac{2}{3} \times 1.51 = 1.01$$

$$S_{D1} = \frac{2}{3} \times 1.14 = 0.76$$

- Determine the SDC.

From IBC Table 1604.5, the occupancy category is II. Since $S_1 > 0.75$, this building is assigned to SDC E (see ASCE/SEI 11.6).

- Determine the seismic base shear using the provisions of the equivalent lateral force procedure (ASCE/SEI 12.8).

1. Check whether the equivalent lateral force procedure can be used. Since this building does not exceed two stories in height, this method is permitted to be used (see ASCE/SEI Table 12.6–1).

2. Determine the response modification coefficient R. From ASCE/SEI Table 12.2–1, $R = 5.5$ for a building frame system with special reinforced-masonry shear walls.

3. Determine the importance factor I. From ASCE/SEI Table 11.5–1, $I = 1.0$ for occupancy category II.

4. Determine the period T. An approximate fundamental period T_a is determined by ASCE/SEI Eq. (12.8–7):

$$T_a = C_t h_n^x$$

From ASCE/SEI Table 12.8–2 for all other structural systems, $C_t = 0.02$ and $x = 0.75$. Thus

$$T_a = 0.02 \times (18)^{0.75} = 0.2 \text{ second}$$

5. Determine the seismic response coefficient C_s. The value of C_s determined by ASCE/SEI Eq. (12.8–3) is

$$C_s = \frac{S_{D1}}{T(R/I)} = \frac{0.76}{0.2 \times 5.5} = 0.69$$

The value of C_s need not exceed that determined by ASCE/SEI Eq. (12.8–2):

$$C_s = \frac{S_{DS}}{R/I} = \frac{1.01}{5.5} = 0.184$$

Also, C_s must not be less than the larger of the following:

- $0.044 S_{DS} I = 0.044$

- 0.01

- $\dfrac{0.5 S_1}{R/I} = \dfrac{0.5 \times 0.76}{5.5} = 0.069$

Thus the value from ASCE/SEI Eq. (12.8–2) governs.

6. Determine the effective seismic weight W. The effective seismic weight W is equal to the weight of the roof plus the weight of the masonry walls tributary to the roof, conservatively assuming that there are no openings in the walls.

$$\text{Weight of roof} = 0.015 \times 140 \times 256 = 538 \text{ kips}$$

$$\text{Weight of walls} = 0.083 \times \frac{(18+5.33)+18}{2\times 2} \times [2\times(256+140)] = 679 \text{ kips}$$

$$W = 538 + 679 = 1,217 \text{ kips}$$

7. Determine the seismic base shear V. The seismic base shear is determined by ASCE/SEI Eq. (12.8–1):

$$V = C_s W = 0.184 \times 1,217 = 224 \text{ kips}$$

Part 2: Design Seismic Force on the Diaphragm   ASCE/SEI Eq. (12.10–1) is used to determine the seismic design force on the diaphragm:

$$F_{px} = \frac{\displaystyle\sum_{i=x}^{n} F_i}{\displaystyle\sum_{i=1}^{n} w_i} w_{px}$$

Since this is a one-story building, ASCE/SEI Eq. (12.10–1) reduces to $F_{px} = 0.184 w_{px}$:

$$\text{Minimum } F_{px} = 0.2 S_{DS} I w_{px} = 0.20 w_{px}$$

$$\text{Maximum } F_{px} = 0.4 S_{DS} I w_{px} = 0.40 w_{px}$$

Thus

$$F_{px} = 0.20 w_{px}$$

The metal deck roof is permitted to be idealized as a flexible diaphragm in accordance with ASCE/SEI 12.3.1.1. Seismic forces are computed from the tributary weight of the roof and the walls oriented perpendicular to the direction of analysis (*Note:* Walls parallel to the direction of the seismic forces typically are not considered in the tributary width because these walls do not obtain support from the diaphragm in the direction of the seismic force).

- N-S direction

$$w_{pN-S} = (0.20 \times 15 \times 140) + \left[0.20 \times 83 \times 2 \times \frac{(18+5.33)+18}{2\times 2} \right] = 420 + 343 = 763 \text{ plf}$$

- E-W direction

$$w_{pE-W} = (0.20 \times 15 \times 256) + \left[0.20 \times 83 \times 2 \times \frac{(18+5.33)+18}{2\times 2} \right] = 768 + 343 = 1,111 \text{ plf}$$

Part 3: Out-of-Plane Design Seismic Forces on the Masonry Wall   Structural walls are to be designed for a force normal to the surface equal to $0.4S_{DS}I$ times the weight of the wall (ASCE/SEI 12.11.1). The minimum normal force is equal to 10 percent of the weight of the wall.

$$W_p = 0.083 \times \frac{(18+5.33)+18}{2} = 1.7 \text{ kips/ft}$$

$$0.1 \times 1.7 = 0.17 \text{ kips/ft}$$

$$0.4S_{DS}I = 0.4 \times 1.01 \times 1.0 \times 1.7 = 0.69 \text{ kips/ft}$$

$$\text{Distributed load} = \frac{0.69}{\frac{(18+5.33)+18}{2}} = 0.03 \text{ kips/ft per foot width of wall}$$

Anchorage of the masonry walls to the flexible diaphragm must develop the out-of-plane force given by ASCE/SEI Eq. (12.11–1):

$$F_p = 0.8S_{DS}IW_p = 1.4 \text{ kips/ft}$$

Example 2.8: Earthquake Loads, Simplified Design Method

For the warehouse building depicted in Fig. 2.17, determine the seismic base shear using the simplified alternative structural design criteria of ASCE/SEI 12.14, assuming the design data given in Example 2.6.

Solution

- Determine whether the simplified design method can be used.

The simplified method is permitted to be used if the following 12 limitations are met:

1. The structure shall qualify for occupancy category I or II in accordance with ASCE/SEI Table 1–1. It was determined in Example 2.6 that the occupancy category is II.
2. The site class shall not be E or F. The site class is given as D.
3. The structure shall not exceed three stories in height. The structure is one story.
4. The seismic force–resisting system shall be either a bearing-wall system or a building frame system in accordance with ASCE/SEI Table 12.14–1. The seismic force–resisting system is a building frame system.
5. The structure has at least two lines of lateral resistance in each of the two major axis directions. Masonry shear walls are provided along two lines at the perimeter in both directions.

6. At least one line of resistance shall be provided on each side of the center of mass in each direction. The center of mass is essentially located at the geometric center of the building, and walls are provided on all four sides of the perimeter.

7. For structures with flexible diaphragms, overhangs beyond the outside line of shear walls or braced frames shall satisfy $a \leq d/5$. The diaphragm does not overhang the line of shear walls at the perimeter.

8. For buildings with diaphragms that are not flexible, the distance between the center of rigidity and the center of mass parallel to each major axis shall not exceed 15 percent of the greatest width of the diaphragm parallel to that axis. The diaphragm in this example is flexible, so this limitation is not applicable.

9. Lines of resistance of the seismic force–resisting system shall be oriented at angles of no more than 15 degrees from alignment with the major orthogonal axes of the building. The shear walls in both directions are parallel to the major axes.

10. The simplified design procedure shall be used for each major orthogonal horizontal axis direction of the building. The simplified design procedure is used in both directions.

11. System irregularities caused by in-plane or out-of-plane effects of the seismic force–resisting elements shall not be permitted. The building does not have any irregularities.

12. The lateral load resistance of any story shall not be less than 80 percent of the story above. Since this building is one story high, this limitation is not applicable.

Since all 12 limitations are satisfied, the simplified design procedure may be used.

- Determine the seismic base shear.

1. Determine S_{DS} from ASCE/SEI 12.14.8.1:

$$S_{DS} = \frac{2}{3} F_a S_s$$

According to ASCE/SEI 14.8.1, F_a is permitted to be determined in accordance with ASCE/SEI 11.4.3. Thus, from Example 3.6, $F_a = 1.0$. Therefore, $S_{DS} = \frac{2}{3} \times 1.0 \times 1.51 = 1.01$.

2. Determine the response modification coefficient R. From ASCE/SEI Table 12.14–1, $R = 5.5$ for a building frame system with special reinforced-masonry shear walls.

3. Determine the seismic base shear V. The seismic base shear is determined by ASCE/SEI Eq. (12.14–11):

$$V = \frac{F S_{DS}}{R} W = \frac{1.0 \times 1.01}{5.5} \times 1,217 = 224 \text{ kips}$$

Note that $F = 1.0$ for a one-story building, and the effective seismic weight W was determined in Example 3.6.

References

1. International Code Council (ICC). 2009. *International Building Code*. ICC, Washington, DC.

2. Structural Engineering Institute (SEI) of the American Society of Civil Engineers (ASCE). 2006. *Minimum Design Loads for Buildings and Other Structures, Including Supplements Nos. 1 and 2*. ASCE/SEI 7–05. ASCE, Reston, VA.

3. Fanella, D. 2009. *Structural Load Determination under 2009 IBC and ASCE/SEI 7–05*. International Code Council (ICC), 500 New Jersey Avenue, NW, 6th Floor, Washington, DC.

4. Structural Engineering Institute (SEI) of the American Society of Civil Engineers (ASCE). 2005. *Flood Resistant Design and Construction*. ASCE/SEI 24–05. ASCE, Reston, VA.

5. National Council of Structural Engineers Associations (NCSEA). 2009. *Guide to the Design of Diaphragms, Chords and Collectors: Based on the 2006 IBC and ASCE/SEI 7–05*. International Code Council (ICC), 500 New Jersey Avenue, NW, 6th Floor, Washington, DC.

6. Mays, T., National Council of Structural Engineers Associations (NCSEA). 2009. *Guide to the Design of Out-of-Plane Wall Anchorage: Based on the 2006 IBC and ASCE/SEI 7–05*. International Code Council (ICC), 500 New Jersey Avenue, NW, 6th Floor, Washington, DC.

CHAPTER 3

Structural Materials

One of the most important ingredients in the design of any structure is thorough understanding of the mechanical properties and behavior of the materials used in its construction. This chapter outlines the critical properties of materials employed in constructing cold-formed steel, reinforced-masonry, structural timber, and open-web joist structural systems.

3.1 Cold-Formed Steel Materials

American and Iron Steel Institute (AISI), *North American Specification for the Design of Cold-Formed Steel Members* (ANSI/AISI S100-07),[4] Sec. A2.1 lists sheet and strip materials applicable for cold-formed steel member design. Other steels can be used for structural members if they meet ductility and other requirements as stipulated in ANSI/AISI S100-07 Sec. A2.2. The basic ductility requirement is for the ratio of tensile strength F_u to yield strength F_y of at least 1.08 and for a total elongation of at least 10 percent in 2 in. If these requirements cannot be met, alternative criteria related to local elongation may be applicable. In addition, certain steels that do not meet the criteria, such as American Society for Testing and Materials (ASTM) A653 SS Grade 80 may be used for multiple-web configurations (i.e., roofing, siding, decking, etc.), provided that the nominal yield strength in design is taken as the lesser of 75 percent of the specified yield strength or 60 ksi (414 MPa) and the nominal tensile strength is taken as the lesser of 75 percent of the specified tensile stress or 62 ksi (428 MPa).

The following 16 ASTM steels are permitted by ANSI/AISI S100-07 for use with cold-formed steel members or connections:

- Carbon structural steel (ASTM A36)
- High-strength low-alloy structural steel (ASTM A242)
- Low and intermediate tensile strength carbon and steel plates (ASTM A283)
- Cold-formed welded and seamless carbon steel structural tubing in round and shapes (ASTM A500)
- High-strength carbon-manganese steel of structural quality (ASTM A529)
- High-strength low-alloy columbium-vanadium structural steel (ASTM A572)
- High-strength low-alloy structural steel with 50 ksi minimum yield point to 4 in. thick (ASTM A588)
- Steel, sheet and strip, high-strength, low-alloy, hot-rolled and cold-rolled, with improved atmospheric corrosion resistance (ASTM A606)

- Steel sheet, zinc-coated (galvanized) or zinc-iron alloy-coated (galvannealed) by the hot-dip process (ASTM A653)

- Steel sheet, 55 percent aluminum–zinc alloy–coated by the hot-dip process (ASTM A792)

- Cold-formed welded and seamless high-strength, low-alloy structural tubing with improved atmospheric corrosion resistance (ASTM A847)

- Steel sheet, zinc–5 percent aluminum alloy–coated by the hot-dip process (ASTM A875)

- Steel sheet, carbon, metallic- and nonmetallic-coated for cold-formed framing members (ASTM A1003)

- Steel, sheet, cold-reduced, carbon, structural, high-strength low-alloy, high-strength low-alloy with improved formability, solution hardened, and bake hardened (ASTM A1008)

- Steel, sheet and strip, hot-rolled, carbon, structural, high-strength low-alloy and high-strength low-alloy with improved formability (ASTM A1011)

- Steel, sheet, hot-rolled, carbon, commercial and structural, produced by the twin-roll casting process (ASTM A1039)

Table 1.2 of the AISI *Cold-Formed Steel Design Manual* (AISI D100-08)[5] provides a summary of the mechanical properties of the 16 approved steels.

There are two general types of stress-strain curves, sharp yielding and gradual yielding. Steels produced by hot rolling usually exhibit sharp-yielding characteristics (Fig. 3.1a). That is, the sharp-yielding type of steel has a yield point that is defined by the level at which the stress-strain curve becomes horizontal. Steels that are cold-reduced or cold-worked yield gradually (Fig. 3.1b). For gradual-yielding steel, the stress-strain curve is rounded out at the "knee" (Fig. 3.2a), and the yield stress generally is determined by the *offset method*, as illustrated in Fig. 3.2b. As shown in Fig. 3.2b, the yield stress is determined by offsetting the elastic portion of the stress-strain curve by 0.01 percent (distance *om*), where the intersection of the offset line *mn* and the stress curve locates the point *r*, thus defining the yield-strength magnitude *R*. The *load method* determination of yield stress is similar to the offset method, except that the offset line *mn* is formed parallel with the stress axis of the stress-strain diagram (Fig. 3.2c).

Ductility is a measure of the ability of the steel to sustain plastic deformation without fracture. For cold-formed steel members, ductility is needed for plastic redistribution of stress in connections and is required in the forming process. The percent elongation and the ratio F_u/F_y typically are used to quantify ductility.

The required ductility for cold-formed steel structural members depends mainly on the type of application and the suitability of the material. The same amount of elongation that is considered necessary for individual framing members may not be needed for such applications such as roof panels and siding, which typically are formed with large radii and are not used in conjunction with highly stressed connections.

The mechanical properties of cold-formed sections sometimes are different from those of the steel sheet or strip before the cold-forming operation. This is so because the cold-forming operation increases the yield stress and tensile strength but decreases the material ductility. Since the material in the corners of a section is cold-worked to a

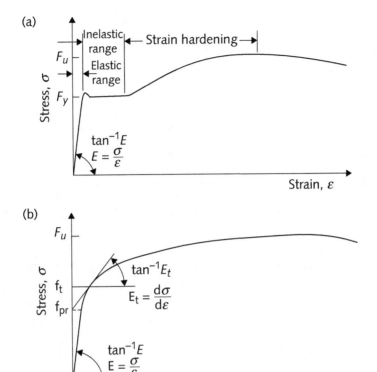

FIGURE 3.1   Stress-strain curves of carbon steel sheet or strip. (*AISI, 2007.*[6])

considerably higher degree than the material in the flat elements, the mechanical properties will be different in various parts of the cross section.

Based on research,[6, 39] the F_u/F_y and inside-radius-to-thickness (R/t) ratios are the most important factors to affect the change in mechanical properties of cold-formed sections. Virgin material with a large F_u/F_y ratio possesses a large potential for strain hardening. Consequently, as the F_u/F_y ratio increases, the effect of cold work on the increase in the yield stress of steel increases. Small inside-radius-to-thickness ratios R/t correspond to a large degree of cold work in a corner, and therefore, for a given material, the smaller the R/t ratio, the larger is the increase in yield stress.

The ANSI/AISI S100-07 Sec. A7.2 permits the use of the increase in material properties that result from a cold-forming operation. The provisions of Sec. A7.2 are applicable only to compact sections that are not subject to local buckling.

For further information on the materials used in fabrication of cold-formed steel members, the reader is encouraged to refer to Yu and LaBoube.[39]

3.2 Reinforced Masonry

Reinforced masonry consists of four primary components—masonry units, mortar, reinforcement, and grout. These components are combined to create structural elements that are highly effective in resisting loads. Figures 3.3 through 3.5 show common types

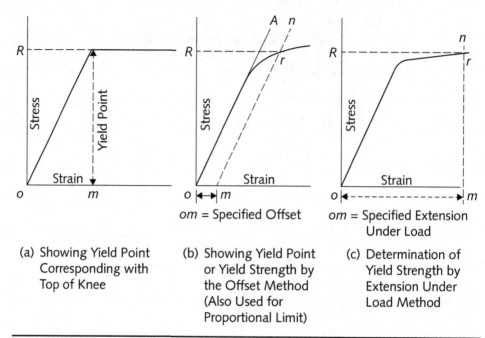

(a) Showing Yield Point Corresponding with Top of Knee

(b) Showing Yield Point or Yield Strength by the Offset Method (Also Used for Proportional Limit)

om = Specified Offset

(c) Determination of Yield Strength by Extension Under Load Method

om = Specified Extension Under Load

Figure 3.2 Stress-strain diagrams showing methods of yield stress and yield strength determination. (*AISI, 2007.*[6])

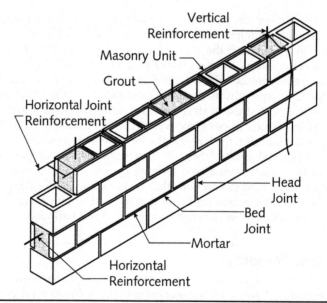

Figure 3.3 Partially grouted reinforced-concrete masonry.

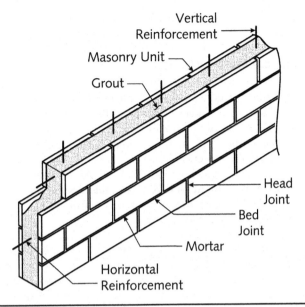

FIGURE 3.4 Fully grouted reinforced-concrete masonry.

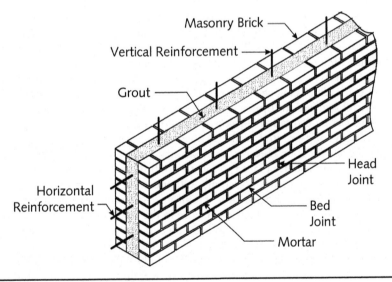

FIGURE 3.5 Fully grouted reinforced-clay masonry.

of reinforced-masonry construction. In partially grouted masonry with concrete masonry units (CMUs), grout is placed only in cells with reinforcement. Horizontal reinforcement typically consists of joint reinforcement placed in the mortar in the bed joints. In fully grouted masonry, all voids are filled with grout to create a solid element with no voids. Clay masonry usually consists of two wythes of clay units. Reinforcement is placed in the space between the wythes, which then is filled completely with grout.

3.2.1 Concrete Masonry Units

Concrete masonry units are manufactured using primarily portland cement, aggregates, and water. The units usually are hollow masonry units, which are defined as units with a net cross-sectional area in every plane parallel to the bearing surface that is less than 75 percent of the gross cross-sectional area in the same plane. Solid units or units with fewer voids sometimes are manufactured, but these are used rarely in modern reinforced-masonry construction.

The ASTM provides specifications that govern the manufacture of masonry units. The primary ASTM document for CMUs is *Standard Specification for Load Bearing Masonry Units* (ASTM C90).[12] This document provides requirements for materials, dimensions, finish, and appearance of masonry units.

The modular nature of masonry construction requires that CMUs be produced with standardized dimensions. The industry standard is to specify the unit dimensions with its width (or thickness) first, the unit's height next, and unit's length last. For example, an 8 × 8 × 16 CMU is nominally 8 in. wide, 8 in. high, and 16 in. long. These are called the *nominal dimensions* of the unit. The *actual dimensions* of masonry units typically are ⅜ in. smaller than the nominal dimensions to allow for mortar joints. This means that an 8 × 8 × 16 unit is actually 7⅝ in. wide, 7⅝ in. high, and 15⅝ in. long. These are called the *specified standard dimensions*. Figure 3.6 shows typical dimensions of 8 × 8 × 16 units. As can be seen from the figure, various kinds of units can be used depending on where the unit is placed in the structure.

Figure 3.6 also shows that CMUs are produced with different shapes within the modular dimension to serve various purposes in the structure. *Bond beam blocks* are hollow units with portions depressed 1¼ in. or more to permit the forming of a continuous channel for reinforcing steel and grout. *Open-end blocks*, or *A-blocks*, are hollow units with one end closed and the opposite end open, forming two cells when laid in a wall. *Double-open-end blocks*, also called *H-blocks*, are hollow units with both ends open. They are used typically in solid-grouted walls because they permit the grout to flow freely between units. *Lintel blocks*, or *U-blocks*, are masonry units with a solid bottom surface and no webs. They are usually placed to form a continuous beam over openings. *Sill blocks* are solid CMUs used for sills or openings. Other commonly used units include *pilaster blocks*, which are used in the construction of reinforced-concrete masonry pilasters and columns; *return blocks* or *L-blocks*, which are for use in the construction of corners for various wall thicknesses; and *sash blocks*, which have an end slot for use in openings to receive metal window frames and premolded expansion-joint material.

Depending on the oven-dry density of the concrete used, ASTM C90 classifies CMUs into lightweight (<105 pcf), medium-weight (105 to 125 pcf), and normal-weight (>125 pcf) units. The linear shrinkage of units at the time of delivery to the purchaser should not be greater than 0.065 percent. CMUs also can be classified based on appearance. *Precision units*, which are the most commonly used CMUs, have a relatively smooth surface. No overall dimension in a precision unit should differ from the specified standard dimensions by more than ⅛ in. *Split-face units* are popular for the rough-textured surface that is obtained by splitting a large unit crosswise to obtain two units. *Slumped units*, or *slumpstone units*, are obtained by applying compression to units before they are completely cured to obtain the distinctive concave shape.

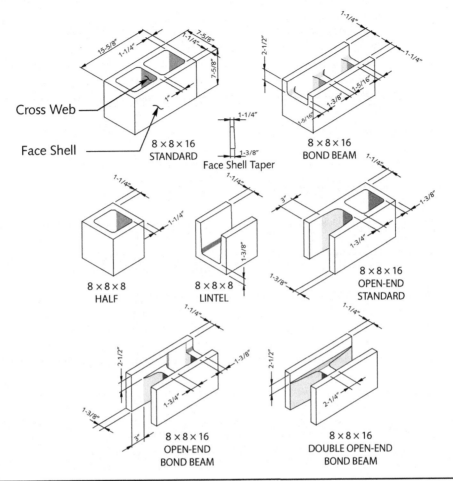

FIGURE 3.6 Typical dimensions for 8-in. concrete masonry units (CMUs).

3.2.2 Clay Masonry Units

Clay masonry units are manufactured with clay, which is found abundantly on the earth's surface. The traditional method of manufacturing clay masonry units, which is still used in some parts of the world, involves the *soft-mud process*. The units are formed using molds, demolded and dried for about 24 hours, and then fired in a kiln at temperatures high enough to cause *vitrification*, a process of ceramic fusion that involves extensive fluxing and hardening of the clay mass. Modern manufacture of clay units usually involves the *stiff-mud process*, in which the units are formed by extruding the clay through a die and then wire cutting to the appropriate size before drying and firing.

Clay masonry units typically are manufactured to comply with *Standard Specification for Building Brick—Solid Masonry Units Made from Clay or Shale* (ASTM C62).[11] When the surface appearance of the brick is a requirement, facing units that comply with *Standard Specification for Facing Brick—Solid Masonry Units Made from Clay or Shale* (ASTM C216)[13] are used. For hollow clay units, *Standard Specification for Hollow Brick—Hollow Masonry Units Made from Clay or Shale* (ASTM C652)[16] applies.

	Minimum Compressive Strength for Brick Flat-Wise Based on Gross Area (psi)		Maximum Water Absorption by 5-Hour Boiling (%)		Maximum Saturation Coefficient	
	Average of Five Bricks	Individual Brick	Average of Five Bricks	Individual Brick	Average of Five Bricks	Individual Brick
Grade SW	3,000	2,500	17.0	20.0	0.78	0.80
Grade MW	2,500	2,200	22.0	25.0	0.88	0.90
Grade NW	1,500	1,250	No limit	No limit	No limit	No limit

TABLE 3.1 Physical Requirements for Solid and Hollow Units

ASTM C62 classifies clay units into grades based on their resistance to damage by freezing when wet. Grade SW (severe weathering) units offer high resistance to frost action and disintegration by weathering. Grade MW (moderate weathering) units are used when a moderate degree of weathering resistance is required and it is unlikely that units will be exposed to freezing temperatures when permeated with water. Grade NW (negligible weathering) units are intended for use as backup material for interior masonry or for interior walls and partitions. Because of poor resistance to weathering, Grade NW is not applicable to facing or hollow brick.

Since compressive strength and absorption are reasonably accurate indicators of the durability of clay masonry units, they are used, in addition with the saturation coefficient, as requirements for the various grades of clay units. Table 3.1 shows the physical requirements for solid and hollow units.

Clay masonry units come in a variety of sizes. While there are no standard sizes, as there are for CMUs, most sizes are based on a 4- × 4-in. module, with the actual sizes of the bricks ⅜ to ½ in. smaller that the specified dimensions. Suppliers should be consulted for actual dimensions because sizes vary depending on the manufacturer. Coring patterns also vary depending on the manufacturer and are assumed to have no impact on the strength of solid units as long as the net cross-sectional area in every plane parallel to the bearing surface is greater than 75 percent of the gross cross-sectional area measured in the same plane.

3.2.3 Mortar

Mortar consists of a mix of cementitious materials and fine aggregates (sand) to which water is added to form a workable paste. The mortar is placed between the joints of the masonry units to bond the individual units into a solid unit. Mortar also serves as a seating material that enables the masonry units to be aligned precisely by sealing the variances in the units and correcting minor inaccuracies in placement.

The cementitious materials used in mortar may be portland cement, masonry cement, or mortar cement. Hydrated lime or lime putty is also used (only with portland cement) to improve the workability of the mortar. In the western United States, portland cement and lime are the most commonly used cementitious materials in mortar. Mortar cement is not used commonly, and masonry cement, which is a proprietary blend of portland cement and plasticizers, typically is not permitted for structural masonry.

Standard Specification for Mortar for Unit Masonry (ASTM C270)[14] provides specifications for mortar used in masonry construction. This standard categorizes mortar in two ways:

1. The mortar can be specified by its properties, which are obtained by testing under laboratory conditions.

2. The mortar can be specified by the volumetric proportions of the constituent materials.

Tables 3.2 and 3.3 show the requirements for specifying mortar by property and proportion, respectively. Mortar is classified as Type M, S, N, or O depending either on the proportion of materials or on its performance.

Each of the principal components of mortar makes a definite contribution to its performance. Portland cement contributes to the mortar's strength, durability, and bond strength. Lime, which sets only on contact with air, improves workability by helping the mortar retain water and elasticity. Sand acts as a filler and contributes to the strength of the mix. Sand also reduces cracking by decreasing the setting time and drying shrinkage of mortar. Sand also enables the unset mortar to retain its shape and thickness under several courses of CMUs. Water is the mixing agent that gives workability and hydrates the cement.

Workable mortar has a smooth, plastic consistency that makes it easy to spread and still support the weight of masonry units. Workable mortar also adheres to vertical masonry surfaces and readily squeezes out of mortar joints. A well-graded, smooth aggregate improves workability. Air entrainment improves workability through the action of miniature air bubbles, which function like ball bearings in the mixture.

Admixtures may be added to improve certain mortar characteristics. In very hot, dry weather, an admixture may be added to retard the setting time to allow a little more time before the mortar must be retempered. On the other hand, admixtures may be added to mortar in cold climates to accelerate the hydration of the cement in the mortar.

Mortar	Type	Average Compressive Strength at 28 Days (psi)	Minimum Water Retention (%)	Maximum Air Content (%)	Aggregate Ratio (Measured in Damp, Loose Condition)
Cement-lime	M	2,500	75	12	Not less than 2¼ and not more than 3½ times the sum of the separate volumes of cementitious materials
	S	1,800	75	12	
	N	750	75	14	
	O	350	75	14	
Mortar cement	M	2,500	75	12	
	S	1,800	75	12	
	N	750	75	14	
	O	350	75	14	
Masonry cement	M	2,500	75	18	
	S	1,800	75	18	
	N	750	75	20	
	O	350	75	20	

TABLE 3.2   Mortar Specification by Properties

Mortar	Type	Proportions by Volume (Cementitious Materials)								Aggregate Measured in a Damp, Loose Condition
		Portland Cement or Blended Cement	Masonry Cement			Mortar Cement			Hydrated Lime or Lime Putty	
			M	S	N	M	S	N		
Cement-lime	M	1	—	—	—	—	—	—	¼	Not less than 2¼ and not more than 3 times the sum of the separate volumes of cementitious materials
	S	1	—	—	—	—	—	—	Over ¼ to ½	
	N	1	—	—	—	—	—	—	Over ½ to 1¼	
	O	1	—	—	—	—	—	—	Over 1¼ to 2½	
Mortar cement	M	1	—	—	—	—	—	1	—	
	M	—	—	—	—	1	—	—	—	
	S	½	—	—	—	—	—	1	—	
	S	—	—	—	—	—	1	—	—	
	N	—	—	—	—	—	—	1	—	
Masonry cement	M	1	—	—	1	—	—	—	—	
	M	—	1	—	—	—	—	—	—	
	S	½	—	—	1	—	—	—	—	
	S	—	—	1	—	—	—	—	—	
	N	—	—	—	1	—	—	—	—	
	O	—	—	—	1	—	—	—	—	

TABLE 3.3 Mortar Specification by Proportions

ASTM C270 provides recommendations for selecting a mortar type for a specific application based on the type of a structural member and its location in reference to grade and exposure. Specifically, for exterior elements at or below grade, Type S mortar is recommended, whereas Type M or N mortars could be used as an alternative depending on the required strength. Type N mortar is recommended for parapets and exterior and interior load-bearing walls, whereas Type O mortar typically is used in non-load-bearing applications.

3.2.4   Reinforcement

Reinforcing steel used in masonry construction consists of deformed bars and joint reinforcement. Deformed bars typically comply with either *Deformed and Plain Carbon-Steel Bars for Concrete Reinforcement* (ASTM A615)[9] or *Low-Alloy Steel Deformed and Plain Bars for Concrete Reinforcement* (ASTM A706).[10] ASTM A615 provides standard specifications for plain carbon-steel reinforcing bars that are used most commonly. ASTM A706 covers low-alloy steel bars, which are used when more restrictive mechanical properties and chemical composition are required to enhance weldability and provide closer control of tensile properties. Deformed bars range from a minimum size of #3 to a recommended maximum size of #9 for strength design and #11 for allowable stress design. The bars are usually Grade 60 with a minimum yield strength of 60,000 psi, although Grade 40 bars with a minimum yield strength of 40,000 psi are sometimes used for #3 and #4 bars.

Joint reinforcement, which is placed horizontally in the mortar joints between the units, is of the ladder or truss type, as shown in Fig. 3.7. The use of truss-type joint reinforcement is discouraged in several jurisdictions because it interferes with the placement of vertical reinforcing steel and the flow of grout in the cells.

3.2.5   Grout

Grout is a mixture of cementitious material, aggregate, and enough water to cause the mixture to flow readily, without segregation, into cells or cavities in the masonry. Grout is always placed in wall spaces containing steel reinforcement. The grout bonds to the masonry units and steel so that they act together to resist imposed loads. In some reinforced load-bearing masonry walls, all cells, including those without reinforcement, are grouted to further increase the strength of the wall.

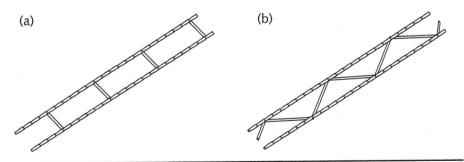

Figure 3.7   Joint reinforcement.

Grout for Masonry (ASTM C 476)[15] provides requirements for grout in masonry construction. The specification identifies two types of grouts—*fine grout* and *coarse grout*—depending on the size of aggregates used. Fine grout typically is used where the grout space is small and consists primarily of 1 part portland cement and 2¼ to 3 parts of sand by volume. Course aggregate contains the same materials as fine aggregate but also includes 1 to 2 parts of pea gravel. Both types of grout are required to contain enough water to have slump of 8 to 11 in.

Admixtures also can be used to improve the properties of grout. Various admixtures can be used to decrease the shrinkage of the grout as it hardens, improve the slump without additional water, reduce the amount of cement required, or improve grout performance in cold or hot climates. However, admixtures must be used with care because they sometimes can diminish certain grout characteristics while improving others. For this reason, admixtures are not typically accepted unless specifically approved by the building official.

Typically, grout is required to have a minimum compressive strength of 2,000 psi. Higher values can be used to obtain a larger overall masonry compressive strength. Lower values are permitted if the required masonry strength is achieved by testing. Standard methods for sampling and testing of grout are given in *Sampling and Testing Grout* (ASTM C1019).[17] The grout must be tested in conditions that are representative of the conditions in the structure.

3.2.6 Masonry Assemblage

Properties of the individual masonry components, such as grout, mortar, and masonry units, affect the compressive strength of the masonry f_m used in design.

Tables 1 and 2 of the Masonry Standards Joint Committee (MSJC), *Specification for Masonry Structures* (TMS 602-08/ACI 530.1-08/ASCE/6-08),[26] relate the compressive strengths of units to f_m for clay and concrete masonry, respectively. The values in these tables are based on mortar Types M, S, or N and may be used when the conditions stipulated in ACI 530.1-08, Articles 1.4 B.2a and 1.4 B.2b for clay and concrete masonry, respectively, are met. When these conditions are not met, or when a more accurate assessment of f_m is desired, prism tests per *Compressive Strength of Masonry Prisms* (ASTM C1314)[18] can be conducted. The user should note, however, that for quality-assurance aspects of construction, the *2009 International Building Code* (2009 IBC)[23] Sec. 2101.2 mandates compliance with the 2009 IBC Sec. 2105 rather than directly referencing TMS 602-08/ACI 530.1-08/ASCE/6-08. The unit-strength method, permitted by 2009 IBC Sec. 2105.2.2.1, uses Table 2105.2.2.1.1, which relates the compressive strengths of clay masonry units and mortar type to f_m. This table is identical to Table 1 in TMS 602-08/ACI 530.1-08/ASCE/6-08. Similarly, the same section also provides Table 2105.2.2.1.2, which is similar to Table 2 in TMS 602-08/ACI 530.1-08/ASCE/6-08, which relates the compressive strengths of CMUs and mortar type to f_m. Section 2105.2.2.2 of 2009 IBC permits the use of a more comprehensive prism test method for determining the value of f_m. The prism test method must be used whenever limitations of 2009 IBC Sec. 2105.2.2.1 are not met or whenever specified in the construction document. Whenever the 2009 IBC is governing the project under consideration, the user must use its Chap. 21 as the first point of reference for determination of f_m as well as for other aspects pertaining to material properties. The minimum value of f_m permitted by the MSJC *Building Code Requirements for Masonry Structures* (TMS 402-08/ACI 530-08/ASCE 5-08)[25] Sec. 3.1.8.1.1 is 1,500 psi. The user should note that when a project is governed by the 2009 IBC, values of f_m as low as 1,000 psi can be specified, as provided for by 2009 IBC Sec. 2105, although an f_m value of less than 1,500 psi is not used commonly in practice.

3.3 Timber Mechanical Properties

Design values for timber are set by the American Lumber Standard Committee (ALSC). This committee is a nonprofit organization incorporated in the state of Maryland that serves as the standing committee for the Voluntary Product Standard (PS 20-10), *American Softwood Lumber Standard*.[28] The ALSC functions as a consensus process organization in accordance with the *Procedures for the Development of Voluntary Product Standards* of the U.S. Department of Commerce. The rules for stress grading lumber are written by several certified rules-writing agencies such as the Southern Pine Inspection Bureau, the National Lumber Grades Authority, the Northern Softwood Lumber Bureau, and the Western Wood Products Association and then are tested before undergoing a review and approval process by the ALSC. The rules are developed under the assumption that the board will be loaded in bending about the strong axis of the member.

Timber is a natural composite that is anisotropic when evaluated at the cellular level. In a log form, timber usually is considered to be orthotropic with the principal axes oriented in the longitudinal, radial, and tangential directions (Fig. 3.8). However, lumber is cut into rectangular cross sections and for design purposes is idealized as having two principal axes, parallel to grain and perpendicular to grain (Fig. 3.8), with

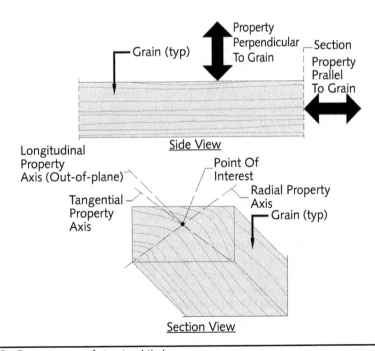

Figure 3.8   Property axes of structural timber.

lumber being weak in the perpendicular-to-grain direction and relatively strong in the parallel-to-grain direction. Due to its high strength-to-weight ratio, timber is suited to and has been used for long-span structures such as arches and domes. Timber is also resistant to slight acids and therefore is used in acidic environments such as cooling towers and pulp mills.

The *National Design Specification for Engineered Wood Construction* (ANSI/AF&PA NDS-2005)[2] provides the design properties for lumber based on species groups and size and divided into dimensional lumber, timbers, decking, and glued-laminated timber categories. Based on PS 20-99,[27] the ANSI/AF&PA NDS-2005 lists nominal design values in the tables for properties that are considered. The ANSI/AF&PA NDS-2005 Sec. 3.8.2 states that tension perpendicular to grain should be avoided whenever possible and suggests mechanical cross-grain reinforcement capable of resisting all applied forces when avoidance of tension perpendicular to grain is not possible. Similarly, the American Society of Civil Engineers (ASCE) *Minimum Design Loads for Buildings and Other Structures* (ASCE/SEI 7-05)[37] Secs. 12.11 and 12.14 prohibit the use of cross-grain tension and bending in ledgers and other framing used in wood diaphragm anchorage. Since timber is a natural viscoelastic material, the nominal properties are based on a set of conditions. These conditions are

- Moisture Content ≤ 19 percent
- Temperature ≤ 100°F
- Duration of Load = 10 years accumulative over a 50-year window of time
- Depth = 12 in.
- The member is placed in bending about the strong axis.

If the end-use conditions are different from these, adjustment factors are used to adjust the nominal design value to the allowable stress design (ASD) or load and resistance factor design (LRFD) values for the environmental and physical conditions of the application.

Timber is graded into visual, mechanically stress rated, and mechanically evaluated grades. Visual graded lumber is the most common grading system used for lumber. There are usually eight different grades possible, although more are possible for certain species groups, and each is assigned a set of mechanical properties for design. The lumber is evaluated by a grader who is trained in the rules specific to each of the grades and depending on the number, size, and location of growth characteristics (e.g., knots, slope of grain, checks, etc.). The piece then is assigned a grade and grouped according to market demands. For instance, construction lumber is most often graded as #2 and better because there is not a sufficient premium to justify grading and inventorying a large number of grades (i.e., select structural and #1 grades in addition to #2). The higher grades can be specially ordered for particular jobs, but the most cost-effective designs will use lumber species and grade groupings that are inventoried by the local lumber yards. Truss and glued-laminated timber manufacturers separate the higher grades to use in the high-stress locations of their products.

Machine-stress-rated (MSR) lumber represents lumber that is graded based on the modulus of elasticity (MOE) of the lumber stresses about the weak axis of the board. There is a correlation between the MOE of the board about the weak axis and the MOE of the board about the strong axis. There is also a correlation between the MOE about the strong axis and the bending strength of the board. There are 37 different grades of MSR lumber that range from 900f-1.0E to 3000f-2.4E. The grade designation represents the design bending stress and the average MOE (i.e., 1200f-1.2E represents 1,200 psi bending strength and 1.2×10^6 psi for MOE). Other mechanical properties are assigned and listed in the NDS.

Machine-evaluated lumber (MEL) is graded using microwaves and other technologies. There are 31 different grades of such lumber, and they are not commonly marketed in most regions of the country.

The adjustments and determination of design values are presented in detail with examples in Chap. 6. The objective of design with wood, as with any material, is to take advantage of the material's strengths while compensating for its weaknesses. As it relates to wood, this results in structural designs that use the material's longitudinal strength while compensating for the perpendicular-to-grain weakness.

3.4 Open-Web Joist Systems

Materials used for noncomposite and composite joists in open-web joist roof and floor systems do not appreciably differ from those used in structural steel and cold-formed steel industries. The main elements of a noncomposite joist system include the steel joist member, steel deck, bridging elements, and deck attachment hardware. Floor applications typically feature a concrete topping slab. The type of steel deck used depends on the purpose for which the joist member is used. Specifically, roof steel deck is used for roof framing, whereas noncomposite or composite floor deck is used for joists supporting concrete slabs. Attachment hardware comes in the form of either welding or mechanical fastening. Furthermore, when concrete topping slabs are used, they include supplemental reinforcement in the form of wire mesh, deformed bars, or fiber reinforcement.

Composite joist members contain a steel joist member, composite steel deck, headed shear studs, bridging elements, deck attachment hardware, and concrete slabs. As is the case with noncomposite systems, concrete slabs include supplemental reinforcement in the form of wire mesh, deformed bars, or fiber reinforcement.

Chapter 7 provides more information on the role and design of each of the constituent elements just listed. The following sections provide an overview of the materials employed in fabrication of these elements.

3.4.1 Steel Joist Members

Permissible materials for fabrication of steel joist members are listed in the following 2009 IBC referenced standards:

- *Standard Specification for Composite Steel Joists, CJ-Series* (SJI-CJ-1.0-06)[33]
- *Standard Specification for Joist Girders* (SJI-JG-1.1-05)[34]
- *Standard Specification for Open-Web Steel Joists, K-Serie*s (SJI-K-1.1-05)[35]
- *Standard Specification for Longspan Steel Joists, LH-Series, and Deep Longspan Steel Joists, DLH-Series* (SJI-LH/DLH-1.1-05)[36]

For fabrication of joist chord and web members, SJI-CJ-1.0-06 Sec. 102.1, SJI-JG-1.1 Sec. 1002.1(a), SJI-K-1.1-05 Sec. 3.1, and SJI-LH/DLH-1.1-05 Sec. 102.1 all permit the use of mild and ductile steels complying with one of the following ASTM material specifications:

- *Carbon Structural Steel* (ASTM A36)
- *High-Strength, Low-Alloy Structural Steel* (ASTM A242)
- *High-Strength Carbon-Manganese Steel of Structural Quality* (ASTM A529, Grade 50)
- *High-Strength Low-Alloy Columbium-Vanadium Structural Steel* (ASTM A572, Grades 42 and 50)
- *High-Strength Low-Alloy Structural Steel with 50 ksi Minimum Yield Point to 4 Inches Thick* (ASTM A588)
- *Steel, Sheet and Strip, High-Strength, Low-Alloy, Hot-Rolled and Cold-Rolled, with Improved Corrosion Resistance* (ASTM A606)
- *Steel, Sheet, Cold-Rolled, Carbon, Structural, High-Strength Low-Alloy and High-Strength Low-Alloy with Improved Formability* (ASTM A1008)
- *Steel, Sheet and Strip, Hot-Rolled, Carbon, Structural, High-Strength Low-Alloy and High-Strength Low-Alloy with Improved Formability* (ASTM A1011)

The SJI-CJ-1.0-06 Sec. 102.1, SJI-JG-1.1 Sec. 1002.2, SJI-K-1.1-05 Sec. 3.2, and SJI-LH/ DLH-1.1-05 Sec. 102.2 all permit the use of steels other than those conforming with the preceding ASTM standards, provided that such steels possess the ductility and strength prescribed in those specification sections and confirmed per *Standard Test Methods and Definitions for Mechanical Testing of Steel Products* (ASTM A370).[8] Ductility so determined must reflect 20 percent elongation over a 2-in. gauge for sheet and strip and 18 percent over an 8-in. gauge for bars, plates, and shapes. In determining the ductility compliance based on percent elongation for plates, shapes, or bars, the preceding elongation limits must be adjusted for member thickness, as stipulated in the ASTM standard that applies to a steel with a similar yield strength; for example, if the tested material indicates a yield of about 36 ksi, ASTM A36 stipulates that elongation adjustments for thickness will apply. When a member is designed using ANSI/AISI S100-07 and the design strength takes into account the strength increase due to cold forming, the entire member represents a test coupon. For cases where the cold-forming effect on strength is not considered, the test coupon can be cut from the as-rolled member, such as a plate or an angle.

The SJI-CJ-1.0-06 Sec. 101, SJI-JG-1.1 Sec. 1001, SJI-K-1.1-05 Sec. 2, and SJI-LH/DLH-1.1-05 Sec. 101 stipulate that the chord design be based on a material with a yield strength of 50 ksi. Materials with yield strength from 36 to 50 ksi are acceptable for the design of web members and bearing elements.

Chord members typically consist of hot-rolled double-angle sections, although hot-rolled members, such as WT sections, or cold-formed members, such as cold-formed double angles, are used in some cases. Web members typically consist of hot-rolled double-angle, single-angle, or round-rod sections. Bearing elements usually consist of double-angle sections welded to top-chord angle vertical legs. Information on the particular role and configuration of each individual element of a steel joist is presented in Chap. 7. Angle sizes for steel joists chord and web elements may be the standard sizes listed in the AISC *Steel Construction Manual* (AISC 325-05)[3] or other proprietary sizes used by an individual manufacturer.

Welding of joint elements typically is achieved using E60 or E70 electrodes and F6XX-EXXX or F7XX-EXXX flux-electrode combinations with weld metal strengths of 60 or 70 ksi, respectively.

3.4.2   Steel Deck

The materials used for fabrication of cold-formed steel decks used in open-web joist systems are listed in the following 2009 IBC–referenced Steel Deck Institute (SDI) standards:

- *Standard for Noncomposite Steel Floor Deck* (ANSI/SDI-NC1.0-06)[30]
- *Standard for Steel Roof Deck* (ANSI/SDI-RD1.0-06)[31]

The *Standard for Composite Steel Floor Deck* (ANSI/SDI-C1.0-06)[29] is used for deck applications in which the deck serves as a permanent support and steel reinforcement for the supported concrete slab. As noted in Chap. 1, although this document is not an IBC-referenced standard, it is generally used to supplement the requirements of 2009 IBC–referenced ASCE *Standard for the Structural Design of Composite Slabs* (ANSI/ASCE 3-91),[7] which is often viewed as outdated. ANSI/ASCE 3-91 simply refers to AISI S100 for the list of permissible material standards.

Section 2.1 of the preceding three SDI standards stipulates the following material standard requirements for roof decks and accessories:

- ASTM A653, *Structural Quality*, Grade 33 or higher—for galvanized deck and structural accessories
- ASTM A1008, or other structural sheet steel, or *High-Strength Low-Alloy Steel* (HSLAS) from Chap. A of AISI S100, Grade 33 or higher—for painted, top-uncoated or phosphatized/painted-bottom, uncoated decks and for structural and nonstructural components
- ASTM A653, *Commercial Quality*, for nonstructural accessories

General characteristics of sheet steels used in cold forming of deck products are covered in Sec. 3.1 and Chap. 4.

Steel decks are manufactured in thicknesses ranging from 26 (0.0179 in.) to 16 (0.0598 in.) gauge with yield strengths ranging from 33 to 80 ksi. The steels corresponding to 80 ksi yield strength typically are used only with Gauges 26 and 24 (0.0235 decks), although the actual thickness may be 95 percent of the nominal value, as permitted by Chap. A of ANSI/AISI S100-07 and Sec. 2.2 of the previously referenced SDI standards. Specifically, the low F_u/F_y ratio associated with Grade 80 steels indicates a very low ductility, creating the possibility of sheet cracking and fracturing during the cold-forming process. Furthermore, as outlined in Sec. 3.1, ANSI/AISI S100-07 Chap. A stipulates a 25 percent reduction in yield strength for nonductile steels to alleviate the possibility of a sudden and premature failure. Therefore, the resulting nominal yield strength for a Grade 80 steel for use in design calculations is 60 ksi.

Steel decks typically are fabricated and shipped in strips up to 36 in. in width, whereas the deck lengths are cut to size to accommodate the actual sheathing surface.

TECHNICAL PRODUCT INFORMATION

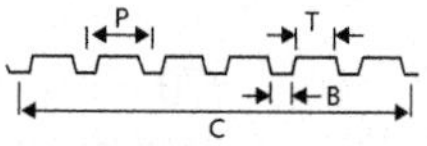

	Deck Type	Gauge	Indiana				Nebraska				South Carolina				Texas				Alabama/New York			
			Approximate Dimensions in Inches																			
			C	P	T	B	C	P	T	B	C	P	T	B	C	P	T	B	C	P	T	B
ROOF	1.5B, 1.5BI, 1.5BA, 1.5BIA 1.5BSV	24 22 20 19 18 16	NA 36 36 36 36 36	6.00	3.50	1.75	30 36 36 36 36 36	6.00	3.50	1.75	36 36 36 36 36 36	6.00	3.50	1.75	NA 36 36 36 36 36	6.00	3.50	1.75	36 36 36 36 36 36	6.00	3.50	1.75
ROOF	1.5F	22 20 18	30	6.00	4.25	0.50	36	6.00	4.25	0.50	36	6.00	4.25	0.50	36	6.00	4.25	0.50	36	6.00	4.25	0.50
ROOF	1.5A	22 20 18	36	6.00	5.00	0.38	36	6.00	5.00	0.38	36	6.00	5.00	0.38	NA	-	-	-	NA	-	-	-
ROOF	3N, 3NI, 3NA, 3NIA	22 20 18 16	24	8.00	5.38	1.88	24	8.00	5.38	1.88	24	8.00	5.38	1.88	24	8.00	5.38	1.88	24	8.00	5.38	1.88
ROOF	1.0E	26 24 22 20	36	4.00	1.13	1.33	32	4.00	1.01	1.25	33	3.67	0.90	0.90	33	3.67	1.00	1.00	36	4.00	1.13	1.13
noncomposite	0.6C and 0.6CSV	28 26 24 22	NA 30 30 NA	2.50	0.62	0.62	NA 36 36 36	3.04	0.63	0.63	35 35 35 35	2.50	0.75	0.75	30 35 35 35	2.50	0.62	0.62	30 30 30 30	2.50	0.75	0.75
noncomposite	1.0C and 1.0CSV	26 24 22 20	36	4.00	1.13	1.13	32	4.00	1.25	1.01	33	3.67	0.90	0.90	33	3.67	1.00	1.00	36	4.00	1.13	1.13
noncomposite	1.3C and 1.3CSV	26 24 22 20	NA	-	-	-	NA	-	-	-	NA	-	-	-	32	4.57	1.06	1.06	NA	-	-	-
noncomposite	1.5C	24 22 20 18	NA 36 36 36	6.00	1.75	3.50	30 36 36 36	6.00	1.75	3.50	36 36 36 36	6.00	1.75	3.50	30 36 36 36	6.00	1.75	3.50	36 36 36 36	6.00	1.75	3.50
noncomposite	2C	22 20 18 16	36	12.0	5.00	5.00	36	12.0	5.00	5.00	36	12.0	5.00	5.00	36	12.0	5.00	5.00	36	12.0	5.00	5.00
noncomposite	3C	22 20 18 16	36	12.0	4.75	4.75	36	12.0	4.75	4.75	36	12.0	4.75	4.75	36	12.0	4.75	4.75	36	12.0	4.75	4.75
COMPOSITE	1.5VL and 1.5VLI	22 20 19 18 16	36	6.00	3.50	1.75	36	6.00	3.50	1.75	36	6.00	3.50	1.75	36	6.00	3.50	1.75	36	6.0	3.50	1.75
COMPOSITE	1.5VLR	22 20 19 18 16	36	6.00	1.75	3.50	36	6.00	1.75	3.50	36	6.00	1.75	3.50	36	6.00	1.75	3.50	36	6.0	1.75	3.50
COMPOSITE	2VLI	22 20 19 18 16	36	12.0	5.00	5.00	36	12.0	5.00	5.00	36	12.0	5.00	5.00	36	12.0	5.00	5.00	36	12.0	5.00	5.00
COMPOSITE	3VLI	22 20 19 18 16	36	12.0	4.75	4.75	36	12.0	4.75	4.75	36	12.0	4.75	4.75	36	12.0	4.75	4.75	36	12.0	4.75	4.75

FIGURE 3.9 Commercial deck profiles and dimensions. (Vulcraft, 2008.[38])

The deck cross-sectional shapes are designed to maximize their usefulness for the purpose for which they are intended. For example, composite deck profiles contain embossments that enhance the concrete-deck bond. Figure 3.9 illustrates an assortment of roof, noncomposite, and composite decks available from Vulcraft.[38]

3.4.3 Concrete

When using noncomposite deck for support of a concrete slab or composite deck as part of a composite slab, Sec. 2.4.B.2 of ANSI/SDI-NC1.0-06 and Sec. 2.4.B.3 of ANSI/SDI-

C1.0-06, respectively, stipulate that concrete must be of compressive strength f_c of at least 3 ksi and that it must meet all applicable provisions of the American Concrete Institute (ACI) *Building Code Requirements for Structural Concrete* (ACI 318-08).[1] Concrete admixtures containing chloride salts are not permitted because they may promote corrosion of the steel deck.

3.4.4 Reinforcement

Composite and noncomposite slabs typically are reinforced using welded-wire reinforcement conforming to *Standard Specification for Steel Welded Wire Reinforcement, Plain, for Concrete* (ASTM A185), with F_y = 65 ksi and F_u = 75 ksi, or *Standard Specification for Steel Welded Wire Reinforcement, Deformed, for Concrete* (ASTM A497), with F_y = 70 ksi and F_u = 80 ksi. Welded wire can be used for both flexural and shrinkage and temperature reinforcement. Current and historic sizes of welded-wire reinforcement are available from the Wire Reinforcement Institute (WRI) at www.wirereinforcementinstitute.org. The current welded wire sizes are reproduced in Appendix E of ACI 318-08. Figure 3.10 illustrates the structure of the designations used to indicate welded-wire reinforcement size and type. For example a 6 × 10–W1.4 × W1.4 designation would indicate a smooth wire, 6 in. width-wise and 10-in. length-wise wire spacing, and a cross-sectional area of 0.028 in.²/ft length-wise and 0.017 in.²/ft width-wise. Deformed wire mats typically are used in composite and noncomposite slabs. The commonly available wire sizes range from D1.4 and W1.4 to D45 and W45, although sizes in excess of W2.1 and W5, and D2.1 and D5, respectively, are seldom used in composite and noncomposite slabs, respectively. Mats with wire spacing of 4 × 4 and 6 × 6 are common for slabs on steel deck. Wire

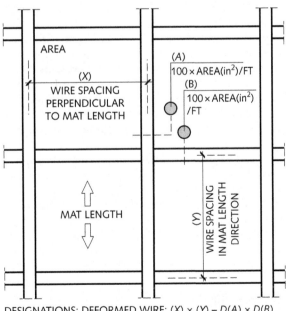

DESIGNATIONS: DEFORMED WIRE: $(X) \times (Y) – D(A) \times D(B)$
PLAIN WIRE: $(X) \times (Y) – W(A) \times W(B)$

FIGURE 3.10 Designation system for welded-wire reinforcement.

reinforcement mat sizes depend on specific manufacturer preference and manufacturing capability.

Mild reinforcement bars conforming to ASTM A615 with F_y = 60 ksi are used occasionally to supplement welded-wire fabric. When used, this reinforcement typically is employed in smaller diameters, such as #3, #4, or #5 bar.

Section 2.4.B.6 of ANSI/SDI-C1.0-06 permits the use of cold-drawn steel fibers conforming to *Standard Specification for Steel Fibers for Fiber-Reinforced Concrete* (ASTM A820) as an alternative to welded-wire reinforcement to satisfy minimum temperature and shrinkage reinforcement requirements.

3.4.5 Shear Connectors

Headed shear studs used to achieve shear connection at the interfaces of concrete slab and steel joist members in composite joist floor systems conform to American Welding Society (AWS) *Structural Welding Code—Steel* (AWS D1.1-04),[20] as required by Sec. 102.2 of SJI-CJ-1.0-06. Section 7.2.5 of AWS D1.1-04 stipulates that the shear connectors be made of material conforming to *Specification for Steel Bars, Carbon, Cold-Finished, Standard Quality* (ASTM A108) Grades 1010 through 1020. For shear connectors in flexural members, Table 7.1 of AWS D1.1-04 requires Type B studs with F_y = 51 ksi and F_u = 65 ksi. To prevent burn-through of the base material, AWS D1.1-04 Sec. 7.2.7 requires a minimum joist top chord thickness of a least one-third the stud diameter when studs are welded directly to the chord. Also, stud diameters may not exceed 2.5 times the joist top chord thickness when they are welded to the chord through cold-formed steel deck.

Stud diameters D_s in Fig. 3.11 of ⅜ in., ½ in., ⅝ in. and ¾ in. are used in composite joist construction, although ⅜-in.-diameter studs are used relatively rarely. The user should note that AWS D1.1-04 did not include ⅜-in.-diameter studs in the Type B classification, but this diameter was added in AWS D1.1-06.[21] As illustrated in Fig. 3.11, shear studs typically are fabricated to a length L_{pr} of ⅛ in. (for ⅜-in. and ½-in. diameter studs) or ³⁄₁₆ in. (for ⅝-in. and ¾-in. diameter studs) longer than the nominal installed length L_{po} to account for the length loss during welding. Stud lengths depend on the deck rib height and overall slab thickness. Lengths ranging from 2½ to 6 in. are typical for composite joist construction. The stud head thickness T_h in Fig. 3.11 is ⅜ in. for ¾-in.-diameter studs, ⁹⁄₃₂ in. for ⅜-in-, ½-in-, and ⅝-in.-diameter studs. Head diameter D_h in Fig. 3.11 is 1¼ in. for ¾-in. and ⅝-in. diameter studs and 1 in. for ½-in. and ⅜-in. diameter studs.

Stud bases can differ somewhat from manufacturer to manufacturer. Base configuration is a function of the welding process. Section 7.2.4 of AWS D1.1-04 requires each configuration of stud base to be qualified prior to use per AWS D1.1-04 Annex IX to ensure weld quality. Two common stud base configurations are illustrate in Fig. 3.11. The first incorporates an edge chamfer and a round tip and also may include a base groove. The second, more basic type includes a cylindrical pointed end. Both configurations facilitate the flow of weld material and subsequent fusion with stud and base metal.

3.4.6 Deck Attachments

Deck attachment to the supporting members is achieved through the use of either welds or mechanical connectors, as provided by Sec. 3.2 of ANSI/SDI-C1.0-06, ANSI/SDI-NC1.0-06, and ANSI/SDI-RD1.0-06. Welded attachments are in the form of arc-spot

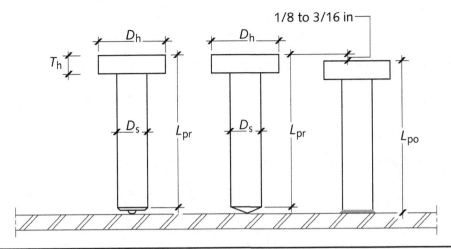

Figure 3.11   Shear connector features and dimensions.

welds, commonly known as *puddle welds*. These welds are executed in accordance with *Structural Welding Code—Sheet Steel* (AWS D1.3-98).[19] The E60 and E70 electrodes, resulting in the weld metal tensile strengths of 60 and 70 ksi, respectively, are used. Effective weld diameter used in computing the weld resistance is computed from the visible weld diameter, as provided by Sec. E2.2 of ANSI/AISI S100-07. A recent study[29] shows that proper weld execution, in particular, the length of welding arc time, is crucial in ensuring that the effective length diameter calculated from the externally visible diameter actually can be achieved. Additional discussion on puddle welds is provided in Chap. 4.

Mechanical fasteners used in deck attachments come in the form of self-tapping screws of power-actuated fasteners[23] (Fig. 3.12). When power-actuated fasteners are employed, they typically are used to attach the deck to the supporting member, whereas screws are used to stitch the side-laps of adjacent deck panels together. Alternatively, the entire deck attachment can be done with screws. The self-tapping screws are capable of cutting their own threads into embedment material as they are drilled into it. The self-tapping screws typically are indicated by number designations that reflect their external diameter, as shown in Fig. 3.13, and by number of threads per inch. For instance, a #12-14 screw from Fig. 3.12 has an out-to-out of thread diameter of 0.216 in. and a density of 14 threads per inch. Furthermore, screws designations also can include length of screw shank. Therefore, a #12-14 × 1½ screw has an out-to-out of thread diameter of 0.216 in. and a density of 14 threads per inch and is 1.5 in. long, not including its head. The #10 and #12 screws typically are used in deck attachments. Screws can be made from various materials, although many are made from metals conforming to *Standard Specification for General Requirements for Wire Rods and Coarse Round Wire, Carbon Steel* (ASTM A510) Grades 1018 to 1022. The screws are subsequently hardened to a Rockwell C hardness (HRC) of up to about 40 to ensure their ability to drill into embedment material. The ability of a screw to drill into a supporting member depends on its hardness, the hardness of the supporting material, the thickness of the supporting material, the depth of the screw drilling point, and the availability of adequate torsional strength, which ensures that a screw can be drilled into a member without fracturing at the interface of screw head and screw shank. The practical member thickness limit for a self-tapping screw application is about ⅜ in. Larger embedment thickness is possible with predrilled holes or with screws specially designed for thicker embedments. In terms of their shape and design, screws are completely nonstandardized, and their specific design features vary from manufacturer to manufacturer. Specific design features of a particular screw are germane to its intended purpose. It is therefore important to follow the manufacturer's application guidelines as they relate to screw selection and installation. Figure E4.4-2 of ANSI/AISI S100-07 indicates typical features of common screw types. Figure 3.12

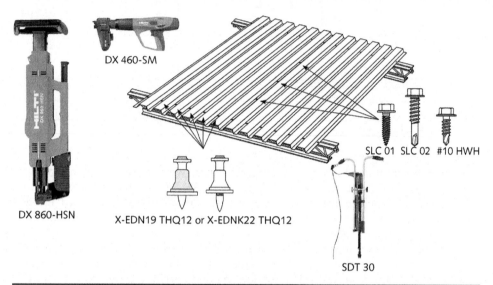

FIGURE 3.12 Screws and power-actuated fasteners in open-web joist systems. (*Hilti, Inc., 2011.*[23])

Number Designation	Nominal Diameter, d	
	in	mm
0	0.060	1.52
1	0.073	1.85
2	0.086	2.18
3	0.099	2.51
4	0.112	2.84
5	0.125	3.18
6	0.138	3.51
7	0.151	3.84
8	0.164	4.17
10	0.190	4.83
12	0.216	5.49
1/4	0.250	6.35

FIGURE 3.13 Screw diameter designations. (*AISI, 2007.*[6])

illustrates a typical deck screw and the associated installation tool from Hilti.[23] Further discussion on screw connections is provided in Chap. 4.

Power-actuated fasteners (PAFs) typically are made from Society of Automotive Engineers (SAE) Grades 1060 through 1080 or other similar high-carbon materials. Power-actuated fasteners made from these materials are hardened subsequently up to about 60 HRC to ensure their ability to penetrate the embedment material. Ultimate tensile strength of the hardened fastener is typically around 260 ksi. Unlike screws, which penetrate material through circular drilling action, PAFs are fired into the embedment member and must possess hardness four to five times higher than the embedment material[22] to penetrate it successfully. The fastener penetration of the embedment material generates large temperatures, causing partial fusion of fastener surface with the base material hole. Furthermore, the base material displaced by fastener penetration has the propensity to partially return in its original position, creating confinement pressure at the fastener surface. Therefore, the pull-out strength of a PAF is derived from partial fusion with the base material and confinement stresses induced by the displaced embedment material. Certain types of PAFs contain knurling patterns which increase the fastener surface roughness and pullout resistance. Power actuated fasteners are highly proprietary connection systems whose successful choice of application and installation are heavily dependent on manufacturer's installation tools, proper power settings for fastener driving, and application recommendations. When properly installed, PAFs can serve as an effective alternative to screws and welds when forming deck attachments to the supporting members. A typical PAF from one of the major manufacturers[23] and the associated installation tools are shown in Fig. 3.12.

References

1. American Concrete Institute (ACI). 2005. *Building Code Requirements for Reinforced Concrete* (ACI 318-05). ACI, Farmington Hills, MI.
2. American Forest and Paper Association (AF&PA). 2005. *National Design Specification for Engineered Wood Construction* (ANSI/AF&PA NDS-2005). AF&PA, Washington, DC.
3. American Institute of Steel Construction (AISC). 2005. *Steel Construction Manual* (AISC 325-05). AISC, Chicago, IL.
4. American Iron and Steel Institute (AISI). 2007. *North American Specification for the Design of Cold-Formed Steel Members* (ANSI/AISI S100-07). AISI, Washington, DC.
5. American Iron and Steel Institute (AISI). 2008. *Cold-Formed Steel Design Manual* (AISI D100-08). AISI, Washington, DC.
6. American Iron and Steel Institute (AISI). 2007. *Commentary on North American Specification for the Design of Cold-Formed Steel Members.* AISI, Washington, DC.
7. American Society of Civil Engineers (ASCE). 1991. *Standard for the Structural Design of Composite Slabs* (ANSI/ASCE 3-91). ASCE, New York, NY.
8. American Society for Testing and Materials (ASTM). 2003. *Standard Test Methods and Definitions for Mechanical Testing of Steel Products* (ASTM A370). ASTM, Philadelphia, PA.
9. American Society for Testing and Materials (ASTM). 2009. *Deformed and Plain Carbon-Steel Bars for Concrete Reinforcement* (ASTM A615). ASTM, Philadelphia, PA.
10. American Society for Testing and Materials (ASTM). 2009. *Low-Alloy Steel Deformed and Plain Bars for Concrete Reinforcement* (ASTM A706). ASTM, Philadelphia, PA.

11. American Society for Testing and Materials (ASTM). 2004. *Standard Specification for Building Brick—Solid Masonry Units Made from Clay or Shale* (ASTM C62). ASTM, Philadelphia, PA.

12. American Society for Testing and Materials (ASTM). 2009. *Standard Specification for Load Bearing Masonry Units* (ASTM C90). ASTM, Philadelphia, PA.

13. American Society for Testing and Materials (ASTM). 2007. *Standard Specification for Facing Brick—Solid Masonry Units Made from Clay or Shale* (ASTM C216). ASTM, Philadelphia, PA.

14. American Society for Testing and Materials (ASTM). 2008. *Standard Specification for Mortar for Unit Masonry* (ASTM C270). ASTM, Philadelphia, PA.

15. American Society for Testing and Materials (ASTM). 2009. *Grout for Masonry* (ASTM C 476). ASTM, Philadelphia, PA.

16. American Society for Testing and Materials (ASTM). 2004. *Standard Specification for Hollow Brick—Hollow Masonry Units Made from Clay or Shale* (ASTM C 652). ASTM, Philadelphia, PA.

17. American Society for Testing and Materials (ASTM). 2009. *Sampling and Testing Grout* (ASTM C1019). ASTM, Philadelphia, PA.

18. American Society for Testing and Materials (ASTM). 2007. *Compressive Strength of Masonry Prisms* (ASTM C1314). ASTM, Philadelphia, PA.

19. American Welding Society (AWS). 1998. *Structural Welding Code—Sheet Steel* (AWS D1.3-98). AWS, Miami, FL.

20. American Welding Society (AWS). 2004. *Structural Welding Code—Steel* (AWS D1.1-04). AWS, Miami, FL.

21. American Welding Society (AWS). 2006. *Structural Welding Code—Steel* (AWS D1.1-06). AWS, Miami, FL.

22. Beck, H., and Reuter, M. 2005. *Powder-Actuated Fasteners in Steel Construction*. Steel Construction Calendar 2005. Ernst & Sohn, Berlin, Germany.

23. Hilti, Inc. 2011. *North American Product Technical Guide: A Guide to Specification and Installation*. Hilti, Tulsa, OK.

24. International Code Council (ICC). 2009. *International Building Code*. ICC, Washington, DC.

25. Masonry Standards Joint Committee (MSJC). 2008. *Building Code Requirements for Masonry Structures* (TMS 402-08/ACI 530-08/ASCE 5-08). The Masonry Society (TMS), Boulder, CO, American Concrete Institute (ACI), Farmington Hills, MI, and American Society of Civil Engineers, Reston, VA.

26. Masonry Standards Joint Committee (MSJC). 2008. *Specification for Masonry Structures* (TMS 602-08/ACI 530.1-08/ASCE 6-08). The Masonry Society (TMS), Boulder, CO, American Concrete Institute (ACI), Farmington Hills, MI, and American Society of Civil Engineers, Reston, VA.

27. National Institute of Standards and Technology (NIST). 1999. *Voluntary Product Standard* (PS 20-99), *American Softwood Lumber Standard*. U.S. Department of Commerce, National Institute of Standards and Technology, Gaithersburg, MD.

28. National Institute of Standards and Technology (NIST). 2010. *Voluntary Product Standard* (PS 20-10), *American Softwood Lumber Standard*. U.S. Department of Commerce, National Institute of Standards and Technology, Gaithersburg, MD.

29. Snow, G. L. 2008. *Strength of Arc Spot Welds Made in Single and Multiple Steel Sheets*. Thesis, Virginia Tech, Blacksburg, VA.

30. Steel Deck Institute (SDI). 2006. *Standard for Composite Steel Floor Deck* (ANSI/SDI-C1.0-06). SDI, Fox River Grove, IL.

31. Steel Deck Institute (SDI). 2006. *Standard for Noncomposite Steel Floor Deck* (ANSI/ SDI-NC1.0-06). SDI, Fox River Grove, IL.

32. Steel Deck Institute (SDI). 2006. *Standard for Steel Roof Deck* (ANSI/SDI-RD1.0-06). SDI, Fox River Grove, IL.

33. Steel Joist Institute (SJI). 2006. *Standard Specification for Composite Steel Joists, CJ-Series* (SJI-CJ-1.0-06). SJI, Forest, VA.

34. Steel Joist Institute (SJI). 2005. *Standard Specification for Joist Girders* (SJI-JG-1.1-05). SJI, Forest, VA.

35. Steel Joist Institute (SJI). 2005. *Standard Specification for Open Web Steel Joists, K-Series* (SJI-K-1.1-05). SJI, Forest, VA.

36. Steel Joist Institute (SJI). 2005. *Standard Specification for Longspan Steel Joists, LH-Series, and Deep Longspan Steel Joists, DLH-Series* (SJI-LH/DLH-1.1-05). SJI, Forest, VA.

37. Structural Engineering Institute (SEI) of the American Society of Civil Engineers (ASCE). 2006. *Minimum Design Loads for Buildings and Other Structures, Including Supplements Nos. 1 and 2* (ASCE/SEI 7-05). ASCE, Reston, VA.

38. Vulcraft. 2008. Vulcraft Steel Roof & Floor Decks. Vulcraft, Norfolk, NE.

39. Yu, W. W., and LaBoube, R. A. (2010). *Cold-Formed Steel Design.*, John Wiley & Sons, Hoboken, NJ.

CHAPTER 4

Design of Cold-Formed Steel Structures

4.1 Introduction

Cold-formed steel members include such products as studs and joists used in light commercial and residential construction, purlins and girts used in metal building construction, and deck and panel profiles used in the construction of floors and roofs. This chapter focuses primarily on the application of cold-formed steel members for light-framed commercial and residential applications. Such applications may be for axial load-bearing wall construction, in-fill curtain-wall systems, and floor or roof joists.

Attributes that have enabled the more widespread use of cold-formed steel construction are high strength-to-weight ratio, ease of mass production, easy adaptability to many cross-sectional geometries, noncombustibility, green construction, favorable performance to seismic loading, and cost-effectiveness. Figure 4.1 illustrates cross sections of typical products.

4.2 Cold-Formed Steel Framing

Cold-formed steel design for building construction is defined by the requirements stipulated by the American Iron and Steel Institute (AISI) *North American Specification for the Design of Cold-Formed Steel Structural Members* (AISI S100-07).[2] The 2009 *International Building Code* (2009 IBC)[15] references the 2007 edition of this standard (AISI S100-07), although AISI S100-07 has since been augmented by *Supplement No. 2 to the North American Specification for the Design of Cold-Formed Steel Structural Members*, AISI S100-07/S2-10.[2a] To facilitate the use of cold-formed steel in residential and light commercial applications, the AISI also has developed a comprehensive series of American National Standards Institute (ANSI)–approved consensus standards. Included are standards for floor and roof systems, wall studs, headers over doors and windows, lateral bracing, and trusses. These framing standards supplement the design criteria provided in AISI S100-07.

The standards may be group as general-use documents, such as

- AISI S200-07, *North American Standard for Cold-Formed Steel Framing—General Provisions*[4]
- AISI S201-07, *North American Standard for Cold-Formed Steel Framing—Product Data*[5]

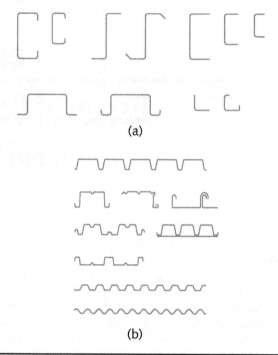

Figure 4.1 Typical cold-formed steel members: (a) structural members; (b) decks and panels.

Other standards are for specific applications of walls, floors, and roofs, such as

- AISI S210-07, *North American Standard for Cold-Formed Steel Framing—Floor and Roof System Design*[6]
- AISI S211-07, *North American Standard for Cold-Formed Steel Framing—Wall Stud Design*[7]
- AISI S212-07, *North American Standard for Cold-Formed Steel Framing Header Design*[8]
- AISI S213-07, *North American Standard for Cold-Formed Steel Framing—Lateral Design*[9]
- AISI S214-07, *North American Standard for Cold-Formed Steel Framing—Truss Design,*[10] and S2-08, *Supplement 2*[10]

For design assistance or prescriptive design,

- AISI S230-07, *Standard for Cold-Formed Steel Framing—Prescriptive Method for One- and Two-Family Dwellings*[11] and S2-08, *Supplement 2*[11]

These standards are the latest available editions. Chapter 35 of 2009 IBC should be consulted to determine the code-adopted edition of each of the standards referenced in the code. Section 2209 of 2009 IBC states that all cold-formed steel structural members must be designed in accordance with AISI S100-07. Section 2210 of 2009 IBC defines correlation of AISI S100-07 with the AISI framing standards as it relates to light-frame applications. For example, 2009 IBC Sec. 2210.4 permits the wall studs to be designed per either AISI S100-07 or AISI S211-07. On the other hand, 2009 IBC Sec. 2210.6 stipulates AISI S213-07 as the sole standard for lateral design. In many instances, as is the case with AISI S213-07, the framing standard provisions will refer to S100 for the design of specific members and elements.

Typical studs for load-bearing walls are C-sections of 33- and 50-ksi yield-stress steel, 3.5 and 6 in. deep by 1.625 in. wide, 0.033 to 0.071 in. thick, with a ½-in. stiffener lip on the edge of the flange. The exterior surfaces of the outside walls often are sheathed with wood or metal structural panels, whereas the interior surfaces receive gypsum-board sheathing. Both surfaces of the interior walls may be sheathed with gypsum. Similar C-sections, 6 to 12 in. deep, are used for floor joists. Roof construction may be with C-section rafters or with cold-formed steel trusses.

The design of wall studs can be based on either an all-steel design, in which discrete braces are provided along the member's length, or a sheathing-braced design. When sheathing-braced design is employed, the wall stud also should be evaluated without the sheathing bracing for the load combination stipulated in AISI S211-07. This provides for the possibility that the sheathing has been removed or has become ineffective accidentally.

Steel stud wall assemblies with sheathing or with diagonal bracing (X-bracing) also serve as in-plane diaphragms and shear walls to brace the structure and resist racking from wind or seismic loads.

4.3 Design Specifications and Materials

Cold-formed members used for light commercial and residential applications are designed in accordance with AISI S100-07. This is the recognized design standard in the United States, Canada, and Mexico. In the United States, this design standard is published by the AISI.

AISI S100-07 applies to members cold-formed to shape from carbon or low-alloy steel sheet, strip, plate, or bar, not more than 1 in. thick, used for load-carrying purposes in buildings. AISI S100-07 provides a list of 16 American Society for Testing and Materials (ASTM) steels that may be used for the fabrication and design of cold-formed steel members. Other steels may be used for structural members if they meet AISI S100-07 requirements such as minimum ductility. The minimum-ductility requirement is defined as a ratio of tensile strength to yield stress not less than 1.08 and a total elongation of at least 10 percent in 2-in. gauge length. For studs and joists used in light commercial and residential applications, the common steel is ASTM A1003 Grades 33 and 50. For deck and panel applications, ASTM A653 steels are used commonly. A summary of the 16 ASTM steels and the associated S100 requirements can be found in Part I of the AISI *Cold-Formed Steel Design Manual* (AISI D100-08).[12]

4.4 Manufacturing Methods and Effects

The cross section of a cold-formed member is achieved by a bending operation at room temperature. The dominant cold-forming process is known as *roll-forming*. In this process, a coil of steel is fed through a series of rolls, each of which bends the sheet progressively until the final shape is reached. Because the steel is fed in coil form, with successive coils weld-spliced as needed, the process can achieve speeds of up to about 300 ft/min and is well suited for large-quantity production. Small quantities may be produced on a press-brake, particularly if the shape is simple, such as an angle or channel cross section. In its simplest form, a press brake consists of a male die that presses the steel sheet into a matching female die.

4.5 Design Methodology

AISI S100-07 defines the nominal strength equations for the various structural members, such as beams and columns, as well as for connections. For allowable strength design (ASD), the nominal strength is divided by a safety factor and compared with the required strength based on nominal loads. For load and resistance factor design (LRFD), the nominal strength is multiplied by a resistance factor and compared with the required strength based on factored loads.

4.5.1 ASD Strength Requirements

A design satisfies the ASD requirements when the allowable strength of a structural component equals or exceeds the required strength expressed as

$$R \leq R_n/\Omega \tag{4.1}$$

where R = required strength

R_n = nominal strength

Ω = safety factor

R_n/Ω = allowable strength

4.5.2 LRFD Strength Requirements

A design satisfies the LRFD requirements when the design strength of a structural component equals or exceeds the required strength determined on the basis of the nominal loads multiplied by the appropriate load factors

$$R_u < \phi R_n \tag{4.2}$$

where R_u = required strength

R_n = nominal strength

ϕ = resistance factor

ϕR_n = design strength

4.6 Section Property Calculations

The properties of common sections used for light commercial and residential construction are summarized in AISI D100-08. Because the cross section of a cold-formed C-section is of a single thickness of steel, computation of section properties is simplified by using the linear method. In this method, the material is considered to be concentrated along the centerline of the steel sheet, and area elements are replaced by straight or curved line elements. For more in-depth discussion of the section property calculation, refer to Yu and LaBoube.[16] To reflect the influence of local buckling, an effective width rather than the full width of a compression element is used.

In addition to the effective-width method, an alternative method, the *direct strength method*, is available for determining nominal axial and flexural strengths directly from the full cross section properties. The direct strength method is particularly useful for members with complex cross-sectional shapes. Refer to the AISI S100-07 and the AISI *Direct Strength Method Design Guide* (AISI CF06-1),[14] for additional information.

4.7 Effective-Width Concept

The design of cold-formed steel members differs from that of hot-rolled steel members because the flat width-to-thickness ratios of compression elements of a cold-formed steel member cross section typically are large. Therefore, because of the large width-to-thickness ratios, the compression elements are subject to local buckling. Figure 4.2 illustrates local buckling in beams and columns.

To account for the effect of local buckling in design, the concept of effective width is employed for elements subject to compression. The equations for calculating effective widths of elements are summarized by Eqs. (4.3) to (4.7), as given by AISI S100-07. These equations are based on theoretical elastic plate buckling but modified as a result of extensive experimental studies.

4.7.1 Maximum Width-to-Thickness Ratio

Because the design equations in AISI S100-07 are empirical, maximum width-to-thickness ratios that reflect the test data from the experimental studies must be adhered

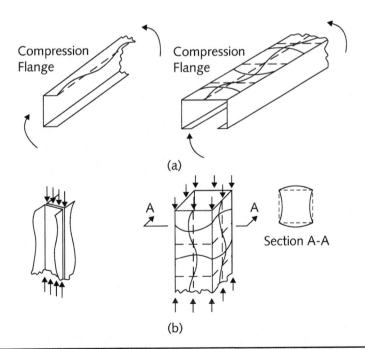

FIGURE 4.2 Local buckling of compression elements: (a) in beams; (b) in columns. (*American Iron and Steel Institute, 2007.*[3])

to. For compression elements, such as in flexural members or columns, the maximum flat width-to-thickness ratio w/t is as follows:

- Stiffened compression element having one longitudinal edge connected to a web or flange element, the other stiffened by a simple lip, 60
- Stiffened compression element with both longitudinal edges connected to other stiffened elements, 500
- Unstiffened compression element, 60
- For web elements of flexural members, the maximum web depth-to-thickness ratio h/t disregarding any intermediate stiffeners is as follows:
  - o Unreinforced webs, 200
  - o Webs with bearing stiffeners only, 260; bearing and intermediate stiffeners, 300

4.7.2 Effective Widths of Stiffened Elements

The effective width for load-capacity determination depends on a slenderness factor λ, defined in terms of the plate elastic buckling stress F_{cr} as

$$\lambda = \sqrt{\frac{f}{F_{cr}}} \tag{4.3}$$

where

$$F_{cr} = \frac{k\pi^2 E}{12(1-\mu^2)}\left(\frac{t}{w}\right)^2 \tag{4.4}$$

where k is the plate buckling coefficient (varies depending on the stress condition and element boundary conditions; for example, $k = 4.0$ for stiffened elements supported by a web along each longitudinal edge); f is the maximum compressive stress in the compression element; E is the modulus of elasticity (29,500 ksi); and μ equals 0.30.

For flexural members, when initial yielding is in compression, $f = F_y$, where F_y is the yield stress; when the initial yielding is in tension, $f =$ the compressive stress determined on the basis of effective section. For compression members, $f =$ column buckling stress F_n.

The effective width is defined as follows:

When $\lambda \le 0.673$, $b = w$ $\tag{4.5}$
When $\lambda > 0.673$, $b = \rho w$ $\tag{4.6}$

where the reduction factor ρ is defined as

$$\rho = \frac{1-0.22/\lambda}{\lambda} \tag{4.7}$$

Figure 4.3 illustrates the effective width of a cross section.

To calculate effective width for deflection determination, use the preceding equations but substitute for f the compressive stress at design loads, $f_d = 0.6F_y$, in Eqs. (4.3) to (4.7).

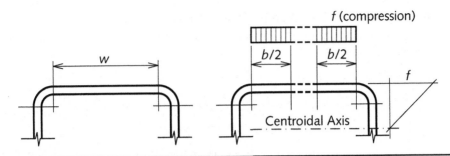

FIGURE 4.3 Illustration of uniformly compressed stiffened element: (a) actual element; (b) stress on effective element. (*American Iron and Steel Institute, 2007.*[2])

For compression elements having other boundary conditions, for example, unstiffened elements, or stress gradients such as webs subjected to compression from bending alone, Eqs. (4.3) through (4.7) are used with the appropriate buckling coefficient as defined by Chap. B of AISI S100-07. For more in-depth discussion of the effective-width concept, refer to Yu and LaBoube.[16]

4.8 Tension Members

The nominal tensile strength T_n of a concentrically loaded tension member is the smallest of three limit states: (1) yielding in the gross section (Eq. 4.8), (2) fracture in the net section away from the connections (Eq. 4.9), or (3) tension rupture in the net section at connections (Art. 4.13.2):

$$T_n = A_g F_y \tag{4.8}$$
$$\Omega = 1.67, \phi = 0.90$$

$$T_n = A_n F_u \tag{4.9}$$
$$\Omega = 2.0, \phi = 0.75$$

where A_g is the gross cross-sectional area, A_n is the net cross-sectional area, and F_y is the design yield stress.

4.9 Flexural Members

When designing a flexural member, the strength-limit states of bending, shear, web crippling, combined bending and shear, and combined bending and web crippling must be considered. Bending strength must consider both yielding and lateral-torsional buckling. For serviceability, deflections at service loads are also an important consideration. When evaluating the bending strength, the influence of local buckling must be considered by application of the effective-width equations. In addition to local or lateral-torsional buckling modes, open cross section members having compression flanges with edge stiffeners also must be evaluated for their distortional buckling strength. This is a buckling mode involving change in cross-sectional shape that differs from that of local buckling. Distortional buckling is more prevalent for cross sections formed using higher-strength steels and with small edge stiffeners. Provisions for distortional buckling are given in the AISI S100-07.

4.9.1   Strength Based on Initiation of Yielding

For an adequately braced beam, the nominal strength M_n is the effective yield moment based on section strength:

$$M_n = S_e F_y \tag{4.10}$$

$\Omega = 1.67$, $\phi = 0.95$ (stiffened and partially stiffened flanges)

$\Omega = 1.67$, $\phi = 0.90$ (unstiffened flanges)

where S_e is the elastic section modulus of the effective section calculated with the extreme fiber at the design yield stress F_y. The stress in the extreme fiber can be compression or tension depending on which is farthest from the neutral axis of the effective section. If the extreme fiber stress is compression, the effective width and the effective section can be calculated directly based on the stress F_y in that compression element. However, if the extreme fiber stress is tension, the stress in the compression element depends on the effective section, and therefore, a trial-and-error solution is required.

4.9.2   Strength Based on Lateral-Torsional Buckling

The nominal strength M_n of laterally unbraced segments of singly, doubly, and point-symmetric sections is given by Eq. (4.11). These provisions apply to I, Z, C, and other singly symmetric sections. The nominal strength is

$$M_n = S_c F_c \tag{4.11}$$

$\Omega = 1.67$, $\phi = 0.90$

where S_c is the elastic section modulus of the effective section calculated relative to the extreme compression fiber at stress F_c.

For $F_e \geq 2.78F_y$,

$$F_c = F_y \tag{4.12}$$

For $2.78F_y > F_e > 0.56F_y$,

$$F_c = \frac{10}{9} F_y \left(1 - \frac{10F_y}{36F_e} \right) \tag{4.13}$$

For $F_e \leq 0.56F_y$,

$$F_c = F_e \tag{4.14}$$

where F_e is the elastic critical lateral-torsional buckling stress calculated as follows.

For Singly and Doubly Symmetric Sections

The elastic buckling stress for bending about the axis of symmetry is

$$F_e = (C_b r_o A \sqrt{\sigma_{ey} \sigma_t})/S_f \tag{4.15}$$

For singly symmetric sections, the x axis is the axis of symmetry oriented such that the shear center has a negative x coordinate. The elastic buckling stress is

$$F_e = C_s A \sigma_{ex} \frac{[j + C_s \sqrt{j^2 + r_o^2 (\sigma_t / \sigma_{ex})}]}{(C_{TF} S_f)} \tag{4.16}$$

for bending about the centroidal axis perpendicular to the axis of symmetry for singly symmetric sections only, and

where

$C_s = +1$ for moment-causing compression on the shear center side of the centroid

$C_s = -1$ for moment-causing tension on the shear center side of the centroid

A = full cross-sectional area

$$\sigma_{ex} = \frac{\pi^2 E}{(K_x L_x / r_x)^2} \tag{4.17}$$

$$\sigma_{ey} = \frac{\pi^2 E}{(K_y L_Y / r_y)^2} \tag{4.18}$$

$$\sigma_t = \frac{1}{A r_o^2} \left[GJ + \frac{\pi^2 E C_w}{(K_t L_t)^2} \right] \tag{4.19}$$

$$C_b = \frac{12.5 M_{max}}{2.5 M_{max} + 3 M_A + 4 M_B + 3 M_C} \tag{4.20}$$

M_{max} = absolute value of maximum moment in the unbraced segment

M_A = absolute value of moment at quarter point of the unbraced segment

M_B = absolute value of moment at centerline of the unbraced segment

M_C = absolute value of moment at three-quarter point of the unbraced segment

E = modulus of elasticity

S_f = elastic section modulus of full, unreduced cross section relative to extreme compression fiber

C_b is permitted to be conservatively taken as unity for all cases. For cantilevers or overhangs where the free end is unbraced, $C_b = 1.0$. For members subject to combined compressive axial load and bending moment, $C_b = 1.0$:

$$C_{TF} = 0.6 - 0.4 \left(\frac{M_1}{M_2} \right) \tag{4.21}$$

where M_1 is the smaller and M_2 the larger bending moment at the ends of the unbraced length in the plane of bending and M_1/M_2, the ratio of end moments, is positive when M_1 and M_2 have the same sign (reverse-curvature bending) and negative when they are of opposite sign (single-curvature bending). When the bending moment at any point within an unbraced length is larger than that at both ends of the length, $C_{TF} = 1.0$.

r_o = polar radius of gyration of the cross section about the shear center.

$$r_o = \sqrt{r_x^2 + r_y^2 + x_o^2}$$

(4.22)

where

r_x, r_y = radii of gyration of the cross section about the centroidal principal axes

G = shear modulus (11,000 ksi)

K_x, K_y, K_t = effective length factors for bending about the x and y axes, as well as for twisting

L_x, L_y, L_t = unbraced length of compression member for bending about the x and y axes, as well as for twisting

x_o = distance from the shear center to the centroid along the principal x axis, taken as negative

J = St. Venant torsion constant of the cross section

C_w = torsional warping constant of the cross section

$$j = \frac{1}{2I_y}\left[\int_A x^3 dA + \int_A xy^2 dA\right] - x_o$$

(4.23)

4.9.3   Shear Strength

AISI S100-07 provides equations for the nominal shear strength of beam webs (1) based on yielding condition, (2) based on inelastic buckling, and (3) for elastic buckling:

1. For $h/t \le \sqrt{Ek_v/F_y}$,

$$V_n = 0.60F_y ht$$

(4.24)

2. For $\sqrt{Ek_v/F_y} < h/t \le 1.51\sqrt{Ek_v/F_y}$,

$$V_n = 0.60t^2\sqrt{k_v F_y E}$$

(4.25)

3. For $h/t > 1.51\sqrt{Ek_v/F_v}$,

$$V_n = \frac{\pi^2 Ek_v t^3}{12(1-\mu^2)h} = 0.904\frac{Ek_v t^3}{h}$$

(4.26)

where V_n = nominal shear strength of beam, t = web thickness, h = depth of the flat portion of the web measured along the plane of the web, Ω = 1.60, ϕ = 0.95, and k_v = shear buckling coefficient determined as follows:

a. For unreinforced webs, k_v = 5.34.

b. For beam webs with transverse stiffeners satisfying AISI S100-07 requirements: When $a/h \le 1.0$,

$$k_v = 4.00 + \frac{5.34}{(a/h)^2}$$

(4.27)

When $a/h > 1.0$,

$$k_v = 5.34 + \frac{4.00}{(a/h)^2} \tag{4.28}$$

where a = shear panel length for unreinforced web element

= clear distance between transverse stiffeners for reinforced web elements.

For a web consisting of two or more sheets, each sheet is considered as a separate element carrying its share of the shear force. For C-section webs with holes, AISI S100-07 provides a shear strength reduction factor.

4.9.4 Combined Bending and Shear

Combinations of bending and shear may be critical at locations, such as near interior supports of continuous beams or at web holes nearest to a support. To guard against this condition, AISI S100-07 provides interaction equations for both ASD and LRFD.

ASD Method

The required flexural strength M and required shear strength V must not exceed M_n/Ω_b and V_n/Ω_v, respectively. For beams with unreinforced webs, the required flexural strength M and required shear strength V also must satisfy the following:

$$\left(\frac{\Omega_b M}{M_{nxo}}\right)^2 + \left(\frac{\Omega_v V}{V_n}\right) \leq 1.0 \tag{4.29}$$

For beams with transverse web stiffeners, when $\Omega_b M/M_{nxo} > 0.5$ and $\Omega_v V/V_n > 0.7$, then M and V also must satisfy the following interaction equation:

$$0.6\left(\frac{\Omega_b M}{M_{nxo}}\right) + \left(\frac{\Omega_v V}{V_n}\right) \leq 1.3 \tag{4.30}$$

where Ω_b = factor of safety for bending

Ω_v = factor of safety for shear

M_n = nominal flexural strength when bending alone exists

M_{nxo} = nominal flexural strength about the centroidal x axis determined in accordance with AISI, excluding lateral buckling

V_n = nominal shear force when shear alone exists

LRFD Method

The required flexural strength M_u and the required shear strength V_u shall not exceed $\phi_b M_n$ and $\phi_v V_n$, respectively. For beams with unreinforced webs, the required flexural strength M_u and required shear strength V_u also must satisfy the following:

$$\sqrt{\left(\frac{M_u}{\phi_b M_{nxo}}\right)^2 + \left(\frac{V_u}{\phi_v V_n}\right)^2} \leq 1.0 \tag{4.31}$$

For beams with transverse web stiffeners, when $M_u/(\phi_b M_{nxo}) > 0.5$ and $V_u/(\phi_v V_n) > 0.7$, then M_u and V_u also must satisfy the following interaction equation:

$$\left(\frac{M_u}{\phi_b M_{nxo}}\right) + \left(\frac{V_u}{\phi_v V_n}\right) \le 1.3 \tag{4.32}$$

where

ϕ_b = resistance factor for bending

ϕ_v = resistance factor for shear

M_n = nominal flexural strength when bending alone exists

4.9.5   Web Crippling

At locations of concentrated loads or reactions, the webs of cold-formed members are susceptible to web crippling. If the web depth-to-thickness ratio h/t is greater than 200, stiffeners must be used to transmit the loads directly into the webs. For unstiffened webs, AISI S100-07 gives an equation with multiple coefficients to calculate the nominal web crippling strength. The coefficients (Tables 4.1 and 4.2) are based on the results of experimental studies and provide for several different conditions for load placement and cross-sectional geometry.

The nominal web crippling strength P_n is determined as follows:

$$P_n = C t^2 F_y \sin\theta \left(1 - C_R\sqrt{\frac{R}{t}}\right)\left(1 + C_N\sqrt{\frac{N}{t}}\right)\left(1 - C_h\sqrt{\frac{h}{t}}\right) \tag{4.33}$$

where

P_n = nominal web crippling strength

C = coefficient

C_h = web slenderness coefficient

C_N = bearing length coefficient

C_R = inside bend radius coefficient

F_y = design yield point

h = flat dimension of web measured in plane of web

N = bearing length

R = inside bend radius

t = web thickness

θ = angle between plane of web and plane of bearing surface

P_n represents the nominal strength for load or reaction for one solid web connecting top and bottom flanges. For webs consisting of two or more webs, P_n should be calculated for each individual web and the results added to obtain the nominal strength for the full section. For C-section webs with holes, AISI S100-07 provides a web crippling reduction factor.

The load cases specified in Tables 4.1 and 4.2 are as follows: One-flange loading or reaction occurs when the clear distance between the bearing edges of adjacent opposite

TABLE 4.1 Web Crippling Coefficients for Built-up Sections

| Support and Flange Conditions | | Load Cases | | C | C_R | C_N | C_h | USA and Mexico | | Canada | |
								ASD Ω_w	LRFD ϕ_w	LSD ϕ_w	Limits
Fastened to support	Stiffened or partially stiffened flanges	One-flange loading or reaction	End	10	0.14	0.28	0.001	2.00	0.75	0.60	$R/t \leq 5$
			Interior	20.5	0.17	0.11	0.001	1.75	0.85	0.75	$R/t \leq 5$
Unfastened	Stiffened or partially stiffened flanges	One-flange loading or reaction	End	10	0.14	0.28	0.001	2.00	0.75	0.60	$R/t \leq 5$
			Interior	20.5	0.17	0.11	0.001	1.75	0.85	0.75	$R/t \leq 3$
		Two-flange loading or reaction	End	15.5	0.09	0.08	0.04	2.00	0.75	0.65	$R/t \leq 3$
			Interior	36	0.14	0.08	0.04	2.00	0.75	0.65	
	Unstiffened flanges	One-flange loading or reaction	End	10	0.14	0.28	0.001	2.00	0.75	0.60	$R/t \leq 5$
			Interior	20.5	0.17	0.11	0.001	1.75	0.85	0.75	$R/t \leq 3<$

Notes: (1) This table applies to I-beams made from two channels connected back to back. (2) These coefficients apply when $h/t \leq 200$, $N/t \leq 210$, $N/h \leq 1.0$, and $\theta = 90$ degrees.

Source: AISI, 2007.[2]

Support and Flange Conditions		Load Cases		C	C_R	C_N	C_h	USA and Mexico ASD Ω_w	USA and Mexico LRFD ϕ_w	Canada LSD ϕ_w	Limits
Fastened to support	Stiffened or partially stiffened flanges	One-flange loading or reaction	End	4	0.14	0.35	0.02	1.75	0.85	0.75	$R/t \leq 9$
			Interior	13	0.23	0.14	0.01	1.65	0.90	0.80	$R/t \leq 5$
		Two-flange loading or reaction	End	7.5	0.08	0.12	0.048	1.75	0.85	0.75	$R/t \leq 12$
			Interior	20	0.10	0.08	0.031	1.75	0.85	0.75	$R/t \leq 12$
Unfastened	Stiffened or partially stiffened flanges	One-flange loading or reaction	End	4	0.14	0.35	0.02	1.85	0.80	0.70	$R/t \leq 5$
			Interior	13	0.23	0.14	0.01	1.65	0.90	0.80	
		Two-flange loading or reaction	End	13	0.32	0.05	0.04	1.65	0.90	0.80	$R/t \leq 3$
			Interior	24	0.52	0.15	0.001	1.90	0.80	0.65	
	Unstiffened flanges	One-flange loading or reaction	End	4	0.40	0.60	0.03	1.80	0.85	0.70	$R/t \leq 2$
			Interior	13	0.32	0.10	0.01	1.80	0.85	0.70	$R/t \leq 1$
		Two-flange loading or reaction	End	2	0.11	0.37	0.01	2.00	0.75	0.65	$R/t \leq 1$
			Interior	13	0.47	0.25	0.04	1.90	0.80	0.65	

Notes: (1) These coefficients apply when $h/t \leq 200$, $N/t \leq 210$, $N/h \leq 2.0$, and $\theta = 90$ degrees. (2) For interior two-flange loading or reaction of members having flanges fastened to the support, the distance from the edge of bearing to the end of the member should be extended at least $2.5h$. For unfastened cases, the distance from the edge of bearing to the end of the member should be extended at least $1.5h$.

Source: AISI, 2007.[2]

TABLE 4.2 Web Crippling Coefficients for Single-Web Channel and C-Sections

concentrated loads or reactions is greater than $1.5h$. Two-flange loading or reaction occurs when the clear distance between the bearing edges of adjacent opposite concentrated loads or reactions is equal to or less than $1.5h$. End loading or reaction occurs when the distance from the edge of the bearing to the end of the member is equal to or less than $1.5h$. Interior loading or reaction occurs when the distance from the edge of the bearing to the end of the member is greater than $1.5h$, except as otherwise noted.

For overhang conditions where the web crippling strength may be greater than predicted for an end-loading condition, AISI S100-07 enables the use of a larger web crippling strength.

4.9.6 Combined Bending and Web Crippling Strength

For beams with unreinforced flat webs, combinations of bending and web crippling near concentrated loads or reactions must satisfy interaction equations given in AISI S100-07. Equations are given for two types of webs, with separate equations for ASD and LRFD. See AISI S100-07 for various exceptions and limitations that may apply. Symbols have common definitions except as noted.

ASD Method

1. For shapes having single unreinforced webs,

$$0.91\left(\frac{P}{P_n}\right)+\left(\frac{M}{M_{nxo}}\right)\le\frac{1.33}{\Omega} \tag{4.34}$$

2. For shapes having multiple unreinforced webs such as I-sections made of two C-sections connected back to back or similar sections that provide a high degree of restraint against rotation of the web (such as I-sections made by welding two angles to a C-section),

$$0.88\left(\frac{P}{P_n}\right)+\left(\frac{M}{M_{nxo}}\right)\le\frac{1.46}{\Omega} \tag{4.35}$$

In Eqs. (4.34) and (4.35),

P = required strength for concentrated load or reaction in the presence of bending moment

P_n = nominal strength for concentrated load or reaction in the absence of bending moment

M = required flexural strength at or immediately adjacent to the point of application of the concentrated load or reaction P

M_{nxo} = nominal flexural strength about the centroidal x-axis, $S_e F_y$

Ω = safety factor for combined bending and web crippling = 1.70

LRFD Method

1. For shapes having single unreinforced webs,

$$0.91\left(\frac{P_u}{P_n}\right) + \left(\frac{M_u}{M_{nxo}}\right) \leq 1.33\phi \tag{4.36}$$

2. For shapes having multiple unreinforced webs such as I-sections made of two C-sections connected back to back or similar sections that provide a high degree of restraint against rotation of the web (such as I-sections made by welding two angles to a C-section),

$$0.88\left(\frac{P_u}{P_n}\right) + \left(\frac{M_u}{M_{nxo}}\right) \leq 1.46\phi \tag{4.37}$$

where ϕ = resistance factor = 0.90

P_u = required strength for the concentrated load or reaction in the presence of bending moment

M_u = required flexural strength at or immediately adjacent to the point of application of the concentrated load or reaction P_u

4.10   Concentrically Loaded Compression Members

These provisions are for members in which the resultant of all loads is an axial load passing through the effective section calculated at the stress F_n as subsequently defined.

The nominal axial strength P_n is

$$P_n = A_e F_n \tag{4.38}$$

where $\Omega = 1.80$, $\phi = 0.85$, and A_e is the effective area at the stress F_n, which is determined as follows:

For $\lambda_c \leq 1.5$,

$$F_n = \left(0.658^{\lambda_c^2}\right) F_y \tag{4.39}$$

For $\lambda_c > 1.5$,

$$F_n = \left(\frac{0.877}{\lambda_c^2}\right) F_y \tag{4.40}$$

where

$$\lambda_c = \sqrt{\frac{F_y}{F_e}} \tag{4.41}$$

F_e is the least of the elastic flexural, torsional, and torsional-flexural buckling stresses. Equation (4.40) is based on elastic buckling, whereas Eq. (4.41) represents inelastic buckling, providing a transition to the yield point stress as the column length decreases.

In addition to the buckling modes just mentioned, open cross-sectional members that have compression flanges with edge stiffeners also must be checked for distortional buckling strength. This is a buckling mode involving a change in cross-sectional shape that differs from that of local buckling. It becomes more important with higher strength steels and with edge-stiffened compression flanges. Provisions for distortional buckling are given in AISI S100-07.

4.10.1 Elastic Flexural Buckling

For doubly symmetric, closed, or any other sections that are not subject to torsional or flexural-torsional buckling, the elastic flexural buckling stress is

$$F_e = \frac{\pi^2}{(KL/r)^2} \tag{4.42}$$

where

E = modulus of elasticity (29,500 ksi)

K = effective-length factor (see AISI S100-07 commentary)[3]

L = unbraced length of member

r = radius of gyration of full, unreduced cross section about axis of buckling

4.10.2 Symmetric Sections Subject to Torsional or Flexural-Torsional Buckling

Singly Symmetric Sections

For singly symmetric sections, such as C-sections subject to flexural-torsional buckling, F_e is the smaller of Eq. (4.42) and that is given by:

$$F_e = \frac{1}{2\beta}\left[\left(\sigma_{ex} + \sigma_t\right) - \sqrt{\left(\sigma_{ex} + \sigma_t\right)^2 - 4\beta\sigma_{ex}\sigma_t} \right] \tag{4.43}$$

In the preceding, σ_{ex} and σ_t are given by Eqs. (4.17) and (4.19), and

$$\beta = 1 - \left(\frac{x_o}{r_o}\right)^2 \tag{4.44}$$

where r_o is given by Eq. (4.22), and x_o is the distance from the shear center to the centroid along the principal x axis taken as negative. For singly symmetric sections, the x axis is assumed to be the axis of symmetry.

Doubly Symmetric Sections

For doubly symmetric sections, such as back-to-back C-sections subject to torsional buckling, F_e is taken as the smaller of Eq. (4.42), and the torsional buckling stress σ_t is given by Eq. (4.19).

4.11 Combined Tensile Axial Load and Bending

Members under combined axial tensile load and bending must satisfy the interaction equations given by AISI S100-07 to prevent yielding. Separate equations are given for ASD and LRFD, but symbols have common definitions except as noted.

ASD Method

To check the tension flange,

$$\frac{\Omega_b M_x}{M_{nxt}} + \frac{\Omega_b M_y}{M_{nyt}} + \frac{\Omega_t T}{T_n} \leq 1.0 \tag{4.45}$$

To check the compression flange,

$$\frac{\Omega_b M_x}{M_{nx}} - \frac{\Omega_b M_y}{M_{ny}} - \frac{\Omega_t T}{T_n} \leq 1.0 \tag{4.46}$$

where

T = required tensile axial strength

M_x, M_y = required flexural strengths with respect to the centroidal axes of the section

T_n = nominal tensile axial strength

M_{nx}, M_{ny} = nominal flexural strengths about the centroidal axes

$M_{nxt}, M_{nyt} = S_{ft} F_y$

S_{ft} = section modulus of the full section for the extreme tension fiber about the appropriate axis

Ω_b = safety factor for bending, 1.67

Ω_t = safety factor for tension member, 1.67

LRFD Method

To check the tension flange,

$$\frac{M_{ux}}{\phi_b M_{nxt}} + \frac{M_{uy}}{\phi_b M_{nyt}} + \frac{T_u}{\phi_t T_n} \leq 1.0 \tag{4.47}$$

To check the compression flange,

$$\frac{M_{ux}}{\phi_b M_{nx}} + \frac{M_{uy}}{\phi_b M_{ny}} - \frac{T_u}{\phi_t T_n} \leq 1.0 \tag{4.48}$$

where

T_u = required tensile axial strength

M_{ux}, M_{uy} = required flexural strengths with respect to the centroidal axes

ϕ_b = 0.90 or 0.95 for bending strength or 0.90 for laterally unbraced beams

ϕ_t = 0.95

4.12   Combined Compressive Axial Load and Bending

Members subject to combined compression axial load and bending generally are referred to as *beam-columns*. Bending in such members may be caused by eccentric loading, lateral loads, or end moments, and the compression load may amplify the bending. These members must satisfy the interaction equations given by AISI S100-07 to prevent both overall buckling and yielding. Separate equations are given for ASD and LRFD, but the symbols have common definitions except as noted. For LRFD only, as an alternative to the interaction equations that follow, AISI S100 permits the use of a second-order analysis to determine required strengths of members.

ASD Method

$$\frac{\Omega_c P}{P_n} + \frac{\Omega_b C_{mx} M_x}{M_{nx}\alpha_x} + \frac{\Omega_b C_{my} M_y}{M_{ny}\alpha_y} \le 1.0 \tag{4.49}$$

$$\frac{\Omega_c P}{P_{no}} + \frac{\Omega_b M_x}{M_{nx}} + \frac{\Omega_b M_y}{M_{ny}} \le 1.0 \tag{4.50}$$

When $\Omega_c P/P_n \le 0.15$, the following equation may be used in lieu of the preceding two equations:

$$\frac{\Omega_c P}{P_n} + \frac{\Omega_b M_x}{M_{nx}} + \frac{\Omega_b M_y}{M_{ny}} \le 1.0 \tag{4.51}$$

where

P = required compressive axial strength

M_x, M_y = required flexural strengths with respect to the centroidal axes of the effective section determined for the required compressive axial strength alone

P_n = nominal axial strength

P_{no} = nominal axis strength with $F_n = F_y$

M_{nx}, M_{ny} = nominal flexural strengths about the centroidal axes

$$\alpha_x = 1 - \frac{\Omega_c P}{P_{Ex}} \tag{4.52}$$

$$\alpha_y = 1 - \frac{\Omega_c P}{P_{Ey}} \tag{4.53}$$

$$P_{Ex} = \frac{\pi^2 E I_x}{(K_x L_x)^2} \tag{4.54}$$

$$P_{Ey} = \frac{\pi^2 E I_y}{(K_y L_y)^2} \tag{4.55}$$

Ω_b = safety factor for bending, 1.67

Ω_c = safety factor for compression, 1.80

I_x = moment of inertia of the full, unreduced cross section about the x axis

I_y = moment of inertia of the full, unreduced cross section about the y axis

L_x = unbraced length for bending about the x axis

L_y = unbraced length for bending about the y axis

K_x = effective length factor for buckling about the x axis

K_y = effective length factor for buckling about the y axis

C_{mx}, C_{my} = coefficients whose values are as follows:

1. For compression members in frames subject to joint translation (sidesway),

$$C_m = 0.85$$

2. For restrained compression members in frames braced against joint translation and not subject to transverse loading between their supports in the plane of bending,

$$C_m = 0.6 - 0.4 \left(\frac{M_1}{M_2} \right) \tag{4.56}$$

where M_1/M_2 is the ratio of the smaller to the larger moment at the ends of that portion of the member under consideration that is unbraced in the plane of bending. M_1/M_2 is positive when the member is bent in reverse curvature and negative when it is bent in single curvature.

3. For compression members in frames braced against joint translation in the plane of loading and subject to transverse loading between their supports, the value of C_m may be determined by rational analysis. However, in lieu of such analysis, the following values may be used:

   a. For members whose ends are restrained, $C_m = 0.85$.

   b. For members whose ends are unrestrained, $C_m = 1.0$.

LRFD Method

$$\frac{P_u}{\phi_c P_n} + \frac{C_{mx} M_{ux}}{\phi_b M_{nx} \alpha_x} + \frac{C_{my} M_{uy}}{\phi_b M_{ny} \alpha_y} \leq 1.0 \tag{4.57}$$

$$\frac{P_u}{\phi_c P_{no}} + \frac{M_{ux}}{\phi_b M_{nx}} + \frac{M_{uy}}{\phi_b M_{ny}} \leq 1.0 \tag{4.58}$$

When $P_u / \phi_c P_n \leq 0.15$, the following equation may be used in lieu of the preceding two equations:

$$\frac{P_u}{\phi_c P_n} + \frac{M_{ux}}{\phi_b M_{nx}} + \frac{M_{uy}}{\phi_b M_{ny}} \leq 1.0 \tag{4.59}$$

where

P_u = required compressive axial strength

M_{ux}, M_{uy} = required flexural strengths with respect to the centroidal axes of the effective section determined for the required compressive axial strength alone

$$\alpha_x = 1 - \frac{P_u}{P_{Ex}} \tag{4.60}$$

$$\alpha_y = 1 - \frac{P_u}{P_{Ey}} \tag{4.61}$$

ϕ_b = 0.90 or 0.95 for bending strength or 0.90 for laterally unbraced beams

ϕ_c = 0.85

4.13 Welded Connections

Various types of welds may be used to join cold-formed steel members such as groove welds in butt joints, fillet welds, flare groove welds, arc spot welds, arc seam welds, and resistance welds. The design strength P_n for the more common of these weld types used in light-steel framing is summarized herein. For an in-depth discussion of the design of welded connections, refer to Yu and LaBoube,[16] and the design provisions may be found in AISI S100-07. The design provisions are applicable for designs where the thickness of the thinnest connected part is $\frac{3}{16}$ in. (4.76 mm) or less. Welds in thicker parts should be designed according to the American Institute of Steel Cnstruction (AISC) *Specification for Structural Steel Buildings* (ANSI/AISC 360-05).[1]

4.13.1 Fillet Welds

Fillet welds may be designed to transmit either longitudinal or transverse loads. For these welds, the nominal strength is the smaller of the limit based on weld strength and that based on the strength of the connected part.

For weld strength (consider only if $t > 0.10$ in. or 2.54 mm),

$$P_n = 0.75 t_w L F_{xx} \tag{4.62}$$

$$\Omega = 2.55, \, \phi = 0.60$$

For strength of connected part, welds longitudinal to the loading,
When $L/t < 25$,

$$P_n = \left(1 - \frac{0.01L}{t}\right) t L F_u \tag{4.63}$$

$$\Omega = 2.55, \, \phi = 0.60$$

When $L/t \, \phi \, 25$,

$$P_n = 0.75 t L F_u \tag{4.64}$$

$$\Omega = 3.05, \, \phi = 0.50$$

For strength of connected part, welds transverse to the loading,

$$P_n = t L F_u \tag{4.65}$$

$$\Omega = 1.35, \, \phi = 0.65$$

where

t = least of t_1 and t_2 (see Fig. 4.4)

t_w = effective throat of weld

= $0.707 w_1$ or $0.707 w_2$, whichever is smaller (see Fig. 4.4)

F_u = tensile strength

4.13.2 Arc Spot Welds

Arc spot welds, often called *puddle welds*, are made to join sheets to thicker members or for joining two thinner sheets. The arc spot weld is used commonly to attached metal

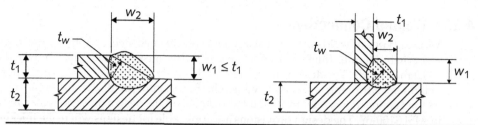

Figure 4.4 Cross sections of fillet welds: (a) at lap joint; (b) at tee joint. (AISI, 2007.[2])

deck to its supporting structural member. They are made typically by using the arc to burn a hole in the top sheet (or sheets) and depositing weld metal to fill the hole, thus fusing the top sheet or sheets to the underlying member. Where the thickness of the sheet is less than 0.028 in. (0.711 mm), a washer is to be used on top of the sheet and the weld made inside this washer. Although arc spot welds are specified by a visible diameter, design is based on the minimum effective diameter of fused area d_e. The design provisions are given by AISI S100-07 for both shear and tension (uplift) loadings.

4.14   Bolted Connections

Bolted connections in cold-formed steel construction are designed as bearing-type connections. Bolt pretensioning is not required. Only bearing-type installations using the snug-tight condition are recognized by AISI S100-07. AISI S100-07 gives applicable provisions when the thickness t of the thinnest connected part is less than $\frac{3}{16}$ in. (4.76 mm). Bolted connections in thicker parts are to be designed according to ANSI/AISC 360-05. For cold-formed steel construction, the most commonly used bolts are A307 carbon steel bolts and A325 high-strength bolts. See AISI S100-07 for the other permitted bolt types and the definition of hole types.

Applicable design limit states for a bolted connection include edge shear strength of sheet (edge distance and spacing effects), tension strength in the connected parts, bearing strength of the sheet, bolt shear strength, and bolt tension strength.

4.14.1   Sheet Shearing (Spacing and Edge Distance)

When bolts are located too close to the end of a member or when the bolts are spaced too closely center to center, the connection may be limited in strength by the shear strength along a line parallel to the application of force. Although the minimum center-to-center spacing of bolts is $3d$ and the minimum center-to-edge distance is $1.5d$, the nominal strength P_n is limited to

$$P_n = teF_u \tag{4.66}$$

$$\Omega = 2.00, \ \phi = 0.70 \ \text{(when } F_u/F_y \geq 1.08)$$

$$\Omega = 2.22, \ \phi = 0.60 \ \text{(when } F_u/F_y < 1.08)$$

where t is the thickness of thinnest part, and e is the distance in line of force from center of hole to the nearest edge of the adjacent hole or end of the connected part.

4.14.2 Tension Rupture in Net Section

The nominal tension strength of a member is determined as discussed in Art. 4.7. The nominal strength in the connection itself for the limit state of fracture, including shear lag effects, is determined as follows:

For flat-sheet connections, the nominal tension strength P_n on the net area of the section A_n of each connected part is

$$P_n = A_n F_t \tag{4.67}$$

Where washers are provided under both the bolt head and the nut, two conditions may apply. For multiple bolts in the line parallel to the force, $F_t = F_u$. For a single bolt or a single row of bolts perpendicular to the line of force,

$$F_t = \left(0.1 + \frac{3d}{s}\right) F_u \leq F_u \tag{4.68}$$

$$\Omega = 2.00, \phi = 0.65 \text{ (double shear)}$$

$$\Omega = 2.00, \phi = 0.70 \text{ (single shear)}$$

Where only one washer or no washers are provided, consider two conditions. For multiple bolts in the line parallel to the force, $F_t = F_u$. For a single bolt or a single row of bolts perpendicular to the line of force,

$$F_t = \left(\frac{2.5d}{s}\right) F_u \leq F_u \tag{4.69}$$

$$\Omega = 2.22, \phi = 0.65$$

where F_u is the tensile strength of sheet, s is the sheet width divided by the number of bolt holes in cross section being analyzed, and d is the nominal bolt diameter. Where holes are staggered, the net area A_n is determined from

$$A_n = 0.90\left[A_g - n_b d_h t + \left(\frac{\Sigma s'^2}{4g}\right)t\right] \tag{4.70}$$

where A_g is the gross cross-sectional area, n_b is the number of bolts in cross section, d_h is the hole diameter, t is the thickness, s' is the longitudinal spacing, and g is the transverse spacing.

For connected parts other than flat sheets, the nominal strength is

$$P_n = A_n U F_u \tag{4.71}$$

$$\Omega = 2.22, \phi = 0.65$$

where U is a factor that reflects the nonuniform distribution of stresses over the cross section (shear lag) and is defined as follows: For angle members having two or more bolts in the line of force,

$$U = 1.0 - \frac{1.20\bar{x}}{L} < 0.9 \text{ but } \geq 0.4 \tag{4.72}$$

For channel members having two or more bolts in the line of force,

$$U = 1.0 - \frac{0.36\bar{x}}{L} < 0.9 \text{ but} \geq 0.5 \tag{4.73}$$

where $\bar{x}$ is the distance from shear plane to the centroid of the cross section, and L is the length of the connected part.

4.14.3 Bearing of Bolt on Connecting Element

The nominal bearing strength on the connected part depends on whether deformation around the bolt holes can be tolerated. When deformation is not a design consideration, the more common design condition, the nominal bearing strength P_n of the sheet for each bolt is

$$P_n = m_f C d t F_u \tag{4.74}$$

$$\Omega = 2.50, \phi = 0.60$$

where C is a bearing factor (Table 4.3), d is the nominal bolt diameter, t is the thickness, F_u is the tensile strength of the sheet, and m_f is the modification factor for the type of connection (Table 4.4).

4.14.4 Shear and Tension in Bolts

The nominal bolt strength resulting from shear, tension, or a combination of shear and tension is defined by the following:

$$P_n = A_b F \tag{4.75}$$

where A_b is the gross cross-sectional area of the bolt. For bolts in shear, $F = F_{nv}$ (nominal shear stress). For bolts in tension, $F = F_{nt}$ (nominal tensile stress). See Table E3.4-1 of AISI S100-07 for F_{nv}, F_{nt}, Ω, and ϕ.

Thickness of Connected Part, t, in. (mm)	Ratio of Fastener Diameter to Member Thickness d/t	C
$0.024 \leq t < 0.1875$	$d/t < 10$	3.0
$(0.61 \leq t < 4.76)$	$10 \leq d/t \leq 22$	$4 - 0.1(d/t)$
	$d/t > 22$	1.8

Source: AISI, 2007.[2]

TABLE 4.3 Bearing Factor C for Bolted Connections

Type of Bearing Connection	m_f
Single-shear and outside sheets of double-shear connection with washers under both bolt head and nut	1.00
Single-shear and outside sheets of double-shear connection without washers under both bolt head and nut or with only one washer	0.75
Inside sheet of double-shear connection with or without washers	1.33

Source: AISI, 2007.[2]

TABLE 4.4 Modification Factor m_f for Bolted Connections

For bolts subject to a combination of shear and tension, $F = F_{nt}'$, as given by the following:

For ASD,
$$F_{nt}' = 1.3F_{nt} - \frac{\Omega F_{nt}}{F_{nv}} f_v \leq F_{nt} \qquad (4.76)$$

For LRFD,
$$F_{nt}' = 1.3F_{nt} - \frac{F_{nt}}{\phi F_{nv}} f_v \leq F_{nt} \qquad (4.77)$$

where

F_{nt}' = nominal tensile stress modified to include effects of required shear stress, ksi

f_v = required shear stress, ksi, $\leq F_{nv}/\Omega$ (ASD) or $\leq \phi F_{nv}$ (LRFD)

4.15   Screw Connections

Screws are most frequently used connector in cold-formed steel-framed construction. This is so because a self-drilling screw can be driven with a hand-held drill without punching or drilling a hole. Section E4 of AISI S100-07 gives the provisions for calculating nominal strength for screws. The screws can be of the thread-forming or thread-cutting type, with or without a self-drilling point.

The distance between the centers of the screw fasteners and the distance from the center of a fastener to the edge of any part should not be less than $3d$. However, if the connection is subjected to shear force in one direction only, the minimum edge distance in the direction perpendicular to the force must be at least $1.5d$.

Nominal strength equations for shear and for tension use the following parameter notation:

P_{ns} = nominal shear strength per screw

P_{ss} = nominal shear strength per screw as reported by manufacturer or as tested

P_{not} = nominal pullout strength per screw

P_{nov} = nominal pullover strength per screw

P_{ts} = nominal shear strength per screw as reported by manufacturer or as tested

t_1 = thickness of member in contact with the screw head

t_2 = thickness of member not in contact with the screw head

F_{u1} = tensile strength of member in contact with the screw head

F_{u2} = tensile strength of member not in contact with the screw head

For screw connections, $\Omega = 3.0$ and $\phi = 0.50$.

4.15.1   Shear

The nominal shear strength per screw P_{ns}, as limited by tilting and bearing, is determined as follows:

For $t_2/t_1 \leq 1.0$, P_{ns} shall be taken as the smallest of

$$P_{ns} = 4.2\left(t_2^3 d\right)^{1/2} F_{u2} \qquad (4.78)$$

$$P_{ns} = 2.7t_1 dF_{u1} \qquad (4.79)$$

$$P_{ns} = 2.7t_2 dF_{u2} \qquad (4.80)$$

For $t_2/t_1 > 2.5$, P_{ns} shall be taken as the smaller of

$$P_{ns} = 2.7t_1 dF_{u1} \qquad (4.81)$$

$$P_{ns} = 2.7t_2 dF_{u2} \qquad (4.82)$$

For $1.0 < t_2/t_1 < 2.5$, P_{ns} should be determined by linear interpolation between the preceding two cases.

Additionally, the limit state based on the strength of the screw itself must be considered:

$$P_{ns} = P_{ss} \qquad (4.83)$$

4.15.2 Tension

For screws subjected to tension force, the diameter of the head of the screw or the diameter of the washer if one is used must be at least $5/16$ in. (7.94 mm). Washers must beat least 0.05 in. (1.27 mm) thick. Two design limit states must be considered: (1) pullout of the screw and (2) pullover of the sheet. In addition, the nominal tensile strength of the screw itself P_{ts} must be considered.

Pullout
The nominal pullout strength P_{not} is calculated as

$$P_{not} = 0.85t_c dF_{u2} \qquad (4.84)$$

where t_c is the lesser of the depth of screw penetration and the thickness t_2.

Pullover
The nominal pullover strength P_{nov} is calculated as

$$P_{nov} = 1.5t_1 d'_w F_{u1} \qquad (4.85)$$

where d'_w is as follows:
For round head, hex head, or hex washer head screws with independent solid steel washers beneath the head,

$$d'_w = d_h + 2t_w + t_1 \le d_w \qquad (4.86)$$

where d_h = screw head or integral hex washer head diameter

t_w = steel washer thickness

d_w = steel washer diameter

For round head, hex head, or hex washer head screw without an independent solid steel washer beneath the head,

$$d'_w = d_h \le \tfrac{1}{2} \text{ in. (12.7 mm)} \qquad (4.87)$$

For a domed washer beneath the screw head, refer to AISI S100-07.

For cases of combined shear and pullover, interaction equations are given in AISI S100-07.

4.16   Other Resources

Additional information pertaining to the application and design of cold-formed steel in construction may be found at the following Web sites:

American Institute of Steel Construction, www.aisc.org
American Iron and Steel Institute, www.steel.org
Canadian Sheet Steel Building Institute, www.cssbi.ca
American Society of Civil Engineers, www.asce.org
Wei-Wen Yu Center for Cold-Formed Steel Structures, http://ccfssonline.org
Cold-Formed Steel Engineers Institute, www.cfsei.org
Metal Building Manufacturers Association, www.mbma.com
Metal Construction Association, www.metalconstruction.org
Rack Manufacturers Institute, www.mhia.org
Steel Deck Institute, www.sdi.org
Steel Framing Alliance, www.steelframing.org
Steel Stud Manufacturers Association, www.ssma.com
Structural Stability Research Council, http:/stabilitycouncil.org

4.17   Example Problems

4.17.1   Example Problem 1: Curtain-Wall Stud

The following example is based on Example 1 in the *Cold-Formed Steel Framing Design Guide* (AISI D110-07).[13] A *curtain-wall stud* is a structural member that resists primarily wind load only. AISI S200-07 defines a *curtain wall* as "a wall that transfers transverse (out-of-plane) loads and is limited to a superimposed vertical load, exclusive of sheathing materials, of not more than 100 lb/ft (1.46 kN/m) or a superimposed vertical load of not more than 200 lb (0.890 kN)."

Given:

Wall height (stud length) = 13 ft

Stud center-to-center spacing = 16 in.

Nominal wind load (component and cladding load per ASCE/SEI 7-05) = 28 psf or uniform stud load = 28(16/12) = 37.33 lb/ft

Stud deflection limit = span/360

Evaluate the design limit states for the curtain-wall stud. In this application, the wall stud resists the applied wind load as a flexural member. Therefore, the appropriate strength limit states are bending, shear, and web crippling. The serviceability also will be evaluated.

Bending

The selection of a wall stud cross section generally is accomplished by using design tables. Such tables may be provided by the Steel Stud Manufacturers Association (SSMA) or AISI D100-08.

Using a load table, a 600S162-43 was chosen. Load tables typically provide design data for studs having either a yield stress of 33 or 50 ksi. For this problem, 33-ksi material was selected from the load table.

Using AISI S100, check 600S162-43 (F_y = 33 ksi):

$$\text{Applied moment} = wL^2/8 = 37.33(13)^2/8 = 789 \text{ ft-lb} = 9.46 \text{ in.-kips}$$

Generally, the sheathing material has sufficient strength and stiffness to assume that the bending member is braced adequately against lateral-torsional buckling. Thus, using the provisions of AISI S100-07 Sec. C3.1.1 or computer software, the allowable ASD moment capacity is determined. Wall studs typically have web punch-outs at 24 in. on center. Because the location of the punch-out is undetermined during design, the wall stud will be evaluated considering both cross-sectional configurations:

$$M_a = 15.2 \text{ in-kips (without punch-out)}$$

$$M_a = 15.1 \text{ in-kips (with 1.5-in. punch-out)}$$

Okay for bending!

Deflection

For architectural wall panels, out-of-plane deflection, serviceability may be the limiting design consideration for a curtain wall.

$$\text{Deflection } \Delta = 5wL^4/(384EI_x) = 0.351 \text{ in.}$$

AISI S100 stipulates that the moment of inertia for deflection I_{xd} be evaluated using the effective-width equations contained in Chap. B. But the stress used when evaluating the effective widths is at design load, not ultimate load. Therefore, the maximum design stress is taken as $0.6F_y$, or 20 ksi.

$$I_x = I_{xd} = \text{full moment of inertia} = 2.316 \text{ in.}^4$$

$$w = 37.33 \text{ lb/ft (components and cladding wind load)}$$

$$L = 13 \text{ ft} \times 12 = 156 \text{ in.}; \quad E = 29,500 \text{ ksi}$$

$$\text{Permissible deflection} = \text{span}/360 = 156/360 = 0.433 \text{ in.}$$

However, Sec. A3.1 of the *Wall Stud Standard* permits the use of 70 percent of components and cladding wind load, which results in a deflection $\delta = 0.7 \times 0.351$ in. = 0.246 in.

Okay for serviceability!

Shear

AISI S100-07 requires that the shear strength of the wall stud be evaluated, but rarely will this limit state control the stud's design. Although some design engineers will check the shear at the location of the nearest punch-out to the end support, this example evaluates the applied end shear only.

$$\text{Applied shear } V = 0.5wL = 0.5(37.33)(13) = 243 \text{ lb} = 0.243 \text{ kips}$$

Using the design provisions of AISI S100-07 Sec. C3.2 or computer software, the allowable ASD shear strength is determined to be

$$V_a = 1.42 \text{ kips (without punch-out)}$$
$$V_a = 1.24 \text{ kips (with 1.5-in. punch-out)}$$

Okay for shear!

Web Crippling

AISI S100-07 Sec. C3.4 contains the design requirements for evaluating the influence of local concentrated loads. The stud-to-track connection will be evaluated for the web crippling limit state:

$$P_n = Ct^2 F_y \sin\theta \left(1 - C_R\sqrt{\frac{R}{t}}\right)\left(1 + C_N\sqrt{\frac{N}{t}}\right)\left(1 - C_h\sqrt{\frac{h}{t}}\right) \qquad \text{AISI S100-07 Eq. (C3.4.1-1)}$$

From Table 4.2 for a single C-section fastened to its support end one-flange loading or reaction,

$$C = 4, \quad C_R = 0.14, \quad C_N = 0.35, \quad C_h = 0.02, \quad \Omega_w = 1.75$$

These values are for a web without a web punch-out near the support. If a punch-out is near the support, AISI S100-07 Sec. C3.4.2 must be considered.

Given that the 600S162-43 is nested in a 600T125-43 track section, it is common practice to use the same thickness for the stud and track sections. However, economy may be achieved by using a thinner track section.

AISI S100-07 Sec. A2.4 stipulates that the minimum delivered thickness be at least $0.95t_{design}$:

$$\text{Design thickness} = 0.043/0.95 = 0.0451 \text{ in.}$$
$$R/t = 0.0712/0.0451 = 1.58$$
$$h/t = 5.767/0.0451 = 128$$

Section C1 of the AISI S211-07 stipulates that for curtain walls, the ends of the wall studs shall be seated squarely in the track with no more than a ¼-in. gap between the end of the stud and the track.

$$N = 1 \text{ in. with a ¼-in. gap}$$
$$N/t = 1/0.0451 = 22.2$$

The nominal web crippling capacity is computed to be

$$P_n = Ct^2 F_y \sin\theta \left(1 - C_R\sqrt{\frac{R}{t}}\right)\left(1 + C_N\sqrt{\frac{N}{t}}\right)\left(1 - C_h\sqrt{\frac{h}{t}}\right)$$

$$P_a = P_n/\Omega = 0.453 \text{ kip}/1.75 = 0.259 \text{ kip}$$

which is greater than the end reaction, 0.243 kip.

Okay for web crippling!

The preceding calculations reflect a web without a punch-out near the support. AISI S100 Sec. C3.4.2 addresses the condition where a web punch-out is near the support. Referring to AISI S100-07 Sec. C3.4.2 for a web opening near the support, a modified web crippling strength is determined.

For an industry-standard wall stud, the distance from the edge of the stud bearing to the nearest edge of the punch-out $x = 9$ in.:

$$R_c = 1.01 - 0.325(d_o/h) + 0.083(x/h) \leq 1.0 \qquad \text{AISI S100-07 Eq. (C3.4.2-1)}$$
$$= 1.01 - 0.325(2.5/5.767) + 0.083(9/5.767)$$
$$= 0.999 \approx 1.0$$

Thus, no reduction is required for a punch-out located 10 in. from the end of the member to the nearest edge of the web punch-out.

The AISI S211-07 permits an increase in web crippling strength when the stud is screw-attached to the track section. The nominal strength equation is the same in both AISI S100-07 and AISI S211-07, but the coefficients are different:

$$P_{nst} = Ct^2F_y\left(1 - C_R\sqrt{\frac{R}{t}}\right)\left(1 + C_N\sqrt{\frac{N}{t}}\right)\left(1 - C_h\sqrt{\frac{h}{t}}\right)$$

From the *Wall Stud Standard*, $C = 3.7$, $C_R = 0.19$, $C_N = 0.74$, $C_h = 0.019$, and $\Omega = 1.70$.

$$P_a = P_{nst}/\Omega = 0.666 \text{ kip}/1.70 = 0.392 \text{ kip}$$

This value is for a web without a web opening near the support. See AISI S100-07 Sec. C3.4.2 for a web punch out near the support.

Note: Using the AISI S211-07 results in a 51 percent increase in web crippling design strength.

4.17.2 Example Problem 2: Web Crippling

Check the web crippling limit state for a curtain-wall stud, 600S162-43. The 600S162-43 is nested in a 600T125-43 track. *Note:* The track thickness does not have to match stud thickness.

Because the section designator 600S162-43 does not stipulate a yield stress, the value of $F_y = 33$ ksi. According to AISI S201-07, the F_y value must be identified in the section designator if the value is other than 33 ksi. Furthermore, if the coating is other than G60, the section designator also must indicate the coating thickness.

AISI S100-07 Sec. A2.4 requires that the minimum delivered thickness must be no less than 95 percent of the design thickness. AISI S201-07 section designator thickness, that is, 43 mils, is the minimum delivered thickness.

$$\text{Design thickness} = 0.043/0.95 = 0.0451 \text{ in.}$$
$$R/t = 0.0712/0.0451 = 1.58$$
$$h/t = 5.767/0.0451 = 128$$

Section C1 of AISI S211-07 stipulates that for curtain walls, the ends of the wall studs shall be seated squarely in the track with no more than a ¼-in. gap between the end of the stud and the track. Therefore, $N = 1$ in. for the curtain-wall stud with a ¼-in. gap between the end of the stud and the track web.

$$N/t = 1/0.0451 = 22.2$$

If the wind load is 28 psf and the curtain-wall studs are 16 in. on center, the applied load is 37.33 lb/ft. The stud height is 13 ft. Thus, the applied end reaction is computed as

$$P = 0.5wL$$

$$= 0.5(37.33)(13)$$

$$= 243 \text{ lb} = 0.243 \text{ kip}$$

Using AISI S100 Sec. C3.4 (Eq. C3.4.1-1), the nominal web crippling capacity is determined by

$$P_n = Ct^2F_y \sin\theta\left(1 - C_R\sqrt{\frac{R}{t}}\right)\left(1 + C_N\sqrt{\frac{N}{t}}\right)\left(1 - C_h\sqrt{\frac{h}{t}}\right)$$

From Table C3.4.1-2, $C = 4$, $C_R = 0.14$, $C_N = 0.35$, $C_h = 0.02$, and $\Omega = 1.75$.

$$P_a = P_n/\Omega = 0.453 \text{ kip}/1.75 = 0.259 \text{ kip}$$

which is greater than 0.243 kip.

Okay!

Note: This value is for a web without a web opening near the location of the applied end reaction. See AISI Sec. C3.4.2 if a web opening is near the applied load.

For the condition of a web opening near the location of the applied end reaction, AISI S100-07 Sec. C3.4.2, Eq. (C4.3.2-1) applies.

If the hole depth d_o is 1.5 in. and $x = 9$ in., which is the nearest distance from the edge of the hole and the edge of bearing,

$$R_c = 1.01 - 0.325(d_o/h) + 0.083(x/h) \leq 1.0$$

$$= 1.01 - 0.325(1.5/5.767) + 0.083(9/5.767)$$

$$= 1.05 > 1.0$$

Thus, no strength reduction is required for a 1.5-in. web hole located 10 in. from the end of the member to the near edge of the web hole.

Reevaluate the web crippling capacity using the AISI S211-07. The nominal strength equation is the same as with AISI S100-07:

$$P_{nst} = Ct^2F_y \sin\theta\left(1 - C_R\sqrt{\frac{R}{t}}\right)\left(1 + C_N\sqrt{\frac{N}{t}}\right)\left(1 - C_h\sqrt{\frac{h}{t}}\right)$$

However, the coefficients given in AISI S211-07 are $C = 3.7$, $C_R = 0.19$, $C_N = 0.74$, $C_h = 0.019$, and $\Omega = 1.70$. These values reflect the increase in web crippling strength achieve

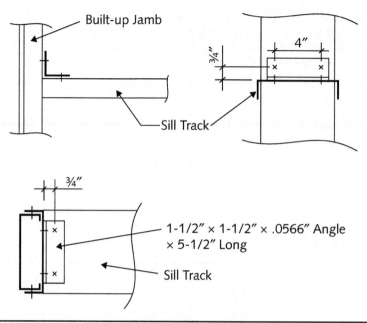

Figure 4.5 Sill track to jamb stud connection. (*AISI, 2007.*[13])

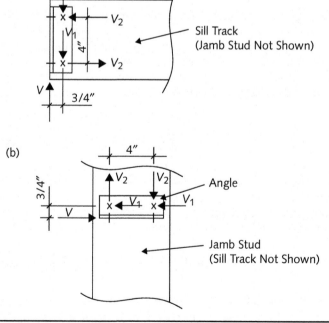

Figure 4.6 Connection forces: (*a*) track-to-clip connection; (*b*) clip-to-jamb connection. (*AISI, 2007.*[13])

by considering the behavior of the stud when nested in the track. AISI S100-07 considers only the strength of the stud:

$$P_a = P_{nst}/\Omega = 0.666 \text{ kip}/1.70 = 0.392 \text{ kip}$$

Therefore, using the AISI S211-07 results in a 51 percent increase in web crippling design strength.

Note: This value is for a web without a web opening near the location of the applied end reaction. See AISI S100-07 Sec. C3.4.2 if a web opening is near the applied load.

4.17.3   Example Problem 3: Sill Track to Jamb Stud Connection

Evaluate the adequacy of the sill track to jamb stud connection (Fig. 4.5).

The eccentricity of the connection will be resisted, as shown by Fig. 4.6. This is the most efficient distribution of the eccentric forces because the fasteners are subjected only to shear. Because of the symmetry of the connection, the required load and the allowable strength for all four screws are the same.

The sill track is assumed to be a simple span beam with a uniform load. The end shear is computed as the end reaction: $V = 296$ lb.

As illustrated by Fig. 4-6,

$$V_1 = 296/2 = 148 \text{ lb}$$

Also, summing forces on the connection,

$$V_2 = Ve/4 = 296(0.75)/4 = 55.5 \text{ lb}$$

V_{req} is the resulting shear force on a screw

$$V_{req} = \sqrt{V_1^2 + V_2^2} = 158 \text{ lb}$$

The screw is a #10-16, which has an outside diameter $d = 0.190$ in.

Connection sheet design values are

$$\text{Angle } t_1 = 0.0566 \text{ in.,} \quad F_{u1} = 65 \text{ ksi}$$

$$\text{Track } t_2 = 0.0451 \text{ in.,} \quad F_{u2} = 45 \text{ ksi}$$

$$\text{Jamb } t_2 = 0.0451 \text{ in.,} \quad F_{u2} = 45 \text{ ksi}$$

The allowable shear capacity for a connection must consider the strength of the connecting elements, that is, angle and track or stud, which are limited by AISI S100-07 Sec. E4.3.1 tilting and bearing and AISI S100-07 Sec. E4.3.2 edge shear.

Tilting and Bearing

Because $t_2/t_1 = 0.797 < 1.0$, therefore, AISI S100-07 Eqs. (E4.3.1-1), (E4.3.1-2), and (E4.3.1-3) will define the governing sheet capacity for tilting and bearing P_{ns}:

$$P_{ns} = 4.2(t^3 d)^{1/2} F_{u2} = 789 \text{ lb} \qquad \text{This governs sheet capacity.}$$

$$P_{ns} = 2.7 t_1 d F_{u1} = 1,887 \text{ lb}$$

$$P_{ns} = 2.7t_2 dF_{u2} = 1{,}041 \text{ lb}$$

$$V_a = P_{ns}/W = 789/3 = 263 \text{ lb} > 158 \text{ lb}.$$

Okay!

Edge Shear

Screw connection allowable shear is limited by AISI S100-07 Sec. E4.3.2 for edge shear. Edge distance $e = \frac{3}{4}$ in. is used for both the angle and the track.

Conservatively assume that the resulting shear acts perpendicular to the end of the sill track:

$$P_{ns} = teF_u$$

$$= 0.0451(0.75)(45) = 1{,}522 \text{ lb}$$

$$V_a = P_{ns}/W = 1{,}522/3 = 507 \text{ lb} > 158 \text{ lb}$$

Okay!

Screw Shear Strength

According to AISI S100-07 Sec. E4.3.3, the screw allowable shear strength must be defined by the manufacturer or by independent tests. For a #10 screw,

$$P_{ns} = P_{ss} = 1{,}400 \text{ lb}$$

$$V_a = P_{ns}/\Omega = 1{,}400/3 = 467 \text{ lb} > 158 \text{ lb}$$

Okay!

The connection is adequate for all limit states.

References

1. American Institute of Steel Construction (AISC). 2005. *Specification for Structural Steel Buildings* (ANSI/AISC 360-05). AISC, Chicago, IL.
2. American Iron and Steel Institute (AISI). 2007. *North American Specification for the Design of Cold-Formed Steel Structural Members* (AISI S100-07). AISI, Washington, DC.
2a. American Iron and Steel Institute (AISI). 2010. *Supplement No. 2 to the North American Specification for the Design of Cold-Formed Steel Structural Members* (AISI S100-07/S2-10). AISI, Washington, DC.
3. American Iron and Steel Institute (AISI). 2007. *Commentary to the North American Specification for the Design of Cold-Formed Steel Structural Members.* AISI, Washington, DC.
4. American Iron and Steel Institute (AISI). 2007. *North American Standard for Cold-Formed Steel Framing—General Provisions* (AISI S200-07). AISI, Washington, DC.
5. American Iron and Steel Institute (AISI). 2007. *North American Standard for Cold-Formed Steel Framing—Product Data* (AISI S201-07). AISI, Washington, DC.
6. American Iron and Steel Institute (AISI). 2007. *North American Standard for Cold-Formed Steel Framing—Floor and Roof System Design* (AISI S210-07). AISI, Washington, DC.

7. American Iron and Steel Institute (AISI). 2007. *North American Standard for Cold-Formed Steel Framing—Wall Stud Design* (AISI S211-07). AISI, Washington, DC.

8. American Iron and Steel Institute (AISI). 2007. *North American Standard for Cold-Formed Steel Framing Header Design* (AISI S212-07). AISI, Washington, DC.

9. American Iron and Steel Institute (AISI). 2007. *North American Standard for Cold-Formed Steel Framing—Lateral Design* (AISI S213-07). AISI, Washington, DC.

10. American Iron and Steel Institute (AISI). 2008. *North American Standard for Cold-Formed Steel Framing—Truss Design* (AISI S214-07), with Supplement No. 2 (S2-08). AISI, Washington, DC.

11. American Iron and Steel Institute (AISI). 2008. *Standard for Cold-Formed Steel Framing—Prescriptive Method for One and Two Family Dwellings* (AISI S230-07), with Supplement No. 2 (S2-08). AISI, Washington, DC.

12. American Iron and Steel Institute (AISI). 2008. *Cold-Formed Steel Design Manual* (AISI D100-08). AISI, Washington, DC.

13. American Iron and Steel Institute (AISI). 2007. *Cold-Formed Steel Framing Design Guide* (AISI D110-07). AISI, Washington, DC.

14. American Iron and Steel Institute (AISI). 2006. *Direct Strength Method Design Guide* (AISI CF06-1). AISI, Washington, DC.

15. International Code Council (ICC). 2009. *2009 International Building Code* (2009 IBC). ICC, Washington, DC.

16. Yu, W. W., and LaBoube, R. A. (2010). *Cold-Formed Steel Design.* John Wiley & Sons, Hoboken, NJ.

CHAPTER 5

Structural Design of Reinforced Masonry

5.1 Introduction

While masonry is used for partitions and other nonstructural components in all types of construction, it is used as the main load-resisting system primarily in low-rise structures. This chapter will provide a summary of topics related to the structural design of low-rise masonry construction. The discussion will be on design in accordance with the 2009 *International Building Code* (2009 IBC),[1] which references Masonry Standards Joint Committee (MSJC) *Building Code Requirements for Masonry Structures* (TMS 402-08/ACI 530-08/ASCE 5-08).[2] Chapter 1 of TMS 402-08/ACI 530-08/ASCE 5-08 provides general requirements for masonry design, whereas Chaps. 2 and 3 provide procedures for the design of masonry using allowable stress design and strength design, respectively. This chapter will discuss both strength design and allowable stress design procedures.

Procedures for the design of flexural members such as beams and lintels, members subjected to design of flexural and axial loads such as columns and piers, and shear walls will be described. In addition, the chapter will discuss the design of slender walls loaded out of plane and connections of masonry structures, which involves the design of anchor bolts, splices, and the development of reinforcement.

5.2 Load Paths and Analysis

5.2.1 Out-of-Plane Loads on Masonry Walls

The design of masonry walls to resist out-of-plane loads is an important aspect in building design and often is the critical phase of the design of large buildings that use masonry walls as the lateral load resisting system. The common masonry warehouse or department store type building with few openings usually has enough walls to resist the in-plane demands generated by wind or earthquake loads. However, the large story height means that the out-of-plane demands can be relatively high. Out-of-plane loads on masonry walls in buildings usually are induced by inertial earthquake forces or wind pressures. In basement walls and other retaining walls, out-of-plane loads also are caused by lateral soil pressures.

Figure 5.1 illustrates the load path for resisting out-of-plane loads. When the walls are loaded out of plane, they are not part of the lateral load resisting system but instead act as elements of the structure or components that support their own direct loads. For satisfactory structural response, walls must be capable of spanning between supports and transfering lateral loads to the floor or roof diaphragm, which, in turn, transfers the loads to the walls that form the lateral load resisting system. The connection between the wall and the supporting diaphragm is a critical link in the load path for out-of-plane response that must be designed carefully.

The design of walls for out-of-plane response is complicated by the fact that because walls usually are slender compared with their height, the deflection induced by the lateral loads sometimes can be comparable with the wall thickness. This means that the usual assumption of small deflections is not valid, and secondary deformation effects (P-δ effects) need to be considered in order to determine the wall demands accurately.

Figure 5.2 shows the effective height H for walls with different support configurations and the effective height for each case. Since response to earthquake and wind loads is a dynamic response driven by the fundamental out-of-plane mode, walls typically are analyzed as simple spans between supports because the connection of the wall to the roof or floor usually does not possess sufficient stiffness or strength to restrain the rotation of the wall.

Walls usually are designed to span vertically between floor or roof levels, as shown in Figs. 5.2 and 5.3a. However, when supported by relatively closely spaced pilasters or cross-walls, it may be more economical to design a wall to span horizontally between the vertical supports, as shown in Fig. 5.3b. In either case, the designer has control over the wall behavior. This is so because after the masonry cracks, loads are transferred in the direction with the larger flexural capacity.

As described in Chapter 2, Structural Engineering Institute (SEI) of the American Society of Civil Engineers (ASCE) *Minimum Design Loads for Buildings and Other Structures* (ASCE/SEI 7-05)[3] provides methods for calculating the out-of-plane wind loads on masonry walls. Using the simplified procedure in ASCE/SEI 7-05, the design

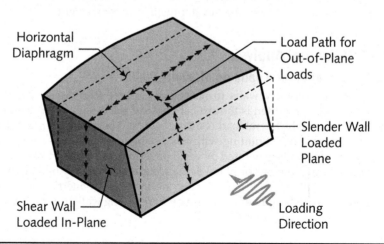

Figure 5.1   Load path for walls loaded out of plane.

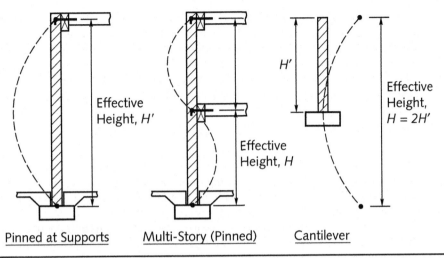

Figure 5.2   Effective height of walls loaded out of plane.

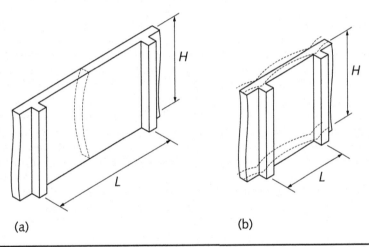

Figure 5.3   Direction of wall span: (a) wall spanning vertically; (b) wall spanning horizontally.

wind pressure p_{net} on masonry walls that are considered as components or cladding for out-of-plane loading is given by

$$p_s = \frac{\lambda K_{zt}}{p_{s30}} \qquad (5.1)$$

ASCE/SEI 7-05 provides values of p_{net30} for various basic wind speeds that are based on 3-s gusts at a height of 30 ft for Exposure B. The adjustment factor λ accounts for building height and exposure, I_w is the importance factor, and K_{zt} is the topographic factor.

Out-of-plane earthquake loads on concrete masonry walls also can be calculated using ASCE/SEI 7-05 and depend on whether the wall is classified as a structural wall, which is part of the vertical or lateral load resisting system, or a nonstructural wall, such as a partition, which is not part of the building's load resisting system. In order to be classified as a nonstructural wall, the connection of a wall to the diaphragm must be detailed such that no in-plane forces are induced during earthquake response. For structural walls, the out-of-plane seismic load is calculated from the following equation:

$$F_p = 0.4S_{DS}I_E W_l \tag{5.2}$$

where S_{DS} is the design spectral response acceleration parameter in the short-period range, as discussed in Chap. 2. I_E is the importance factor, and W_p is the weight of the structural wall (usually in lb/ft^2). The out-of-plane force used in design should not be less than 10 percent of the weight of the structural wall.

For nonstructural walls, which are treated as architectural components, the following equation calculates the seismic design force F_p on the wall, which is to be distributed relative to the wall mass distribution:

$$F_p = \frac{0.4a_p S_{DS}W_P}{R_P/I_P}\left(1+2\frac{z}{h}\right) \tag{5.3}$$

where S_{DS} is the short-period 5 percent damped spectral response acceleration at the building location. The value $0.4S_{DS}$ represents the effective ground acceleration at the site. And a_p is the amplification factor that represents the dynamic amplification of the wall relative to the fundamental period of the structure. For most masonry walls, $a_p = 1.0$, except for parapets and unbraced walls, for which $a_p = 2.5$. I_p is the importance factor, which is equal to 1.0 or, alternately, 1.5 for components that are required for life-safety purposes, contain hazardous materials, or are needed for continued operation of Occupancy Category IV structures. W_p is the wall weight. R_p is the response modification factor that represents the wall overstrength and ductility or energy-absorbing capability. For reinforced-masonry walls, $R_p = 2.5$, whereas for unreinforced-masonry walls, $R_p = 1.5$. Also, z is the height of point of the wall attachment with respect to the base, and h is the average roof height of the structure with respect to the base.

The seismic force need not exceed

$$F_{p,\max} = 1.6S_{DS}I_p W_P \tag{5.4}$$

and should not be less than

$$F_{p,\min} = 0.3S_{DS}I_p W_P \tag{5.5}$$

Designing masonry to resist out-of-plane loads, as shown in Fig. 5.4, around openings involves the design of the masonry above the opening (and below the opening for windows) and the design of the jamb reinforcement on either side of the opening. It is generally more convenient to design the masonry above the opening to span horizontally because all the reinforcement above the opening can be used to resist the out-of-plane forces.

The jamb reinforcement on either side of the opening also must be designed to support the reaction from the masonry above and below the opening. A simplifying approach that is often used when designing jambs is to ignore any reduction in out-of-plane loading due to the opening. Then the jamb reinforcing is determined using additional load from a tributary width equal to half the width of the opening. This approach is reasonably accurate for wind loading and typically conservative for earthquake loading because the weight of the door or window in the opening is usually much less than that of the masonry assumed in the calculations. A more rigorous procedure, in which the actual load imposed on the jamb by the masonry above the opening is calculated, may be used to reduce the conservatism in the design.

One approach is to use equations similar to those for the moment magnifier procedure described in American Concrete Institute (ACI) *Building Code Requirements for Structural Concrete* (ACI 318-08).[4] This method accounts for secondary moment (P-δ) effects and the fact that the out-of-plane loads on the jamb are not uniformly distributed over the jamb height. The maximum moment M_u calculated without P-δ effects is magnified to account for P-δ effects as follows:

$$M_c = \frac{M_u}{1 - \dfrac{P_u}{0.75P_e}} \tag{5.6}$$

where P_u is the factored axial load, and P_e is the critical buckling load, which is given by

$$P_e = \frac{\pi^2 E_m I_{\text{eff}}}{H^2} \tag{5.7}$$

where E_m is the modulus of elasticity of the concrete masonry, and H is the effective height of the wall. The effective moment of inertia I_{eff} depends on whether the jamb is cracked or uncracked:

$$I_{\text{eff}} = I_g \quad \text{if} \quad M_c \leq M_{cr}$$
$$I_{\text{eff}} = I_{cr} \quad \text{if} \quad M_c > M_{cr}$$

where M_{cr} is the cracking moment, I_g is the gross moment of inertia, and I_{cr} is the cracked moment of inertia. Details on how to calculate these variables are discussed in Sec. 5.6.

Walls that retain soil such as basement walls or free-standing retaining (cantilever) walls are also subjected to soil-induced out-of-plane loads. The project's geotechnical engineer usually provides the lateral soil loads on retaining walls. The 2009 IBC Sec. 1610 provides default soil lateral load values that may be used when no site-specific geotechnical information is available. Cantilever retaining walls are designed to resist the active soil pressure that develops as the wall deflects. Basement retaining walls usually are restrained from displacement at the top and bottom by foundation and floor levels and do not deflect enough to mobilize the shear strength of the soil. Such walls must be designed for the at-rest soil pressures, which can be significantly larger than the active soil pressures.

5.2.2 Out-of-Plane Anchorage Loads

The damage that occurred to out-of-plane connections between roofs and concrete or masonry walls during past earthquakes demonstrated that earlier codes may not have provided buildings with sufficient protection against out-of plane forces during strong ground shaking. The most significant failures occurred in large box-type structures with high walls and flexible diaphragms. Recordings from instrumented buildings showed that in such buildings, the acceleration at the center of the diaphragm is often over three times the acceleration at the ground level. This led to changes in the code requirements for out-of-plane anchorage of concrete and masonry walls, particularly when they are supported by flexible diaphragms.

As shown in Fig. 5.4, masonry walls that are subjected to lateral loads from earthquake and wind demands typically span between floors and/or roof levels. This means that adequate reactions, or anchorages, must be developed at floor and roof supports to resist the tendency of the wall to pull away and collapse during earthquakes or high winds.

When concrete masonry walls are supported by diaphragms that are not flexible, the out-of-plane anchorage forces are calculated in a manner similar to the out-of-plane forces used for wall design. For structural walls, the anchorage force is given by Eq. (5.2), where W_p is the weight of the wall tributary to the anchorage. In addition, the anchorage force must not be less than the greater of the following:

$400S_{DS}I_E$ lb/ft

280 lb/ft

Structural walls are to be designed to resist bending between anchors that are spaced more than 4 ft apart. Anchorage forces in non-structural walls connected to nonflexible diaphragms are calculated using Eq. (5.3), where W_p is the weight of the wall tributary to the anchorage, as shown in Fig. 5.5. The following additional

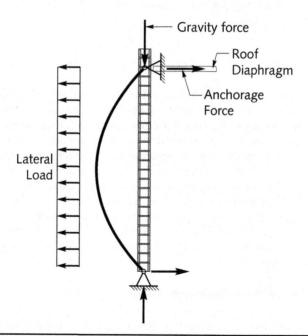

FIGURE 5.4   Out-of-plane anchorage forces on a masonry wall.

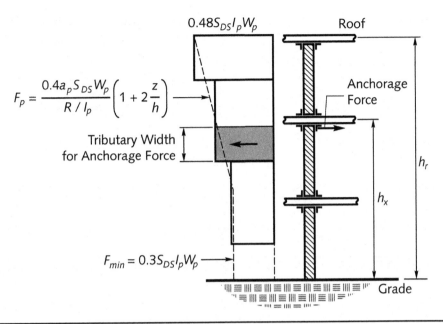

$$F_p = \frac{0.4a_p S_{DS} W_p}{R/I_p}\left(1 + 2\frac{z}{h}\right)$$

$0.48S_{DS}I_p W_p$ Roof

Anchorage
Force

Tributary Width
for Anchorage Force

$F_{min} = 0.3S_{DS}I_p W_p$

h_r

h_x

Grade

FIGURE 5.5   Anchorage forces in non-structural walls.

requirements are applicable to the anchorage of nonstructural components to concrete masonry walls:

1. Anchors shall be designed to resist the smaller of 1.3 times the calculated forces or the maximum force that can be transmitted to the anchor.

2. The value of R_p shall not exceed 1.5 unless the anchor is designed to be governed by the strength of a ductile steel element.

For buildings with flexible diaphragms, the anchorage forces are increased significantly because of amplification of the building acceleration by the diaphragm, as shown in Fig. 5.6. Earlier codes required that the flexible diaphragm effect be included only at the diaphragm mid-span. However, recent research indicates that since a flexible diaphragm is a shear-yielding beam and that because of secondary mode effects, the amplification should be applied uniformly across the length of the diaphragm. Consequently, ASCE/SEI 7-05 uses the following equation for determining the anchorage forces in structural and nonstructural masonry walls supported by flexible diaphragms:

$$F_p = 0.8S_{DS}I_E W_l \tag{5.8}$$

Comparison with Eq. (5.2) shows that anchorage forces for walls supported by flexible diaphragms are assumed to be twice as large as those for nonflexible diaphragms as a result of diaphragm amplification.

When designing buildings to resist earthquakes, it is important that all parts of the structure are tied together with struts, collectors, or ties that are capable of transmitting

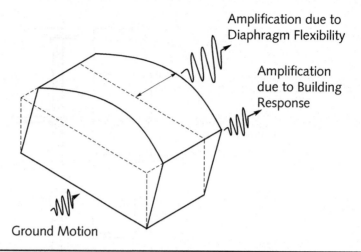

FIGURE 5.6 Amplification of diaphragm forces in flexible diaphragms.

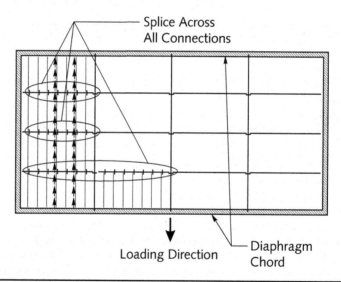

FIGURE 5.7 Continuous ties without subdiaphragms.

the earthquake-induced forces to the primary lateral load resisting system. Bearing in mind that this is particularly important for diaphragms that provide out-of-plane support for relatively heavy concrete or masonry walls, ASCE/SEI 7-05 states that diaphragms must have continuous ties or struts between diaphragm chords to distribute anchorage forces into the diaphragms.

One way of interpreting this requirement would mean that each joist in the north-south direction in Fig. 5.7 would have to transfer the load from one end of the building to the other. Each joist-to-girder connection would have to be capable of transmitting the anchorage force and require splices with the capacity to resist the horizontal loads.

Such a design would be quite expensive and clearly is not the most effective method of providing continuity.

Fortunately, ASCE/SEI 7-05 also states that added chords may be used to form subdiaphragms to transmit the anchorage forces to the main continuous cross-ties. The maximum length-to-width ratio of the structural subdiaphragm shall be 2.5:1. This concept of subdiaphragms is extremely useful in providing continuous ties, as illustrated in Fig. 5.8. While each subdiaphragm chord must be designed to resist axial chord forces, the number of splice connections required to transmit the anchorage load is reduced significantly.

5.2.3 Concentrated Loads on Masonry Walls

Concentrated loads on masonry walls cannot be distributed over a length greater than the bearing width plus the length determined by assuming that the load is dispersed at a slope of 2 vertical to 1 horizontal, as shown in Fig. 5.9. The dispersion of the concentrated load must stop at the wall mid-height. In addition, concentrated loads cannot be dispersed across movement joints, openings, or beyond the end of a wall. The length over which concentrated loads are distributed also should not exceed the center-to-center distance between concentrated loads. For walls that are not laid in running bond, concentrated loads cannot be dispersed beyond head joints unless a bond beam is used to distribute the loads. This means that loads typically can be dispersed across head joints in solid-grouted walls that are laid in stack bond if there is sufficient reinforcement. This is usually the case for buildings assigned to Seismic Design Categories D, E, and F because the minimum reinforcement requirements for masonry walls provide enough reinforcement to distribute concentrated loads across head joints in stack bond construction.

In addition to designing walls to resist the distributed concentrated loads, the bearing on the masonry also must be checked. Bearing stresses are calculated using the

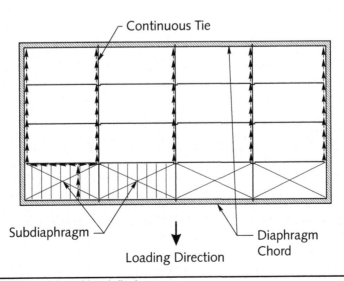

FIGURE 5.8   Continuous ties with subdiaphragms.

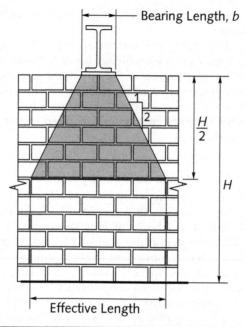

FIGURE 5.9   Distribution of concentrated loads on masonry.

bearing area A_{br}, which is equal to the smaller of the areas obtained from the following equations:

$$A_{br} = A_1 \sqrt{\frac{A_2}{A_1}}$$ (5.9)

$$A_{br} = 2A_1$$ (5.10)

where A_1 is the loaded area, and A_2 is the area of the base of the frustum of a pyramid or a cone with the loaded area at the top and sides that slope at 45 degrees that is wholly contained within the support. The bearing areas are shown in Fig. 5.10.

5.2.4   Lintels

The portion of the wall above an opening that resists gravity loads typically is called a *lintel*. Lintels can be constructed with numerous materials, including concrete (precast or cast in place) or structural steel. However, most lintels, particularly those in solid-grouted masonry, are constructed homogeneously with the rest of the wall using concrete masonry.

The determination of gravity loads on a lintel is based on a number of principles. Arching action above the opening enables loads to span around the opening. When the height of masonry above an opening is sufficient (greater than about half the lintel span), the structure has a natural tendency to span across openings by forming an arch, as shown in Fig. 5.11. The masonry lintel may be assumed to support only the loads

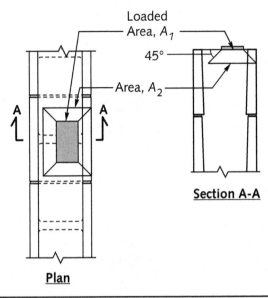

FIGURE 5.10   Areas for calculating bearing loads.

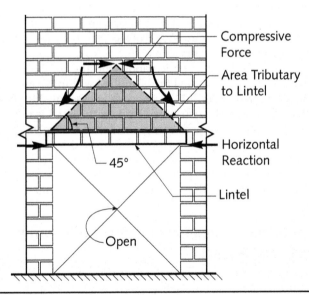

FIGURE 5.11   Arching of masonry above an opening.

within the triangle created by two lines inclined at 45 degrees and extending from the ends of the span. Wall gravity loads and distributed loads outside this area are resisted by the surrounding wall and by a horizontal reaction at the ends of the arch. For arching to occur, there must be sufficient lateral resistance at the ends of the arch to resist the thrust. This lateral resistance may be provided by the mass of masonry adjacent to the opening or by specifically designing an element to resist the imposed lateral load. Note

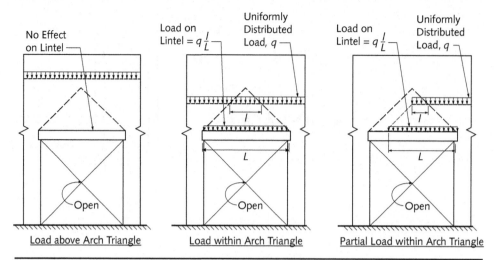

Figure 5.12   Lintel loading from uniformly distributed loads.

that arching cannot occur if the ends of the arch are near wall ends, corners, or control joints or when stack bond is used. There also must be enough masonry above the arch to resist the compressive forces generated by arching action. Arching therefore cannot occur if the height of masonry above an opening is less than half the lintel span. If arching is not possible, the entire weight of the wall above the opening and all distributed loads are applied to the masonry lintel.

Distributed loads that are applied above the apex of the arch triangle are assumed to arch over the lintel. Distributed loads applied within the arch triangle are supported by the lintel, as shown in Fig. 5.12. Such loads are assumed to spread out at a 45 degree angle so that the uniformly distributed load on the lintel is reduced by the ratio of the width of the load within the arch triangle and the lintel span. Distributed loads that do not extend entirely across the arch triangle may be distributed to only a portion of the lintel.

As described in the preceding section, concentrated loads are assumed to be dispersed at a slope of 2 vertical to 1 horizontal, as illustrated in Fig. 5.13.

5.3 Design of Masonry Members Subjected Primarily to Flexure (and Shear)

Masonry members that are subjected primarily to flexural loads (with negligible axial loads) include beams and lintels. Section 3.3.4.2.1 of TMS 402-08/ACI 530-08/ASCE 5-08 states that the axial loads in beams must not exceed $0.05 A_n f'_m$. This is usually taken as the value for which axial loads can be neglected in the design of a member. Reinforced-masonry flexural members may be designed using either strength design or allowable-stress design procedures. In either case, the compression face of beams should be supported laterally at a maximum spacing of 32 times the beam thickness. The nominal depth of beams must be no less than 8 in., and no more than two bar sizes should be used as longitudinal reinforcement. In addition, the variation in reinforcing bar size should not exceed one size.

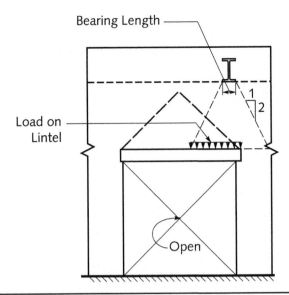

FIGURE 5.13   Lintel loading from concentrated loads.

5.3.1   Strength Design of Flexural Members

Strength design of masonry to resist flexure is similar to the strength design of reinforced concrete and is based on the following assumptions:

1. Plane sections before bending remain plane during and after bending. This means that strain varies linearly across the depth of the cross section and that strain in the reinforcement and masonry is assumed to be directly proportional to the distance from the neutral axis.

2. There is strain continuity among the reinforcement, grout, and concrete masonry such that applied loads are resisted in a composite manner.

3. Tensile strength of masonry is neglected when calculating flexural strength but considered when calculating deflections.

4. The stress in reinforcement is taken as the steel strain multiplied by the steel modulus of elasticity E_s but not greater than the steel yield stress f_y.

5. The maximum usable strain ε_{mu} at the extreme masonry compression fiber is equal to 0.0025 for concrete masonry and 0.0035 for clay masonry.

6. A masonry compressive stress of $0.8f'_m$ is assumed to be uniformly distributed over an equivalent compression zone that is bounded by the edges of the cross section and a line parallel to the neutral axis located at a distance $0.8c$ from the extreme compression fiber, where c is the distance between the extreme compression fiber and the neutral axis.

Figure 5.14 illustrates the assumptions used in the calculation of flexural capacity for a singly reinforced beam, which has only one layer of reinforcement. As can be seen in the figure, an equivalent rectangular compression stress block is used to represent

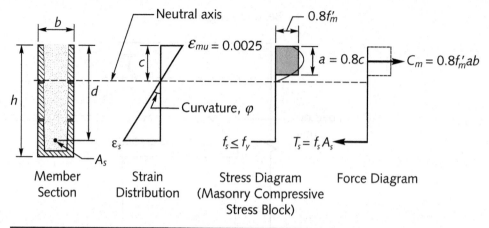

Figure **5.14** Stress and strain distributions for strength design of masonry.

the actual distribution of masonry compressive strain. This assumption greatly simplifies the calculation of flexural capacity because the area and centroid can be obtained much more easily for a rectangular compression block than for a shape defined by the actual stress-strain relationship of masonry.

Reinforcement is not considered effective in compression unless adequate lateral support in the form of ties is provided to prevent buckling of the reinforcing steel bars. Section 5.4 describes the specific code requirements that must be satisfied in order for reinforcement to be used in resisting compression. Most flexural members such as beams and lintels do not possess sufficient lateral reinforcement, so compression steel typically is ignored when calculating flexural strength.

The basic design equation for flexural members is based on the limit-state equation:

$$\phi M_n \geq M_u \tag{5.11}$$

where M_n is the nominal moment strength, ϕ is the strength-reduction factor, and M_u is the factored load demand. The capacity-reduction factor for flexure, axial load, or combinations of flexure and axial load is 0.9.

The principles of strength of materials can be used to determine the flexural capacity of a singly reinforced member such as the beam shown in Fig. 5.14. Assuming that the steel reinforcement yields (which is a code requirement), the tensile force in the reinforcement is equal to the product of the steel yield stress and the area of reinforcement, that is,

$$T_s = f_y A_s \tag{5.12}$$

The compression force in the masonry is given by

$$C_m = 0.8 f_m' ab \tag{5.13}$$

Since the tension and compression force on the cross section must be equal,

$$T_s = C_m \tag{5.14}$$

$$f_y A_s = 0.8 f'_m ab \tag{5.15}$$

which means that the depth of the compression block is given by

$$a = \frac{f_y A_s}{0.8 f'_m b} \tag{5.16}$$

The nominal moment capacity can be determined by taking moments about the centroid of the compression force:

$$M_n = f_y A_s \left(d - \frac{a}{2} \right) \tag{5.17}$$

and substituting Eq. (5.16) into Eq. (5.17):

$$M_n = f_y A_s \left(d - \frac{f_y A_s}{1.6 f'_m b} \right) \tag{5.18}$$

In addition to comparing the flexural strength of a member to the moment demand, the code also requires that the nominal flexural strength of a beam not be less than 1.3 times the flexural cracking moment strength. This requirement is to help ensure that a beam's flexural capacity is greater than the cracking moment and reduce the possibility of sudden reduction in stiffness when a beam cracks. The cracking moment M_{cr} is obtained from

$$M_{cr} = S_n f_r \tag{5.19}$$

where S_n is the section modulus of the net cross section of the beam, and f_r is the modulus of rupture. Values for f_r are provided in Chap. 3.

For a singly reinforced cross section with given dimensions, the amount of tensile reinforcement required can be derived from Eq. (5.18):

$$\frac{M_u}{\phi} = f_y A_{s,\mathrm{req}} \left(d - \frac{f_y A_{s,\mathrm{req}}}{1.6 f'_m b} \right) \tag{5.20}$$

Solving for $A_{s,\mathrm{req}}$, we obtain

$$A_{s,\mathrm{req}} = \frac{d - \sqrt{d^2 - \dfrac{4M_u}{\phi 1.6 f'_m b}}}{\dfrac{2 f_y}{1.6 f'_m b}} \tag{5.21}$$

Or in terms of the reinforcement ratio, ρ_s:

$$\rho_{s,\mathrm{req}} = \frac{A_{s,\mathrm{req}}}{bd} = \frac{1 - \sqrt{1 - \dfrac{2.5}{\phi} \left(\dfrac{M_u}{bd^2 f'_m} \right)}}{1.25 \left(\dfrac{f_y}{f'_m} \right)} \tag{5.22}$$

Equations (5.21) and (5.22) may be used to obtain the amount of reinforcement required in a singly reinforced cross section directly. However, during the design process, it is often more practical to develop an initial design using the following relationship:

$$A_{s,\mathrm{req}} = \frac{M_u}{\phi f_y \left(d - \dfrac{a}{2}\right)} \tag{5.23}$$

and selecting an estimate of the reinforcing steel required by assuming that the moment arm for forces on the cross section is given by

$$d - \frac{a}{2} \approx 0.9 - 0.95d \tag{5.24}$$

Equation (5.18) then can be used to check whether the selected steel provides sufficient flexural strength.

One of the goals in the design of masonry members is to ensure that the steel reinforcement yields before the masonry reaches the maximum usable compressive strain ε_{mu} at the ultimate limit state. This is necessary for a number of reasons. While the failure of masonry at the maximum usable strain is brittle and sudden, reinforcing steel typically possesses substantial deformation capacity after yielding. This means that yielding of reinforcing steel can provide a warning before a dangerous failure occurs. In addition, the ductile behavior that results when reinforcing steel yields prior to the ultimate limit state is important when performance in the nonlinear range is required, such as during response to severe earthquake loading.

The postyield curvature or rotation capacity of a cross section can be evaluated by the strain that exists in the reinforcing steel when the masonry is at the maximum usable compressive strain. The code stipulates that in flexural members the strain in the reinforcing steel must be at least 1.5 times the yield strain when the masonry attains the maximum usable strain. Figure 5.15 shows a cross section with acceptable and unacceptable strain profiles.

Since a larger amount of reinforcing steel will require a smaller strain to maintain equilibrium, it is apparent that the code-required ductility can be achieved by limiting

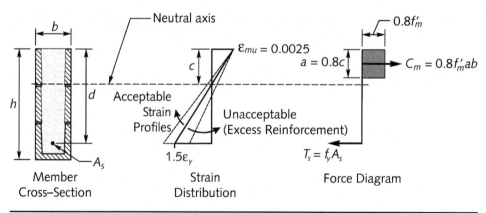

FIGURE 5.15 Maximum reinforcement ratio in beams.

the amount of reinforcement that is placed in a cross section. The maximum amount of reinforcement that will satisfy the code requirements can be determined from Fig. 5.15. With the maximum permissible amount of reinforcement, the distance from the extreme compression fiber to the neutral axis is given by

$$c = \frac{d\varepsilon_{mu}}{\alpha\varepsilon_y + \varepsilon_{mu}} \tag{5.25}$$

where α is the multiple of the steel yield strain needed to achieve the required ductility. Since the total compressive force on the cross section must be equal to the total tensile force, the equilibrium equation for a member with no axial load is given by

$$A_s f_y = 0.8 f'_m ab = 0.64 f'_m cb \tag{5.26}$$

Substituting Eq. (5.25) into Eq. (5.26), we obtain

$$A_{s,\max} = 0.64 \frac{f'_m}{f_y} bd \left(\frac{\varepsilon_{mu}}{\alpha\varepsilon_y + \varepsilon_{mu}} \right) \tag{5.27}$$

Or in terms of the maximum reinforcement ratio

$$\rho_{\max} = \frac{A_{s,\max}}{bd} = 0.64 \frac{f'_m}{f_y} \left(\frac{\varepsilon_{mu}}{\alpha\varepsilon_y + \varepsilon_{mu}} \right) \tag{5.28}$$

For masonry beams with no axial load,

$$\alpha = 1.5$$

$$\varepsilon_{mu} = 0.0025$$

The maximum reinforcement ratio then is given by

$$\rho_{\max} = 0.64 \frac{f'_m}{f_y} \left(\frac{0.0025}{1.5 f_y / E_s + 0.0025} \right) \tag{5.29}$$

Note that the requirements for shear walls are more stringent than those for beams and other flexural members because they form part of the lateral load resisting system. This is so because shear walls typically require ductility to resist earthquake loads adequately. A discussion on the maximum reinforcement ratio in shear walls will be presented in Sec. 5.6.

Flexural members also must be designed to resist the shear forces that correspond to the flexural demands. When using strength design, the shear strength of masonry members is a combination of the shear resistance provided by the masonry V_m and the shear resistance provided by shear reinforcement V_s. The nominal shear strength therefore is equal to

$$V_n = V_m + V_s \tag{5.30}$$

And the basic limit-state design equation for shear is given by

$$\phi V_n \geq V_u \tag{5.31}$$

where V_u is the factored shear demand on the member. The strength-reduction factor ϕ for shear is equal to 0.8. Shear strength is highly dependent on the shear span-to-depth ratio of the member, which is defined by the term $M_u / V_u d_v$, in which M_u is the moment demand corresponding to the factored shear demand and d_v is the actual depth of the masonry cross section in the direction the shear is being considered. Squat or short span members tend to have higher shear strengths so that when $M_u / V_u d_v \leq 0.25$,

$$V_n \leq 6 A_n \sqrt{f'_m}$$

and when $M_u / V_u d_v \geq 1.0$,

$$V_n \leq 4 A_n \sqrt{f'_m}$$

where A_n is the net area of the cross section. The maximum value of V_n for $M_u / V_u d_v$ values between 0.25 and 1.0 is interpolated. Note that for most beams and lintels subjected to primarily flexural loads, the lower limit of V_n will govern.

The shear strength contributed by masonry is given by

$$V_m = \left[4 - 1.75 \left(\frac{M_u}{V_u d_v} \right) \right] A_n \sqrt{f'_m} + 0.25 P_u \tag{5.32}$$

The value of $M_u / V_u d_v$ must be positive and need not exceed 1.0. As can be seen from Eq. (5.32), masonry shear strength increases with smaller shear span-to-depth ratios. In addition, the presence of axial load results in increased masonry shear capacity as a result of improved aggregate interlock. This beneficial impact of axial load is not applicable to beams and lintels, however, because such members are subjected to negligible axial loads.

The shear contributed by the steel reinforcement can be derived from Fig. 5.16, which shows a free body diagram of a beam subjected to shear forces. Shear reinforcement typically is placed parallel to the direction of the applied shear. At locations of high shear demand, such as near the supports, diagonal tension cracks form at an angle of approximately 45 degrees to the longitudinal axis. The free body diagram in Fig. 5.16 shows that the number of reinforcing bars that intersect an inclined crack depends on the spacing of the shear reinforcement, which is parallel to the direction of applied shear. Research has shown that at the ultimate shear limit state, shear reinforcing is effective only in part in resisting shear. Ignoring the resistance due to dowel action provided by the horizontal reinforcement and assuming that the vertical steel reinforcement is 50 percent effective, the shear strength based on the strength of the steel reinforcement can be computed as provided by Eq. (5.33) by summing the vertical forces from the free body diagram shown in Fig. 5.16:

$$V_s = 0.5 \sum A_v f_y \tag{5.33}$$

where A_v is the area of each shear reinforcing bar, and f_y is the steel yield stress. Since the shear reinforcing bars are at a spacing s and the member has an effective

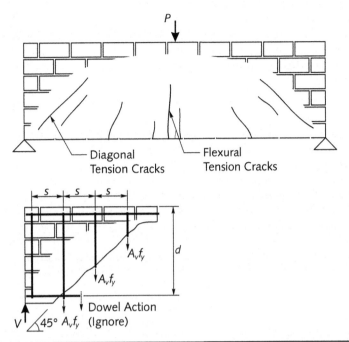

FIGURE 5.16 Shear strength provided by steel reinforcement.

depth d, the number of bars intersecting a 45-degree crack can be incorporated in Eq. (5.33):

$$V_s = 0.5 A_v f_y \frac{d}{s} \tag{5.34}$$

The transverse reinforcement used to resist shear must satisfy the following requirements:

1. Shear reinforcement must consist of a single bar with a 180-degree hook at each end.
2. Shear reinforcement must be hooked around longitudinal reinforcing bars.
3. The minimum area of transverse reinforcement should be $0.0007 bd_v$.
4. The first transverse bar should be placed no more than one-fourth the beam depth d_v from the end of the beam.
5. The maximum spacing of shear reinforcement must be no greater than one-half the beam depth or 48 in. .

5.3.2 Allowable Stress Design of Flexural Members

When using allowable stress design, the stresses in the materials from external loads must not exceed prescribed allowable stresses. The allowable stresses are established as a fraction of the strength or capacity of the materials. Code-calculated dead, live, wind, and seismic loads, without load factors, are used to determine the induced stresses.

The allowable stresses in masonry (i.e., masonry units, mortar, and grout) are expressed as a fraction of the specified compressive strength of the masonry at the age of 28 days f'_m. When a masonry element is subjected to flexural loads, the allowable compressive stress in the masonry is given by

$$F_b = \frac{1}{3} f'_m \qquad (5.35)$$

The allowable stresses in reinforcing steel depend on whether the reinforcement is in tension or compression. For bars in tension, Table 5.1 provides values for the allowable steel tensile stress F_s.

The code does not permit the use of reinforcement for resisting compression stresses unless adequate lateral support in the form of ties is provided to prevent buckling of the reinforcing steel bars. Section 5.4.1 describes the specific code requirements that must be satisfied if the designer intends to use reinforcement to resist compression stresses. Typically, compression reinforcement is effective only in columns and boundary elements of walls with ties that satisfy the minimum lateral reinforcement requirements for columns. When the lateral reinforcement requirements are satisfied, the allowable compressive stress in reinforcing steel is given by

$$F_{sc} = 0.4 f_y \leq 24{,}000 \text{ psi} \qquad (5.36)$$

where f_y is the specified yield stress of the reinforcing steel.

Note that Sec. 1605.3.2 of the 2009 IBC permits the allowable stresses in masonry and reinforcing steel to be increased by one-third if the alternate basic load combinations that include wind or seismic forces are used. No increase in allowable stresses can be used if the basic load combinations in 2009 IBC Sec. 1605.3.1 are used.

The procedures for allowable stress design are based on the following assumptions:

1. Plane sections before bending remain plane during and after bending; strain varies linearly across the depth of the section.

2. All stresses are in the elastic range, and stresses are proportional to strain.

3. The modulus of elasticity of the masonry, mortar, and grout assemblage E_m remains constant.

4. Tensile strength of masonry is neglected.

5. External and internal moments and forces are in equilibrium.

By using these assumptions for a singly reinforced cross section subjected to flexural loads (negligible axial loads), we can obtain the stress, strain, and equilibrium diagrams shown in Fig. 5.17. As can be seen in the figure, plane sections before bending remain plane after bending. This means that the strain at any location on a cross section is

Grade 40 or Grade 50 reinforcement	20,000 psi
Grade 60 reinforcement	24,000 psi
Wire joint reinforcement	30,000 psi

TABLE 5.1 Allowable Tensile Stresses in Reinforcement

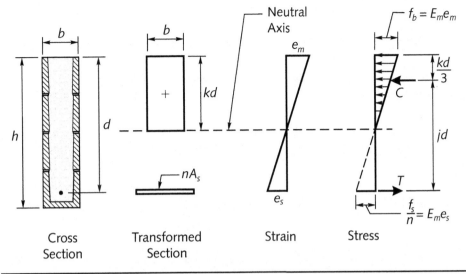

FIGURE **5.17**   Stress and strain distributions for allowable stress design of masonry.

proportional to the distance from the neutral axis and can be determined by similar triangles. In addition, since stress is proportional to strain, the stress at any location on the cross section can be calculated from the strain diagram, bearing in mind that the masonry does not resist tension.

From Fig. 5.17, the following variables can be defined:

b = width of the cross section

A_s = effective cross-sectional area of reinforcement

d = effective depth of the flexural member, which is the distance from the extreme compression fiber to the centroid of the extreme tensile reinforcement

kd = effective depth of the compression area, which is the distance from the extreme compression fiber to the neutral axis

jd = distance between the centroid of flexural compressive force and the centroid of the tensile force

e_m = maximum compressive strain in the masonry

e_s = maximum tensile strain in the steel

n = E_s/E_m = ratio of modulus of elasticity of steel to modulus of elasticity of masonry in compression, also known as the *modular ratio*

f_b = maximum calculated flexural compressive stress in the masonry

f_s = maximum calculated stress in reinforcement

By using similar triangles, the strain in the reinforcement can be expressed in terms of the maximum compressive strain in the masonry:

$$e_s = e_m \left(\frac{d-kd}{kd} \right) = e_m \left(\frac{1}{k} - 1 \right) \tag{5.37}$$

and the stress in the reinforcement is equal to

$$\frac{f_s}{n} = e_s E_m = e_m \left(\frac{1}{k} - 1 \right) E_m \qquad (5.38)$$

The maximum compressive stress in the masonry is also given by

$$f_b = e_m E_m \qquad (5.39)$$

In order to satisfy equilibrium, the total compressive force on the cross section must be equal to the tensile force on the cross section. That is,

$$T - C = 0 \qquad (5.40)$$

Using the stress distribution from Fig. 5.17,

$$\frac{f_s}{n} nA_s - \frac{1}{2} bkdf_b = 0 \qquad (5.41)$$

We can substitute the expressions for steel and masonry stress given in Eqs. (5.38) and (5.39) into Eq. (5.41) to obtain

$$e_m \left(\frac{1}{k} - 1 \right) E_m n A_s - \frac{1}{2} bkde_m E_m = 0 \qquad (5.42)$$

and dividing by $e_m E_m$ leads to

$$\left(\frac{1}{k} - 1 \right) n \frac{A_s}{bd} - \frac{1}{2} k = 0 \qquad (5.43)$$

Since the reinforcement ratio is given by

$$\rho = \frac{A_s}{bd} \qquad (5.44)$$

rearranging Eq. (5.43) results in the following equation:

$$\frac{1}{2} k^2 + \rho nk - \rho n = 0 \qquad (5.45)$$

Equation (5.45) is a second-degree equation with the following solution for k:

$$k = \sqrt{(\rho n)^2 + 2\rho n} - \rho n \qquad (5.46)$$

Given the triangular distribution of the compressive stress, the ratio of the distance between the centroid of the compressive force and the tensile force and the effective depth of the cross section is equal to

$$j = 1 - \frac{k}{3} \qquad (5.47)$$

The external moment from loading is equal to the internal moment in the cross section. Therefore, the allowable moment on the assumed cracked masonry section shown in Fig. 5.17 may be determined by calculating the moments about the centroid of the compressive force:

$$M = f_s A_s jd \tag{5.48}$$

or in terms of the masonry compressive stress,

$$M = \frac{1}{2} f_b jkbd^2 \tag{5.49}$$

Note also that for a given stress in the masonry and reinforcing steel, the depth of the neutral axis can be determined by similar triangles from Fig. 5.17:

$$k = \frac{1}{1 + f_s/nf_b} \tag{5.50}$$

For a given cross section and design moment M, the required area of tensile steel can be determined by developing a closed-form solution for A_s from the relationships in Eqs. (5.45) to (5.47). However, a common approach is to estimate the required steel area by assuming that

$$j \approx 0.85 \text{ to } 0.95$$

And that the required area of steel is given by

$$A_{s,req} = \frac{M}{f_s jd}$$

A steel area that exceeds the estimated requirement then can be selected. The cross section then is checked using the appropriate equations.

When the allowable-stress design method is used to design masonry, shear resistance is provided either completely by the masonry or completely by shear reinforcement. This is consistent with research that indicates that shear reinforcement in masonry is fully effective only if it is designed to resist the entire shear demand. This means that one first should check to see if the masonry alone can resist the entire shear. If the allowable shear force provided by the masonry is insufficient, the masonry should be ignored, and shear reinforcement should be provided to resist the entire shear.

The allowable shear provided by masonry depends on whether flexural tensile stresses exist on the cross section. When flexural tensile stresses do not exist, the cross section is assumed to be uncracked, and maximum shear stress f_v is calculated using the classic equation from strength of materials:

$$f_v = \frac{VQ}{I_n b} \tag{5.51}$$

where

V = shear demand
Q = first moment about the neutral axis of a section of that portion of the cross section lying between the neutral axis and extreme fiber
I_n = moment of inertia of the net sectional area
b = width of section

The allowable shear stress F_v then is given by the smallest of the following:

1. $1.5\sqrt{f'_m}$

2. 120 psi

3. $60 \text{ psi} + 0.45 N_v / A_n$ (for solid-grouted running bond masonry) or $37 \text{ psi} + 0.45 N_v / A_n$ (for partially grouted running-bond masonry and solid-grouted stack-bond masonry with open-end units)

4. 15 psi (for stack-bond other-than-open-end units)

where N_v is the compressive force acting normal to the shear surface, and A_n is the net cross-sectional area. In lieu of the preceding procedure, elements not subjected to flexural tension may be designed by assuming that reinforcing steel resists the entire shear demand, as described in the following paragraphs.

When a cross section is subjected to flexural tension, the shear stress on the cross section is based on the average shear stress, which is equal to

$$f_v = \frac{V}{bd} \tag{5.52}$$

The allowable shear stress on flexural members subjected to flexural tension is given by

$$F_v = \sqrt{f'_m} \le 50 \text{ psi} \tag{5.53}$$

When the masonry is not capable of resisting the shear demand without the allowable shear stress being exceeded, reinforcing steel must be provided and designed to withstand the entire shear. Since the masonry allowable shear stress has been exceeded, it is assumed that the flexural member cracks, as shown in Fig. 5.16, as was the case for strength design.

If we ignore the dowel action provided by the longitudinal flexural reinforcement, summing the vertical forces on the free-body diagram in Fig. 5.16 (and replacing the yield stress of the reinforcement f_y with the allowable stress F_s) results in the following equation:

$$V = \sum A_v F_s \tag{5.54}$$

where A_v is the area of each shear reinforcing bar. Since the shear reinforcing bars are at a spacing s and the member has an effective depth d, the number of bars intersecting a 45-degree crack can be incorporated in Eq. (5.54):

$$V = A_v F_s \frac{d}{s} \tag{5.55}$$

which means that the area of shear reinforcement required at a spacing s is given by

$$A_v = \frac{sV}{F_s d} \tag{5.56}$$

or that the spacing of reinforcement is equal to

$$s = \frac{A_v F_s d}{V} \tag{5.57}$$

The code requirements stipulate that when shear reinforcement is used, the maximum shear stress should not exceed

$$F_{v,\text{max}} = 3.0\sqrt{f'_m} \le 150 \text{ psi} \tag{5.58}$$

To ensure that each 45-degree crack is intersected by at least two bars, it is recommended that the shear reinforcement be spaced no more than $d/2$ apart.

Since flexural tension cracks often penetrate deeply into the body of a cross section, the shear reinforcement must be placed as close to the compression face (and tension face) as permitted by cover and other bar placement requirements. The reinforcement must be anchored at each end in order to develop the tensile stress required by design. The ends of single-leg or U-stirrups must be anchored with a standard hook plus an effective embedment length of $l_d/2$, where l_d is the required development length. The development length is measured from the mid-depth of the member and the tangency point of the hook. For #5 bars and smaller, the shear reinforcement may be bent around the longitudinal reinforcement with a standard 135-degree hook, and the minimum effective embedment may be equal to $l_d/3$.

5.3.3 Examples

Example 5.3.1

Design the reinforcement for the 32-in.-deep beam shown in Fig. 5.18. The compressive strength of the masonry is 1,500 psi. Grade 60 steel is used for the longitudinal reinforcement, and Grade 40 steel is used for the shear reinforcement.

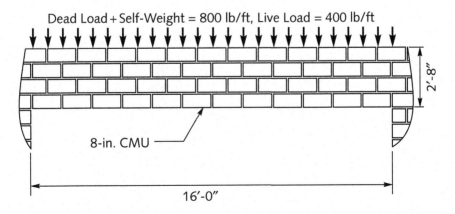

Dead Load + Self-Weight = 800 lb/ft, Live Load = 400 lb/ft

8-in. CMU

2'-8"

16'-0"

FIGURE 5.18 Reinforced-masonry beam.

Strength Design Solution

Assuming that lateral support for the compression area is provided only at the supports,

$$\frac{l}{b} = \frac{16(12)}{7.63} = 25.2 < 32$$ *Okay!*

For dead and live loads, the applicable load combination is $1.2D + 1.6L$:

$$w_u = 1.2(800) + 1.6(400) = 1600 \text{ lb/ft}$$

$$V_u = \frac{w_u l}{2} = \frac{1,600(16)}{2} = 12,800 \text{ lb}$$

$$M_u = \frac{w_u l^2}{8} = \frac{1,600(16)^2}{8} = 51,200 \text{ lb-ft}$$

Assuming that $d = 28$ in., the required area of steel can be estimated:

$$A_{sreq} \approx \frac{M_u}{\phi f_y 0.9d} = \frac{51,200(12)}{0.9(60,000)(28)} = 0.41 \text{ in.}^2$$

Try two #5 bars ($A_s = 0.62$ in.^{2}). Then, from Eq. (5.18),

$$\phi M_n = \phi f_y A_s \left(d - \frac{f_y A_s}{1.6 f_m' b} \right)$$

$$= 0.9(60,000)(0.62) \left(28 - \frac{60,000(0.62)}{1.6(1,500)7.63} \right) \frac{1}{12}$$

$$= 72,452 \text{ lb-ft} > M_u$$ *Okay!*

Compare with the cracking moment:

$$1.3M_{cr} = 1.3S_n f_r = \frac{1.3(7.63)(32)^2}{6} 200 \frac{1}{12}$$

$$= 28,214 \text{ lb-ft} < M_n$$ *Okay!*

From Eq. (5.29), the maximum reinforcement ratio is equal to

$$\rho_{max} = 0.64 \left(\frac{1.5}{60} \right) \left(\frac{0.0025}{1.5(60)/29,000 + 0.0025} \right) = 0.0071$$

$$\rho = \frac{A_s}{bd} = \frac{0.62}{7.63(24)} = 0.0034 < \rho_{max}$$ *Okay!*

Check the maximum shear strength:

$$V_{n,max} = 4A_n \sqrt{f_m'} = 4(7.63)(32)\sqrt{1,500}$$

$$= 37,825 \text{ lb} > V_u$$ *Okay!*

The shear strength provided by the masonry is given by

$$\phi V_m = 0.8(2.25)A_n\sqrt{f'_m}$$

$$= 0.8(2.25)(7.63 \times 32)\sqrt{1,500} = 17,021 \text{ lb} > V_u$$

The shear strength provided by the masonry is sufficient to resist the demand, and no shear reinforcement is required.

Allowable Stress Design Solution

$$V = \frac{wl}{2} = \frac{1,200(16)}{2} = 9,600 \text{ lb}$$

$$M = \frac{wl^2}{8} = \frac{1,200(16)^2}{8} = 38,400 \text{ lb-ft}$$

Allowable stresses:

$$F_b = \frac{1}{3}f'_m = \frac{1,500}{3} = 500 \text{ psi}$$

$$F_s = 20,000 \text{ psi} \qquad \text{Grade 40 steel}$$

$$F_s = 24,000 \text{ psi} \qquad \text{Grade 60 steel}$$

The modulus of elasticity of the masonry is equal to

$$E_m = 900 f'_m = 900(1,500) = 1,350,000 \text{ psi}$$

and the modular ratio is equal to

$$n = \frac{29,000,000}{1,350,000} = 21.5$$

For $d = 28$ in., the amount of reinforcement can be estimated:

$$A_{s,req} = \frac{M}{f_s jd} \approx \frac{38,400(12)}{24,000(0.8)(28)} = 0.86 \text{ in.}^2$$

Try two #7 bars:

$$\rho n = \frac{2(0.60)21.5}{7.63(28)} = 0.121$$

$$k = \sqrt{(\rho n)^2 + 2\rho n} - \rho n = \sqrt{(0.121)^2 + 2(0.121)} - 0.121 = 0.386$$

$$j = 1 - \frac{k}{3} = 1 - \frac{0.386}{3} = 0.871$$

The moment corresponding to allowable masonry compressive stress is given by

$$M = \frac{1}{2} f_b j k b d^2 = \frac{1}{2}(500)(0.871)(0.386)(7.63)(28)^2 \frac{1}{12} = 41,899 \text{ lb-ft}$$

The moment corresponding to the allowable steel tensile stress is given by

$$M = f_s A_s j d = 24,000(2 \times 0.60)(0.871)(28)\frac{1}{12} = 58,531 \text{ lb-ft}$$

The allowable moment is determined by the masonry compressive stress and is greater than the moment demand of 38,400 lb-ft.

Since there is no axial load on the beam, it is subjected to flexural tension. The shear stress therefore is given by Eq. (5.52):

$$f_v = \frac{V}{bd} = \frac{9600}{7.63(28)} \quad = 44.9 \text{ psi}$$

$$> \sqrt{f'_m} = \sqrt{1,500} \quad = 38.7 \text{ psi}$$

$$< 3\sqrt{f'_m} = 3\sqrt{1,500} = 116.1 \text{ psi}$$

Therefore, shear reinforcement is required to resist the entire shear demand. If #4 bars are used for shear reinforcement, the minimum required spacing is given by Eq. (5.57):

$$s = \frac{A_v F_s d}{V} = \frac{0.2(20,000)28}{9,600} = 11.7 \text{ in.}$$

Use #4 bars spaced at 8 in. for shear reinforcement.

5.4 Design of Masonry Members Subjected to Axial Loads and Flexure

In addition to the flexural loads described in the preceding section, members such as piers, columns, and walls are also subjected to axial loads that must be considered. Axial loads may be induced by gravity dead and live loads, whereas flexural loads may occur as a result of lateral wind or seismic forces, lateral soil pressures, or the eccentric application of gravity loads, among other effects. The techniques for designing members subjected to axial loads in combination with flexural loads differ only slightly from the procedures described in the preceding section. The assumptions listed in Sec. 5.3 still apply, and the mechanics of materials must be used to determine the stresses induced on the masonry and reinforcing steel.

Examples of members subjected to combined bending and axial loads include the following:

- Columns
- Piers
- Shear walls (loaded in plane)
- Slender walls (loaded out of plane)
- Coupling beams (axial drag forces due to seismic loads)

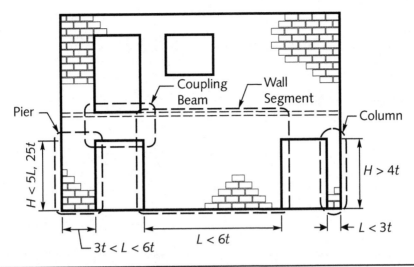

FIGURE 5.19 Members of a masonry wall subjected to axial loads and flexure.

Figure 5.19 shows members in a typical wall that may be subjected to flexural and axial loads. As can be seen in the figure, columns and piers are subject to specific dimensional limits.

The TMS 402-08/ACI 530-08/ASCE 5-08 defines a *column* as an isolated vertical member with a horizontal dimension measured at right angles to its thickness that does not exceed three times its thickness and a height that is greater than four times its thickness. It should be noted that when used in this context, the term *isolated* does not mean that the column is isolated from the lateral load resisting system. Specific dimensional requirements that must be satisfied in order for a member to be classified as a column include the following:

1. The nominal width (or thickness) must be at least 8 in..

2. The distance between lateral supports of a column must not exceed 30 times its nominal width.

3. The nominal depth of a column must not exceed three times the nominal width.

Figure 5.20 illustrates the dimensional limits for columns. All columns must be grouted solid with longitudinal reinforcement that satisfies the following requirements:

1. Longitudinal reinforcement must consist of a minimum of four bars, with at least one bar in each corner of the column.

2. The minimum reinforcement area shall be $0.0025A_n$, where A_n is the net cross-sectional area of the column.

3. The maximum area of longitudinal steel reinforcement area must not exceed $0.04A_n$.

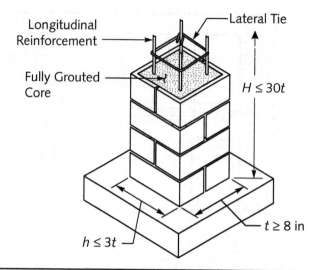

Longitudinal Reinforcement

Lateral Tie

Fully Grouted Core

$H \leq 30t$

$t \geq 8$ in

$h \leq 3t$

FIGURE 5.20 Dimensional limits for columns.

Lateral ties meeting the following specifications must be provided in all columns, which also allow longitudinal reinforcement to be used in resisting compression stresses:

1. Lateral ties must be at least ¼ in. in diameter. (For structures in Seismic Design Categories D, E, and F, lateral ties must be at least ⅜ in. in diameter.)

2. The vertical spacing of lateral ties shall not exceed 16 longitudinal bar diameters, 48 lateral tie diameters, or the minimum cross-sectional dimension of the member. In addition, for structures in Seismic Design Categories D, E, and F, lateral ties must be spaced no more than 8 in. on center.

3. Lateral ties shall be arranged such that every corner and alternate longitudinal bar shall have lateral support provided by the corner of a lateral tie with an included angle of not more than 135 degrees. No bar shall be farther than 6 in. clear on each side along the lateral tie from such a laterally supported bar.

4. Lateral ties shall be located vertically not more than one-half a lateral tie spacing above the top of a footing or a slab in any story and shall be placed not more than one-half a lateral tie spacing below the lowest horizontal reinforcement in a beam, girder, slab, or drop panel.

5. Where beams or brackets frame into a column from four directions, lateral ties shall be terminated not more than 3 in. below the lowest reinforcement in the shallowest of such beams or brackets.

A vertical member that is not isolated from the surrounding masonry, as shown in Fig. 5.21, is called a *pilaster*. A pilaster may be designed as a column if it satisfies the preceding dimensional limits and reinforcement requirements. Otherwise, it should be designed as part of the adjoining wall.

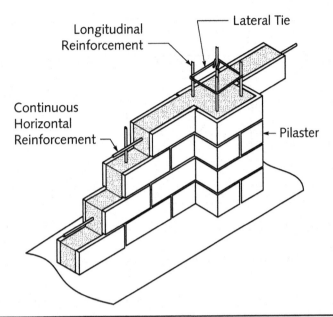

Figure 5.21 Reinforced-masonry pilaster.

Columns and pilasters must be designed to withstand all applied axial and lateral loads. Axial loads usually result from the self-weight of the member and from concentrated reactions from beams, girders, or trusses. The vertical component of earthquake or wind loads sometimes may contribute significantly to the axial load on a column or a pilaster and therefore should be included in the design process.

The eccentricity of the applied axial load results in bending moments that must be accounted for in the design. Columns must be designed to resist loads with a minimum eccentricity of 0.1 times the side dimension. Lateral loads typically are caused by earthquakes or high winds. In addition, lateral loads are imposed on columns and pilasters as they undergo the deformations required to maintain deformation compatibility with the rest of the structure. Figure 5.22 shows that the design loads on a pilaster also must include the out-of-plane load from the area of wall tributary to the pilaster.

As with columns and pilasters, piers are members that usually are subjected to both flexural and axial loads. As shown in Fig. 5.19, piers often are created by openings in walls for doors and windows. The dimensional limits for piers stipulated by the code include the following:

1. The nominal thickness must be no greater than 16 in..

2. The nominal length of a pier (measured perpendicular to its thickness) must be not less than three times or greater than six times its nominal thickness. However, if the factored axial load at the location of maximum moment is less than $0.05 f'_m A_g$, the length of the pier may be equal to the thickness of the pier.

3. The distance between lateral supports of a pier must not exceed 25 times its nominal thickness unless the pier is designed to resist out-of-plane demands

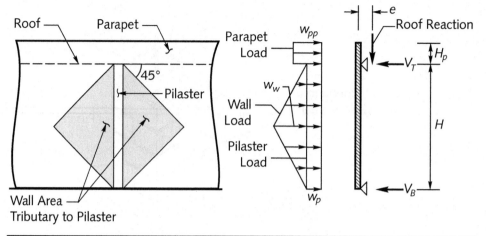

FIGURE 5.22 Out-of-plane loads on a pilaster.

using a procedure that accounts for the effects of displacement and cracking on response.

4. The clear height of a pier must not exceed five times its nominal length.

In addition to dimensional limits for piers, TMS 402-08/ACI 530-08/ASCE 5-08 also provides requirements for longitudinal reinforcement:

1. Piers subjected to load reversals must be reinforced symmetrically.

2. The minimum area of longitudinal reinforcement is $0.0007bd$, where b is the thickness of pier and d is the effective depth.

3. Longitudinal reinforcement must be uniformly distributed throughout the depth of a pier.

Note that the maximum reinforcement in a pier should satisfy the requirements of the walls in the same line of resistance. This means that if a pier is part of a special reinforced-masonry shear wall, the strain in the extreme tensile reinforcement must attain a strain of at least four times the yield strain at the ultimate limit state. This will be discussed in more detail in Sec. 5.6. In the out-of-plane direction, the strain in the extreme tensile reinforcement must achieve a strain of at least 1.5 times the steel yield strain.

Unlike columns, the longitudinal reinforcement in piers is not typically used in compression, and there are no specific requirements for lateral ties. Nevertheless, lateral reinforcement must be provided to resist the shear demands on a pier when required.

5.4.1 Strength Design of Members Subjected to Axial Loads and Flexure

As with columns, piers, and other members subjected to a combination of axial and flexural loads, strength design of piers is based on the same assumptions used for flexural members outlined in Sec. 5.3.1. Figure 5.23 shows strain and stress distributions on a member cross section subjected to axial and flexural loads.

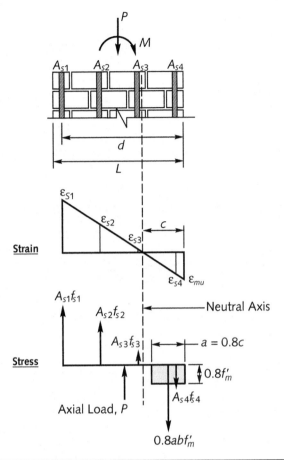

FIGURE 5.23 Stress and strain distribution on a cross section subjected to axial loads and flexure for strength design.

Since a larger compression block is required to resist higher compressive loads, the distribution of stress and strain will vary depending on the level of axial load on the cross section. This means that the flexural strength, which results from equilibrium of the cross section, also will vary with axial load. Design of columns, piers, and other members subjected to combined axial and flexural loads typically involves the development of a relationship between axial load and flexural strength. Such a relationship is called an *axial load versus flexural strength interaction diagram*.

An *interaction diagram* is a plot of the change in flexural strength of a member with variations in axial load. As shown in Fig. 5.24, combinations of factored axial load and bending moment that fall within the interaction diagram are acceptable, and the member can resist such loads adequately. If any combination of axial load and bending moment falls outside the interaction diagram, the cross section is inadequate, and the member needs to be redesigned. Interaction diagrams are developed using the basic assumptions for strength design.

To obtain an accurate interaction diagram, the flexural strength is determined for several values of applied axial load to obtain the smooth curve shown in Fig. 5.24. Such

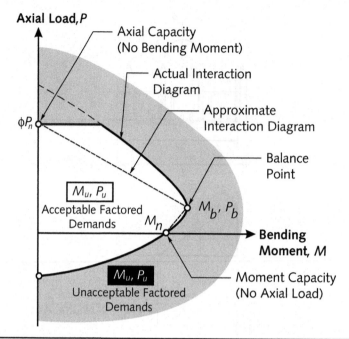

FIGURE 5.24 Interaction diagram.

an analysis can be extremely rigorous and is best performed with a specialized computer program. However, an approximate interaction diagram curve can be developed using selected points on the interaction diagram:

1. The axial load capacity when there is no bending moment on the member
2. The flexural strength when there is no axial load
3. The balanced strain condition when the strain in the extreme-compression fiber is equal to the maximum usable strain and the strain in the extreme-tension steel is equal to the steel yield strain

Note that the strength-reduction factor for all combinations of axial and flexural load is equal to 0.9.

When there is no bending moment on a cross section, the axial strength is determined by the compressive strength of the masonry and longitudinal reinforcement, with a modification factor to account for the effect of slenderness on axial capacity. For relatively squat members with $H/r \leq 99$,

$$P_n = 0.8 \left[0.8 f'_m \left(A_n - A_s \right) + f_y A_s \right] \left[1 - \left(\frac{H}{140r} \right)^2 \right] \tag{5.59}$$

For members with $H/r > 99$,

$$P_n = 0.8 \left[0.8 f'_m \left(A_n - A_s \right) + f_y A_s \right] \left(\frac{70r}{H} \right)^2 \tag{5.60}$$

In Eqs. (5.59) and (5.60), only steel that is confined by lateral ties that satisfy the requirements described earlier may be used to resist compressive stresses. When lateral ties that meet the requirements do not exist, compression reinforcement must be ignored. In addition to the modification factors that account for slenderness effects, a modification factor of 0.8 is used to account for a minimum eccentricity of axial load, which must be considered in the design.

Flexural strength is obtained in the same manner as for beams and lintels. The difference from the procedures described for flexural members in the preceding section lies in the fact that axial load will need to be included in the equilibrium calculations. In addition, more than one layer of reinforcement often will need to be considered, and when adequate lateral ties exist, reinforcement may be used to resist compression. This often makes determination of the neutral axis an iterative process in which a guess at the location of the neutral axis is made, and the equilibrium of forces on the cross section is checked. If the selected neutral axis location does not result in equilibrium of forces on the cross section, the location of the neutral axis is modified until equilibrium exists within acceptable limits. From Fig. 5.23, the strain in each reinforcing bar ε_{si} is given by

$$\varepsilon_{si} = \varepsilon_{mu}\left(\frac{c-d_i}{c}\right) \tag{5.61}$$

where d_i is the distance from the extreme-compression fiber to the reinforcing bar, and c is the depth of the neutral axis. The force in each reinforcing bar then is given by

$$f_{si}A_{si} = \varepsilon_{si}E_sA_{si} = \varepsilon_{mu}\left(\frac{c-d_i}{c}\right)E_sA_{si} \leq f_yA_{si} \tag{5.62}$$

where E_s is the steel modulus of elasticity, A_{si} is the area of the reinforcing bar, f_{si} is the stress in the bar, and f_y is the steel yield stress. The compression force in the masonry is given by

$$C_m = 0.64cbf'_m \tag{5.63}$$

and for equilibrium of forces on the cross section,

$$\sum f_{si}A_{si} + C_m = P \tag{5.64}$$

This means that

$$\sum \max\left[\min\left\{\varepsilon_{mu}\left(\frac{c-d_i}{c}\right)E_sA_s;\ f_yA_s\right\};\ -f_yA_s\right] + 0.64cbf'_m - P = 0 \tag{5.65}$$

The location of the neutral axis is determined from Eq. (5.65) using trial and error or by solution of the closed-form equation. Then the flexural strength corresponding to the axial load is obtained by taking moments about any location on the cross section. If moments are taken about the center of the cross section,

$$M = C_m\left(\frac{h}{2} - \frac{0.8c}{2}\right) + \sum T_{si}\left(d_i - \frac{h}{2}\right) \tag{5.66}$$

where L is the length of the member. When no axial load exists on the cross section, the moment capacity can be determined from Eqs. (5.64) and (5.65) with $P = 0$.

A balanced strain condition occurs when the strain in the extreme-compression fiber is equal to the maximum usable strain, and the strain in the extreme-tension steel is equal to the steel yield strain. The point on the interaction diagram that corresponds to a balance strain condition is called the *balance point*. The balance point can be determined easily because, from Fig. 5.25, the depth of the neutral axis is given by

$$c_b = \frac{d}{\left(\dfrac{f_y}{E_s \varepsilon_{mu}} + 1\right)} \tag{5.67}$$

The axial load at the balance point is the difference between the compression and tension forces on the cross section. The flexural strength at the balanced condition then may be determined from Eq. (5.66).

For singly reinforced cross sections, it sometimes may be more convenient to calculate the flexural strength for a specific axial load instead of developing an interaction diagram. Figure 5.26 shows the forces on a singly reinforced cross section (the steel in compression is ignored because it is not laterally supported). From the figure, and taking into account the strength-reduction factor,

$$A_s f_y + \frac{P_u}{\phi} = 0.8abf'_m \tag{5.68}$$

Therefore, the depth of the masonry compression block is given by

$$a = \frac{A_s f_y + P_u / \phi}{0.8bf'_m} \tag{5.69}$$

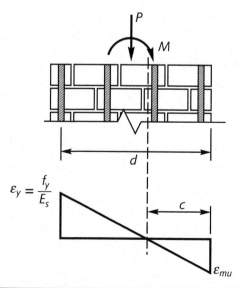

FIGURE 5.25   Strain distribution at balanced condition.

Taking moments about the centroid of the compression block, the nominal flexural strength for a singly reinforced cross section then is equal to

$$M_n = A_s f_y \left(d - \frac{a}{2} \right) + \frac{P_u}{\phi} \left(\frac{h}{2} - \frac{a}{2} \right)$$ (5.70)

Note that this equation is similar to Eq. (5.65) but includes the increase in flexural strength due to axial load.

The shear strength of members subjected to combined axial and flexural loads is calculated in a manner identical to the procedure used for flexural elements such as beams. The basic limit-state design equation for shear is given by

$$\phi V_n \geq V_u$$ (5.71)

where V_u is the factored shear demand on the member, f is the strength-reduction factor for shear (which is equal to 0.8), and V_n is the nominal shear strength. The nominal shear strength is equal to the sum of the shear strength contributed by the masonry and the shear strength contributed by the steel:

$$V_n = V_{nm} + V_{ns}$$ (5.72)

The shear strength contributed by masonry is given by

$$V_{nm} = \left[4 - 1.75 \left(\frac{M_u}{V_u d_v} \right) \right] A_n \sqrt{f_m'} + 0.25 P_u$$ (5.73)

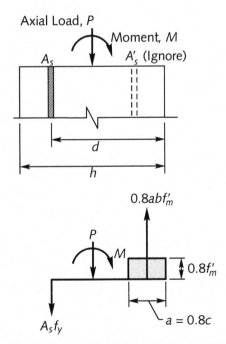

FIGURE 5.26 Singly reinforced cross section with axial load.

where A_n is the net area of the cross section. The value of M_u/V_ud_v used in Eq. (5.73) must be positive and need not exceed 1.0. The shear strength contributed by the steel is equal to

$$V_{ns} = 0.5A_v f_y \frac{d}{s} \qquad (5.74)$$

In addition, the shear strength must not exceed the following values:

When $M_u/V_ud_v \leq 0.25$,

$$V_n \leq 6A_n\sqrt{f_m'}$$

When $M_u/V_ud_v \geq 1.0$,

$$V_n \leq 4A_n\sqrt{f_m'}$$

5.4.2 Allowable-Stress Design of Members Subjected to Axial Loads and Flexure

The techniques for designing members subjected to axial loads in combination with flexural loads differ only slightly from the procedures described in Sec. 5.3.1 for allowable-stress design of members subjected to primarily flexural loads. The assumptions listed in Sec. 5.3.2 still apply, and the mechanics of materials must be used to determine the stresses induced on the masonry and reinforcing steel. The allowable compressive force when due to only the axial component of the load is given by

$$P_a = \left(0.25 f_m'A_n + 0.65 A_{st}F_s\right)\left[1 - \left(\frac{H}{140r}\right)^2\right] ; \quad \frac{H}{r} \leq 99 \qquad (5.75)$$

$$P_a = \left(0.25 f_m'A_n + 0.65 A_{st}F_s\right)\left(\frac{70r}{H}\right)^2 ; \quad \frac{H}{r} > 99 \qquad (5.76)$$

where H is the effective height of the member, and r is the radius of gyration. As with flexural elements, the maximum compressive stresses due to the combination of axial and flexural loads f_b must satisfy the following relationship:

$$f_b \leq F_b = \frac{1}{3}f_m' \qquad (5.77)$$

The allowable tensile stresses in reinforcing steel are listed in Table 5.1. Reinforcing steel is not considered effective in resisting compression unless it is supported adequately by lateral reinforcement, as described in Sec. 5.4.1. When the lateral reinforcement requirements are satisfied, the allowable compressive stress in reinforcing steel is given by

$$F_{sc} = 0.4f_y \leq 24{,}000 \text{ psi} \qquad (5.78)$$

The design of members subjected to a combination of axial and flexural loads depends on whether flexural tension is induced on the member cross section. When there is no flexural tension on a cross section, the compressive maximum stress caused by axial load is less than the maximum tensile stress due to flexural loads alone. This means that

$$f_a = \frac{P}{A_n} \ge \left(f_b = \frac{M}{S_n} \right) \tag{5.79}$$

where P and M are the axial load and bending moment on the cross section, respectively, whereas A_n and S_n are the net cross-sectional area and section modulus.

When there is no flexural tension, a cross section is designed as an uncracked member. This does not mean that there are no cracks in the member because cracking may occur as a result of several other non-load-related factors such as shrinkage. Instead, it means that since, theoretically, only compression stresses are applied, tensile reinforcement is not required, and the member is adequate if Eqs. (5. 75) to (5.77) are satisfied. Reinforcing steel only needs to be used to satisfy minimum reinforcing and detailing requirements.

When flexural tension does exist, tensile reinforcement is required, and the stresses in the masonry and steel are calculated by considering the assumptions in Sec. 5.3.2. Figure 5.27 illustrates the relationships between stress and strain on a singly reinforced

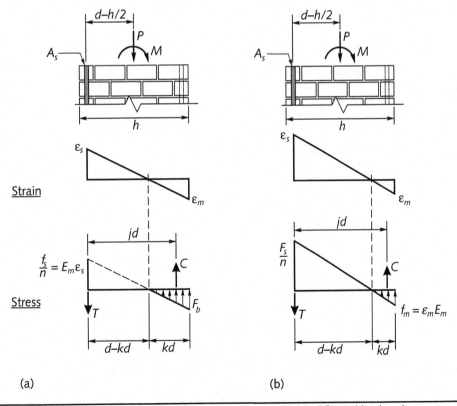

(a) (b)

FIGURE 5.27 Stress and strain distributions for combined axial and flexural loads using allowable-stress design: (a) at allowable compressive stress; (b) at allowable steel tensile stress.

cross section with no compression steel and a given axial load P. Figure 5.27a shows the cross section with moment capacity governed by the allowable compression stress in the masonry, whereas Fig. 5.27b shows the strain governed by the allowable tensile stress in the reinforcing steel.

From Fig. 5.27a, the compression force in the masonry is equal to

$$C = \frac{1}{2} F_b bkd \qquad (5.80)$$

and the tensile force in the reinforcing steel is given by

$$T = f_s A_s = e_s n E_m A_s \qquad (5.81)$$

And since the steel strain is given by

$$e_s = e_m \left(\frac{d - kd}{kd} \right) = \frac{F_b}{E_m} \left(\frac{d - kd}{kd} \right) \qquad (5.82)$$

then

$$T = F_b \left(\frac{d - kd}{kd} \right) n A_s \qquad (5.83)$$

For equilibrium,

$$P + T - C = 0 \qquad (5.84)$$

Substituting Eqs. (5.80) and (5.83) for C and T, respectively, and multiplying by kd, we obtain

$$Pkd + F_b (d - kd) n A_s - \frac{F_b}{2} b(kd)^2 = 0 \qquad (5.85)$$

Multiplying by $2/(bd^2 F_b)$,

$$k^2 + k \left(\frac{2n A_s}{bd} - \frac{2P}{bd F_b} \right) - \frac{2n A_s}{bd} = 0 \qquad (5.86)$$

Since the reinforcement ratio ρ is given by

$$\rho = \frac{A_s}{bd} \qquad (5.87)$$

Equation (5.86) becomes

$$k^2 + k \left(2n\rho - \frac{2P}{bd F_b} \right) - 2n\rho = 0 \qquad (5.88)$$

Solving the preceding second-order equation for k, we get

$$k = \sqrt{\left(n\rho - \frac{P}{bd F_b} \right)^2 + 2n\rho} - \left(n\rho - \frac{P}{bd F_b} \right) \qquad (5.89)$$

Note that this equation is similar to Eq. (5.16), which is the equation for the ratio of the depth of the neutral axis to the effective depth when there is no axial load.

Taking moments about the tensile reinforcing steel on the cross section in Fig. 5.27a, the moment on the cross section is given by

$$M = \frac{1}{2} F_b jkbd^2 - P\left(d - \frac{h}{2}\right) \tag{5.90}$$

where h is the total depth of the cross section. Equations (5.89) and (5.90) provide similar relationships for determining the neutral axis and moment when the cross section is governed by the allowable tensile stress in the reinforcing steel:

$$k = \sqrt{\left(n\rho + \frac{nP}{F_s bd}\right)^2 + 2\left(n\rho + \frac{nP}{F_s bd}\right)} - \left(n\rho + \frac{nP}{F_s bd}\right) \tag{5.91}$$

$$M = A_s F_s jd + P\left(\frac{h}{2} - \frac{kd}{3}\right) \tag{5.92}$$

As with cross sections subjected only to flexural loads, Eqs. (5.90) and (5.92) may be used to calculate the moments that are determined by the allowable masonry compressive stress and allowable steel tensile stress, respectively. The smaller of the two values determines the allowable moment on the cross section. Procedures for allowable stress design of members to resist shear are described in Sec. 5.3.

5.4.3 Examples

Example 5.4.1

Verify that a 20-ft-high, 16-in. concrete masonry column can support the axial load and bending moment from the loads shown in Fig. 5.28. The masonry compressive strength is 1,500 psi. Grade 60 steel is used for the longitudinal reinforcement, and the ties are Grade 40 steel. $S_{DS} = 1.21$ g.

Verify the column dimensions:

$t = 16$ in. > 8 in.	*Okay!*
$L = 16$ in. $< 3t = 48$ in.	*Okay!*
$H = 20(12) = 240$ in. $< 30t = 480$ in.	*Okay!*

Check the ratio of longitudinal reinforcement:

$A_s = 4(0.79) = 3.16$ in.^{2}	
$> 0.0025 A_s = 0.0025(15.63)^2 = 0.61$ in.^{2}	*Okay!*
$< 0.04 A = 0.04(15.63)^2 = 9.8$ in.^{2}	*Okay!*

Check lateral ties:

$d_{bh} = 0.375$ in. ≥ 0.375 in.   *Okay!* (for Seismic Design Categories D, E, and F)

$s = 8$ in. $\leq 16d_b = 16(1.0) = 16$ in.   *Okay!*

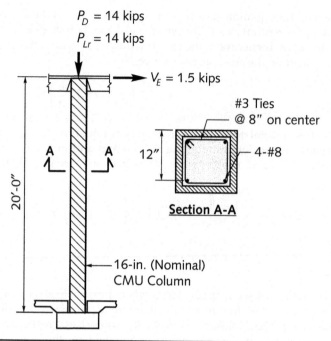

P_D = 14 kips

P_{Lr} = 14 kips

V_E = 1.5 kips

#3 Ties
@ 8" on center

12"

4-#8

Section A-A

20'-0"

16-in. (Nominal)
CMU Column

A   A

FIGURE 5.28 Reinforced-masonry column.

$$\le 48d_{bh} = 48(0.375) = 18 \text{ in.}$$ *Okay!*

$$\le t = 16 \text{ in.}$$ *Okay!*

This also means that longitudinal reinforcement can be used to resist compression.

Strength Design Solution
The factored loads for the applicable load combinations are as follows:

$1.2D + 1.6L_r$:

$$P_u = 1.2(14) + 1.6(14) = 39.2 \text{ kips}$$
$$M_u = 0$$

$1.2D + 1.0E$:

$$P_u = [1.2 + 0.2(1.21)](14) = 20.2 \text{ kips}$$
$$M_u = 1.0(1.5 \times 20) = 30 \text{ kip-ft}$$

$0.9D + 1.0E$:

$$P_u = [0.9 - 0.2(1.21)](14) = 9.2 \text{ kips}$$
$$M_u = 1.0(1.5 \times 20) = 30 \text{ kip-ft}$$

First, we will calculate the axial strength of the column when no bending moment is present. If no modification factors are used, the axial load capacity is given by

$$P_n = \left[0.8 f_m' (A_n - A_s) + f_y A_s\right]$$
$$= \left[0.8(1.5)(244 - 3.16) + 60(3.16)\right] = 479 \text{ kips}$$

$$\phi P_n = 0.9(479) = 431 \text{ kips}$$

However, the column axial strength must be limited to account for slenderness effects and unavoidable eccentricity:

$$r = \frac{t}{\sqrt{12}} = \frac{15.63}{\sqrt{12}} = 4.51 \text{ in.}$$

$$\frac{H}{r} = \frac{20(12)}{4.51} = 53.2 < 99$$

Therefore, Eq. (5.59) applies:

$$P_n = 0.8\left[0.8 f_m' (A_n - A_s) + f_y A_s\right]\left[1 - \left(\frac{H}{140r}\right)^2\right]$$

$$= 0.8\left[0.8(1.5)(244 - 3.16) + 60(3.16)\right]\left[1 - \left(\frac{53.2}{140}\right)^2\right] = 328 \text{ kips}$$

$$\phi P_n = 0.9(328) = 295 \text{ kips}$$

Next, we will calculate the moment strength when there is no axial load. A first estimate in the trial-and-error procedure can be made by equating the compression force in the masonry with the tensile force in the reinforcing steel

$$c = \frac{A_s f_y}{0.64 b f_m'} = \frac{(1.58)(60,000)}{0.64(15.63)(1,500)} = 6.32 \text{ in.}$$

Try $c = 6.0$ in. The compression force in the masonry is given by Eq. (5.63):

$$C_m = 0.64 f_m' bc = 0.64(1.5)(15.63)(6) = 90 \text{ kips}$$

The strain in the tensile reinforcement is calculated from Eq. (5.61):

$$\varepsilon_s = \varepsilon_{mu}\left(\frac{c - d}{c}\right) = 0.0025\left(\frac{6 - 12}{6}\right) = -0.0025$$

$$< \left(-\varepsilon_y = -\frac{f_y}{E_s} = -\frac{60}{29,000} = -0.00207\right)$$

Thus the steel stress is equal to the yield stress, and the force in the tensile reinforcement is given by

$$T_s = f_y A_s = -60(1.58) = -94.8 \text{ kips}$$

For the compression steel,

$$\varepsilon_{sc} = \varepsilon_{mu}\left(\frac{c-d'}{c}\right) = 0.0025\left(\frac{6-3.63}{6}\right) = 0.0010$$

and the steel compression force is obtained from Eq. (5.62):

$$C_s = \varepsilon_{sc} E_s A_{si} = 0.0010(29,000)(1.58) = 45.8 \text{ kips}$$

Then the sum of forces on the cross section is equal to

$$C_m + T_s + C_s = 90 - 94.8 + 45.8 = 41 \text{ kips}$$

The net compression force on the cross section means that the depth of the neutral axis needs to be reduced. Try $c = 4.65$ in.:

$$C_m = 0.64 f_m' bc = 0.64(1.5)(15.63)(4.65) = 69.8 \text{ kips}$$

$$\varepsilon_s = \varepsilon_{mu}\left(\frac{c-d}{c}\right) = 0.0025\left(\frac{4.65-12}{4.65}\right) = -0.0040$$

$$\rightarrow T_s = f_y A_s = -60(1.58) = -94.8 \text{ kips}$$

For the compression steel,

$$\varepsilon_{sc} = \varepsilon_{mu}\left(\frac{c-d'}{c}\right) = 0.0025\left(\frac{4.65-3.63}{4.65}\right) = 0.0005$$

$$T_{sc} = \varepsilon_{sc} E_s A_{si} = 0.0005(29,000)(1.58) = 22.9 \text{ kips}$$

Then sum of axial forces on the cross section is given by

$$C_m + T_s + C_s = 69.8 - 94.8 + 22.9 = -2.1 \text{ kips} \approx 0 \qquad \textit{Okay!}$$

Taking moments about the center of the cross section,

$$M_n = -94.8\left(\frac{15.63}{2} - 12\right) + 22.9\left(\frac{15.63}{2} - 3.63\right) + 69.8\left(\frac{15.63}{2} - \frac{0.8 \times 4}{2}\right)$$

$$= 908 \text{ kip-in.}$$

$$\phi M_n = 0.9(908) = 817 \text{ kip-in.} = 68.1 \text{ kip-ft}$$

At the balance condition, the depth of the neutral axis is given by Eq. (5.67):

$$c_b = \frac{d}{\left(\dfrac{f_y}{E_s \varepsilon_{mu}} + 1\right)} = \frac{d}{\left[\dfrac{60}{29,000(0.0025)} + 1\right]} = 0.547d$$

$$= 6.56 \text{ in.}$$

The sum of axial forces on the cross section then is given by

$$P_b = 0.64 f_m' bc - f_y A_s + \varepsilon_{mu}\left(\frac{c - d'}{c}\right) E_s A_{si}$$

$$= 0.64(1.5)(15.63)(6.56) - 60(1.58) + 0.0025\left(\frac{6.56 - 3.63}{6.56}\right)(29,000)(1.58)$$

$$= 98.4 - 94.8 + 51.2 = 54.8 \text{ kips}$$

$$\phi P_b = 0.9(54.8) = 49.3 \text{ kips} \qquad \text{(compression)}$$

If we take moments about the center of the cross section, we obtain

$$M_b = 98.4\left(\frac{15.63}{2} - \frac{0.8 \times 6.56}{2}\right) - 94.8\left(\frac{15.63}{2} - 12\right) + 51.2\left(\frac{15.63}{2} - 3.63\right)$$

$$= 1122 \text{ kip-in.}$$

$$\phi M_b = 0.9(1,122) = 1,010 \text{ kip-in.} = 84.2 \text{ kip-ft}$$

Figure 5.29 shows the interaction diagram for the column and compares the column strength with the demand due to the three load combinations. All load combinations

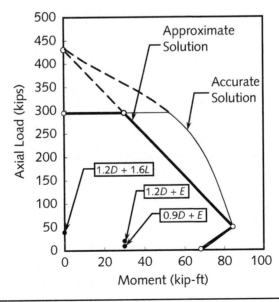

FIGURE 5.29 Interaction diagram for masonry column.

result in axial load and bending moment points that lie within the interaction diagram. This means that the column is capable of resisting the applied loads.

The shear-span ratio for the column is equal to

$$\frac{M}{Vd} = \frac{1.5(20 \times 12)}{1.5(12)} = 20 > 1.0; \text{ use } 1.0.$$

Using the load combination with the minimum axial load ($0.9D - 0.2S_{DS}D = 9.2$ kips), the shear strength contributed by the masonry is given by Eq. (5.72):

$$V_{nm} = \left[4 - 1.75 \left(\frac{M_u}{V_u d_v} \right) \right] A_n \sqrt{f_m'} + 0.25 P_u$$

$$= [4 - 1.75(1.0)] \frac{(15.63 \times 15.63)\sqrt{1,500}}{1,000} + 0.25(9.2)$$

$$= 23.6 \text{ kips}$$

From Eq. (5.74), the shear strength contributed by the shear reinforcement is equal to

$$V_{ns} = 0.5 A_v f_y \frac{d}{s} = 0.5(0.11)(40) \left(\frac{12}{8} \right) = 3.3 \text{ kips}$$

Therefore,

$$V_n = V_{nm} + V_{ns} = 23.6 + 3.3 = 26.9 \text{ kips}$$

Since $M_u / V_u d_v \geq 1.0$, the maximum shear strength is given by

$$V_n \leq 4 A_n \sqrt{f_m'} = \frac{4(15.63 \times 15.63)\sqrt{1,500}}{1,000} = 37.8 \text{ kips}$$

Thus the nominal shear strength is given by

$$\phi V_n = 0.8(26.9) = 21.5 \text{ kips} > 1.5 \text{ kips} \qquad \qquad Okay!$$

Allowable-Stress Design Solution

Using the alternate basic load combinations of Sec. 1605.3.2 of the 2009 IBC, the loads for the applicable load combinations are as follows:

$D + L_r$:

$$P_u = 14 + 14 = 28 \text{ kips}; \quad M_u = 0$$

$D + E/1.4$:

$$P = 14 \text{ kips}; \quad V = \frac{1.5}{1.4} = 1.07 \text{ kips}; \quad M = 1.07(20) = 21.4 \text{ kip-ft}$$

$0.9D + E/1.4$:

$$P = 0.9(14) = 12.6 \text{ kips}; \quad V = 1.07 \text{ kips}; \quad M = 21.4 \text{ kip-ft}$$

Check for flexural tension by assuming an uncracked section:

$$\frac{P}{A} - \frac{M}{S_n} = \frac{12.6(1,000)}{15.63(15.63)} - \frac{21.4(6)(12,000)}{15.63(15.63)^2} = -352 \text{ psi}$$

Since there is net tension on the column, the assumption of an uncracked member is incorrect. Reinforcement must be used to resist all tension, and the stresses must be calculated assuming a cracked section.

Check the stress on the column considering only the axial component of the load:

$$r = \frac{15.63}{\sqrt{12}} = 4.51 \text{ in.}$$

$$\frac{H}{r} = \frac{20(12)}{4.51} = 53.2 < 99$$

Therefore, from Eq. (5.75), and including the one-third increase in allowable stresses permitted by the 2009 IBC when using the alternate basic load combinations for allowable-stress design,

$$P_a = \left(0.25 f_m' A_n + 0.65 A_{st} F_s\right)\left[1 - \left(\frac{H}{140r}\right)^2\right]; \quad \frac{H}{r} \le 99$$

$$P_a = \left(0.25 \times 1.5 \times 15.63^2 + 0.65 \times 3.16 \times 24\right)\left[1 - \left(\frac{53.2}{140}\right)^2\right]\frac{4}{3}$$

$$= 160.7 \text{ kips} > 28 \text{ kips} \qquad\qquad\qquad\qquad\qquad \textit{Okay!}$$

For the combination of axial and flexural loads, the allowable stresses in the masonry and steel are as follows:

$$F_b = \frac{1}{3} f_m' \left(\frac{4}{3}\right) = \frac{1,500}{3}\left(\frac{4}{3}\right) = 667 \text{ psi}$$

$$F_s = 24,000\left(\frac{4}{3}\right) = 32,000 \text{ psi}$$

We may conservatively ignore the contribution of the compression steel and design the column as a singly reinforced member. For the load combination $0.9D + E/1.4$,

$$n = \frac{E_s}{E_m} = \frac{29,000,000}{900 \times 1,500} = 21.5$$

$$\rho = \frac{2(0.79)}{15.63(12)} = 0.0084$$

$$np = (21.5)0.0084 = 0.181$$

$$\frac{P}{bd} = \frac{12.6(1,000)}{15.63(12)} = 67.2 \text{ psi}$$

$$\frac{F_s}{n} = \frac{32,000}{21.5} = 1,488.4 \text{ psi}$$

First, we will calculate the allowable moment based on the masonry compressive stress. The neutral axis for this case is given by Eq. (5.89):

$$k = \sqrt{\left(np - \frac{P}{bdF_b}\right)^2 + 2n\rho} - \left(np - \frac{P}{bdF_b}\right)$$

$$= \left[\sqrt{\left(0.181 - \frac{67.2}{667}\right)^2 + 2(0.181)} - \left(0.181 - \frac{67.2}{667}\right)\right] = 0.527$$

$$j = \left(1 - \frac{k}{3}\right) = \left(1 - \frac{0.527}{3}\right) = 0.824$$

From Eq. (5.90), the allowable moment is given by

$$M = \frac{1}{2}F_b jkbd^2 - P\left(d - \frac{h}{2}\right)$$

$$= \frac{1}{2}(667)(0.824)(0.527)(15.63)(12)^2 - 12,600\left(12 - \frac{15.63}{2}\right)$$

$$= 273,222 \text{ lb-in.} = 22.8 \text{ kip-ft}$$

If the cross section is governed by the steel tensile stress,

$$k = \sqrt{\left(np + \frac{nP}{F_s bd}\right)^2 + 2\left(np + \frac{nP}{F_s bd}\right)} - \left(np + \frac{nP}{F_s bd}\right)$$

$$= \left[\sqrt{\left(0.181 + \frac{67.2}{1,488.4}\right)^2 + 2\left(0.181 + \frac{67.2}{1,488.4}\right)} - \left(0.181 + \frac{67.2}{14,88.4}\right)\right] = 0.483$$

$$j = \left(1 - \frac{k}{3}\right) = \left(1 - \frac{0.483}{3}\right) = 0.839$$

and from Eq. (5.91),

$$M = A_s F_s jd + P\left(\frac{h}{2} - \frac{kd}{3}\right)$$

$$= 1.58(32,000)(0.839)(12) + 12,600\left(\frac{15.63}{2} - \frac{0.483 \times 12}{3}\right)$$

$$= 583,164 \text{ lb-in.} = 48.6 \text{ kip-ft}$$

Therefore, the allowable moment for the given axial load is determined by the allowable masonry stress and is equal to

$$M = 22.8 \text{ kip-ft} > 21.4 \text{ kip-ft}$$ *Okay!*

The preceding calculations can be repeated to show that for the load combination $D + E/1.4$ ($P = 14$ kips), the allowable moment is equal to 22.7 kip-ft, which is also acceptable.

Since the column is subjected to flexural tension, the shear stress is given by

$$f_v = \frac{V}{bd} = \frac{1.07(1,000)}{(15.63 \times 12)} = 5.7 \text{ psi}$$

The allowable shear stress in the masonry alone is given by

$$F_v = \sqrt{f'_m} = \sqrt{1,500} = 38.7 \text{ psi}$$

No shear reinforcement is required other than the minimum lateral reinforcement requirements for columns.

Example 5.4.2

Determine the longitudinal (vertical) reinforcement required to resist the loads on the basement retaining wall shown in Fig. 5.30. The masonry compressive strength is 1,500 psi, and the steel is Grade 60.

Strength Design Solution

The maximum moment occurs at a location 6.7 ft from the top of the wall. The factored loads for the applicable load combinations are as follows:

$1.2D + 1.6L + 0.5L_r + 1.6H$:

$$P_u = 1.2(4,725 + 124 \times 6.7) + 1.6(1,800) + 0.5(300) = 9,697 \text{ lb/ft}$$

$$M_u = (1.6)0.128\left(55 \times \frac{12^3}{2}\right) = 9,732 \text{ lb-ft/ft}$$

$0.9D + 1.6H$:

$$P_u = 0.9(4,725 + 124 \times 6.7) = 5,000 \text{ lb/ft}$$
$$M_u = 9,732 \text{ lb-ft/ft}$$

If we try #5 bars spaced at 16 in. on center, the effective depth is given by

$$d = 11.63 - 2.5 - \frac{0.625}{2} = 8.8 \text{ in.}$$

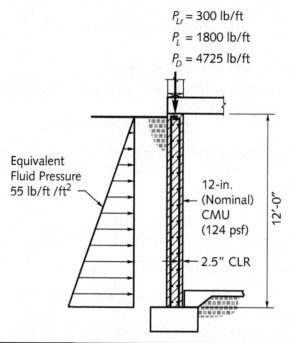

P_{Lr} = 300 lb/ft
P_L = 1800 lb/ft
P_D = 4725 lb/ft

Equivalent
Fluid Pressure
55 lb/ft /ft²

12-in.
(Nominal)
CMU
(124 psf)

2.5" CLR

12'-0"

Figure 5.30 Basement retaining wall.

$$A_s = 0.31\left(\frac{12}{160}\right) = 0.233 \text{ in.}^2$$

Using Eq. (5.69) for a singly reinforced cross section with yielding steel, the depth of the compression block for the first load combination is given by

$$a = \frac{A_s f_y + P_u/\phi}{0.8 f'_m b} = \frac{0.233(60,000) + 9,732/0.9}{0.8(1,500)12} = 1.72 \text{ in.}$$

And from Eq. (5.70), the nominal flexural strength for a singly reinforced cross-section then is equal to

$$M_n = A_s f_y\left(d - \frac{a}{2}\right) + \frac{P_u}{\phi}\left(\frac{h}{2} - \frac{a}{2}\right)$$

$$= 0.233 \times 60\left(8.8 - \frac{1.72}{2}\right) + \frac{9697}{0.9(1,000)}\left(\frac{11.63}{2} - \frac{1.72}{2}\right)$$

$$= 164.4 \text{ kip-in./ft}$$

$$\phi M_n = 0.9\left(\frac{164.4}{12}\right) = 12.3 \text{ kip-ft/ft} > 9.7 \text{ kip-ft/ft} \qquad \textit{Okay!}$$

For the second load combination,

$$a = \frac{A_s f_y + P_u / \phi}{0.8 f_m' b} = \frac{0.233(60,000) + 5,000/0.9}{0.8(1,500)12} = 1.36 \text{ in.}$$

$$M_n = A_s f_y \left(d - \frac{a}{2} \right) + \frac{P_u}{\phi} \left(\frac{h}{2} - \frac{a}{2} \right)$$

$$= 0.233 \times 60 \left(8.8 - \frac{1.36}{2} \right) + \frac{5,000}{0.9(1,000)} \left(\frac{11.63}{2} - \frac{1.36}{2} \right)$$

$$= 142 \text{ kip-in./ft}$$

$$\phi M_n = 0.9 \left(\frac{142}{12} \right) = 10.7 \text{ kip-ft /ft} > 9.7 \text{ kip-ft /ft}$$

Okay!

Thus #5 bars spaced at 16 in. are sufficient to resist the axial and flexural loads on the wall.

Allowable-Stress Design Solution
The loads that correspond to the allowable-stress design load combinations are as follows:

$D + L + H$:

$$P = 1.0(4,725 + 124 \times 6.7) + 1.0(1,800) = 7356 \text{ lb/ft}$$

$$M = 0.128 \left(55 \times \frac{12^3}{2} \right) = 6,083 \text{ lb-ft/ft}$$

$D + 0.75L + 0.75L_r + H$:

$$P = 1.0(4,725 + 124 \times 6.7) + 0.75(1,800) + 0.75(300) = 7,131 \text{ lb/ft}$$
$$M = 6,083 \text{ lb-ft/ft}$$

$0.6D + H$:

$$P = 0.6(4,725 + 124 \times 6.7) = 3,333 \text{ lb/ft}$$
$$M = 6,083 \text{ lb-ft/ft}$$

The allowable stresses in the masonry and steel are given by

$$F_b = \frac{1}{3} f_m' = 500 \text{ psi}$$

$$F_s = 24,000 \text{ psi}$$

We can try #6 bars spaced at 8 in. on center,

$$n = \frac{E_s}{E_m} = \frac{29,000,000}{900 \times 1,500} = 21.5$$

$$\rho = \frac{A_s}{bd} = \frac{0.44(12/8)}{12(8.8)} = 0.0063$$

$$n\rho = (21.5)0.0063 = 0.135$$

For the load combination $D + L + H$,

$$\frac{P}{bd} = \frac{7,356}{12(8.8)} = 69.7 \text{ psi}$$

$$\frac{F_s}{n} = \frac{24,000}{21.5} = 1,116.3 \text{ psi}$$

For the allowable moment based on the masonry compressive stress, the neutral axis for this case is given by Eq. (5.89):

$$k = \sqrt{\left(n\rho - \frac{P}{bdF_b}\right)^2 + 2n\rho} - \left(n\rho - \frac{P}{bdF_b}\right)$$

$$= \left[\sqrt{\left(0.135 - \frac{69.7}{500}\right)^2 + 2(0.135)} - \left(0.135 - \frac{69.7}{500}\right)\right] = 0.524$$

$$j = \left(1 - \frac{k}{3}\right) = \left(1 - \frac{0.524}{3}\right) = 0.825$$

From Eq. (5.90), the allowable moment is given by

$$M = \frac{1}{2}F_b jkbd^2 - P\left(d - \frac{h}{2}\right)$$

$$= \frac{1}{2}(500)(0.825)(0.524)(12)(8.8)^2 - 7,356\left(8.8 - \frac{11.63}{2}\right)$$

$$= 78\,\text{in}.539 \text{ lb-ft / ft}$$

If the cross section is governed by the steel tensile stress,

$$k = \sqrt{\left(n\rho + \frac{nP}{F_s bd}\right)^2 + 2\left(n\rho + \frac{nP}{F_s bd}\right)} - \left(n\rho + \frac{nP}{F_s bd}\right)$$

$$= \left[\sqrt{\left(0.135 + \frac{69.7}{1,116.3}\right)^2 + 2\left(0.135 + \frac{69.7}{1,116.3}\right)} - \left(0.135 + \frac{69.7}{1,116.3}\right)\right] = 0.461$$

$$j = \left(1 - \frac{k}{3}\right) = \left(1 - \frac{0.461}{3}\right) = 0.846$$

and from Eq. (5.91)

$$M = A_s F_s jd + P\left(\frac{h}{2} - \frac{kd}{3}\right)$$

$$= 1.58(24,000)(0.846)(8.8) + 7,356\left(\frac{11.63}{2} - \frac{0.461 \times 8.8}{3}\right)$$

$$= 150,829 \text{ lb-in.}/\text{ft} = 12,569 \text{ lb-ft}/\text{ft}$$

Therefore, for the load combination $D + L + H$, the allowable moment is determined by the masonry compressive stress and

$$M = 6,539 \text{ lb-ft}/\text{ft} > 6,083 \text{ lb-ft}/\text{ft} \qquad\qquad \textit{Okay!}$$

For the load combination $D + 0.75L + 0.75L_r + H$,

$$\frac{P}{bd} = \frac{7,131}{12(8.8)} = 67.5 \text{ psi}$$

For the allowable moment based on the masonry compressive stress,

$$k = \sqrt{\left(n\rho - \frac{P}{bdF_b}\right)^2 + 2n\rho} - \left(n\rho - \frac{P}{bdF_b}\right) = 0.519$$

$$j = \left(1 - \frac{k}{3}\right) = 0.827$$

From Eq. (5.90), the allowable moment is given by

$$M = \frac{1}{2}F_b jkbd^2 - P\left(d - \frac{h}{2}\right) = 6,536 \text{ lb-ft}/\text{ft}$$

If the cross section is governed by the steel tensile stress,

$$k = \sqrt{\left(n\rho + \frac{nP}{F_s bd}\right)^2 + 2\left(n\rho + \frac{nP}{F_s bd}\right)} - \left(n\rho + \frac{nP}{F_s bd}\right) = 0.459$$

$$j = \left(1 - \frac{k}{3}\right) = 0.847$$

$$M = A_s F_s jd + P\left(\frac{h}{2} - \frac{kd}{3}\right) = 12,494 \text{ lb-ft}/\text{ft}$$

Therefore, for the load combination $D + 0.75L + 0.75L_r + H$,

$$M = 6536 \text{ lb-ft}/\text{ft} > 6,083 \text{ lb-ft}/\text{ft} \qquad\qquad \textit{Okay!}$$

For the load combination $0.6D + H$,

$$\frac{P}{bd} = \frac{3,333}{12(8.8)} = 31.6 \text{ psi}$$

For the allowable moment based on the masonry compressive stress,

$$k = \sqrt{\left(n\rho - \frac{P}{bdF_b}\right)^2 + 2n\rho} - \left(n\rho - \frac{P}{bdF_b}\right) = 0.452$$

$$j = \left(1 - \frac{k}{3}\right) = 0.849$$

From Eq. (5.90), the allowable moment is given by

$$M = \frac{1}{2}F_b jkbd^2 - P\left(d - \frac{h}{2}\right) = 6,602 \text{ lb-ft/ft}$$

If the cross section is governed by the steel tensile stress,

$$k = \sqrt{\left(n\rho + \frac{nP}{F_s bd}\right)^2 + 2\left(n\rho + \frac{nP}{F_s bd}\right)} - \left(n\rho + \frac{nP}{F_s bd}\right) = 0.430$$

$$j = \left(1 - \frac{k}{3}\right) = 0.857$$

$$M = A_s F_s jd + P\left(\frac{h}{2} - \frac{kd}{3}\right) = 11,214 \text{ lb-ft/ft}$$

Therefore, for the load combination $0.6D + H$,

$$M = 6,602 \text{ lb-ft/ft} > 6,083 \text{ lb-ft/ft} \qquad\qquad Okay!$$

The selected reinforcement of #6 bars spaced at 8 in. on center satisfies the axial and flexural load requirements.

Note that as for strength design, an interaction diagram can be developed for a cross section using allowable-stress design procedures to determine the allowable moment for various values of axial load. The interaction for the retaining wall in the example is shown in Fig. 5.31.

5.5 Shear Walls

Masonry shear walls are the lateral load resisting elements of most masonry buildings. The code stipulates that when masonry is used as the lateral load resisting system, shear walls must provide at least 80 percent of the lateral load resistance if a response-modification factor R of greater than 1.5 is used in the design. If an R of no greater than 1.5 is used, piers and columns may be used to resist earthquake loads. When resisting lateral loads, shear wall deflection can be a result of any one of the four deformation modes shown in Fig. 5.32 for cantilever shear walls. In general, rocking and sliding

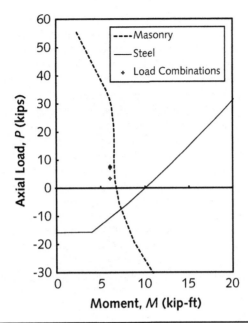

Figure 5.31 Interaction diagram for allowable-stress design of retaining wall.

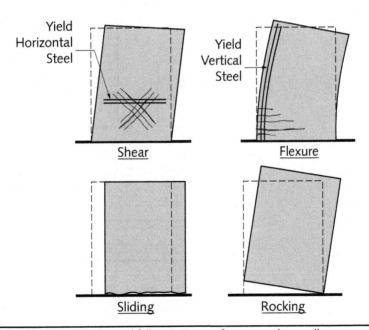

Figure 5.32 Deformation modes and failure patterns of masonry shear walls.

along joints is not an acceptable response, and designers should prevent such response by providing adequate dowels into foundations and across construction surfaces. The amount of deflection due to shear or flexural deformation depends on the ratio of the wall height to its length, which is known as the *aspect ratio*. Short, squat walls tend to be dominated by shear deformation, whereas tall, slender walls are dominated by flexural deformation. Figure 5.33 shows typical deformation modes of walls with various aspect ratios. Whenever possible, designers should endeavor to use walls with predominantly flexural response to resist earthquake loads. This is so because flexural behavior can be significantly more stable under cyclic motions than shear-dominated response.

Designers strive to design walls with adequate *ductility*, which can be defined as the ability of a shear wall to deform beyond the linear elastic range. The early stage of shear wall response typically is characterized by relatively stiff elastic behavior. As the lateral displacement increases and damage occurs, the wall loses stiffness, and its behavior becomes nonlinear and inelastic. The start of nonlinear behavior sometimes can be identified as a specific point in the load-deformation relationship. The displacement at this point is called the *yield displacement* because it usually corresponds to the yielding of flexural reinforcement. Quite often, however, walls do not exhibit a clearly defined yield point, and an approximate yield displacement must be determined.

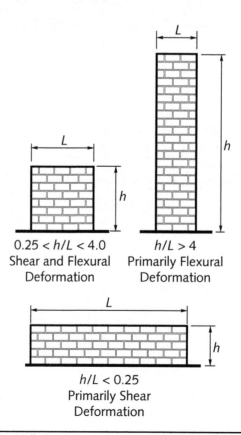

$0.25 < h/L < 4.0$
Shear and Flexural
Deformation

$h/L > 4$
Primarily Flexural
Deformation

$h/L < 0.25$
Primarily Shear
Deformation

FIGURE 5.33   Contribution to displacements of walls with various aspect ratios.

Ductility μ is defined as the ratio of displacement at failure to the yield displacement such that

$$\mu = \frac{\Delta_u}{\Delta_y} \qquad (5.93)$$

where Δ_y is the yield displacement and Δ_u is the maximum allowable displacement or the displacement at failure. Figure 5.34 illustrates the difference in lateral response of a ductile wall and a nonductile or brittle wall. Ductile walls are able to sustain lateral deformations in the nonlinear range, whereas brittle walls fail soon after the yield displacement is reached. The ductility of shear walls is particularly important during large earthquakes because most buildings will respond to intense ground motion in the nonlinear range.

As stated earlier, rocking and sliding generally are unacceptable responses and should be prevented because they are inherently brittle failure modes. Masonry walls that fail in shear develop diagonal shear cracks and have limited ductility capacity. Walls that fail in flexure can have high ductility, provided that masonry compression crushing does not occur at the toe of the wall and the longitudinal reinforcement does not buckle. Therefore, it is preferable to design walls that fail in flexure because this results in a higher ductility capacity. The masonry strength design requirements provide specific procedures to ensure that walls achieve the ductility required for satisfactory performance during earthquakes. The code requirements for allowable-stress design, on the other hand, do not specifically address the ductility of masonry shear walls.

For the purposes of seismic design, masonry shear walls are classified as either unreinforced-masonry shear walls (also called *plain masonry shear walls*) or reinforced-masonry shear walls. Unreinforced-masonry walls have little or no reinforcing steel, and the resistance of reinforcement, if present, is neglected in design. The tensile strength of masonry is taken into account, and walls are designed to ensure that the masonry does not crack by exceeding its tensile strength. The TMS 402-08/ACI 530-08/

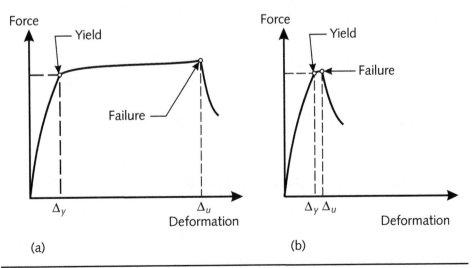

Figure 5.34 Comparison of (a) ductile response and (b) brittle response

ASCE 5-08 defines two types of unreinforced-masonry shear walls. Ordinary plain masonry shear walls do not need to have any reinforcing steel and are designed in accordance with the requirements for unreinforced masonry contained in Secs. 2.2 and 3.2 of the code for allowable stress design and strength design, respectively. Detailed plain masonry shear walls are also designed in accordance with the requirements for unreinforced masonry. However, a wall must contain the following minimum amount of reinforcing steel to be categorized as a detailed plain masonry shear wall:

1. Vertical reinforcement of at least 0.2 in.² in cross-sectional area at corners, within 16 in. of each side of openings, within 8 in. of each side of movement joints, within 8 in. of wall ends, and at a maximum spacing of 10 ft on center

2. Horizontal reinforcement of at least 0.2 in.² in cross-sectional area at a maximum spacing of 10 ft on center (the requirement for horizontal reinforcement may be satisfied by the use of joint reinforcement with at least two wires of W1.7 spaced no more that 16 in. on center)

3. Horizontal reinforcement at the bottom and top of openings that extends at least 24 in. or 40 bar diameters past the opening

4. Continuous horizontal reinforcement at structurally connected roof and floor levels and within 16 in. of the tops of walls

Figure 5.35 illustrates the minimum reinforcement requirements for detailed plain masonry shear walls. Reinforcement adjacent to openings does not need to be provided for openings smaller than 16 in. in either the horizontal or vertical direction unless the spacing of distributed reinforcement is interrupted by such openings.

Reinforced-masonry walls are designed in accordance with Secs. 2.3 and 3.3 of the code for allowable-stress design and strength design, respectively. When designing reinforced masonry, the tensile capacity of the masonry is ignored, and all tension at a

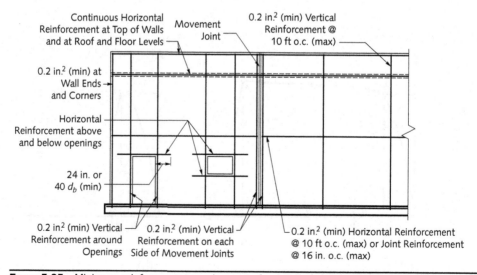

FIGURE 5.35 Minimum reinforcement requirements for detailed plain and ordinary reinforced-masonry shear walls.

cross section is assumed to be resisted by reinforcing steel. Ordinary reinforced-masonry shear walls are designed using reinforced-masonry design requirements and must satisfy the same minimum detailing requirements as detailed plain masonry walls, as shown in Fig. 5.35. Intermediate reinforced-masonry shear walls must satisfy the requirements for ordinary reinforced-masonry shear walls, except that the spacing of vertical reinforcement must not exceed 48 in. Figure 5.36 shows the detailing requirements for intermediate reinforced-masonry shear walls.

In addition to the requirements for intermediate reinforced-masonry shear walls, special reinforced-masonry shear walls must satisfy the following requirements:

1. The sum of the cross-sectional area of horizontal and vertical reinforcements must not be less than 0.002 times the gross cross-sectional area of the wall.

2. The minimum sum of the cross-sectional area of reinforcement in either direction must not be less than 0.0007 times the gross cross-sectional area of the wall using specified dimensions.

3. The minimum cross-sectional area of vertical reinforcement shall be one-third of the required shear reinforcement.

4. Reinforcement should be uniformly distributed, and the maximum spacing of vertical and horizontal reinforcing shall be the smaller of one-third the length of the shear wall, one-third the height of the shear wall, or 48 in..

5. Shear reinforcement should be anchored around vertical reinforcing bars with a standard hook.

6. When the masonry is laid in stack bond, the maximum spacing of reinforcement in special reinforced masonry shear walls shall be 24 in., and fully grouted open-ended masonry units must be used.

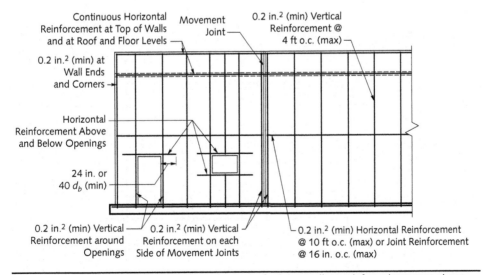

FIGURE 5.36   Minimum reinforcement requirements for intermediate reinforced-masonry shear walls.

The reinforcing requirements for special reinforced-masonry shear walls are illustrated in Fig. 5.37. The detailing requirements are intended to ensure that walls have sufficient ductility to deform during large earthquakes without failure. Special reinforced-masonry shear walls can be used as lateral load resisting elements in all seismic design categories. Intermediate reinforced-masonry shear walls have less ductility than special reinforced masonry shear walls, and ordinary reinforced masonry shear walls have limited ductile capacity. The difference in the seismic response of the different wall types is indicated by the values of the response modification coefficient R. Table 5.2 shows which types of reinforced-masonry shear walls are permitted in the various seismic design categories. Only special reinforced-masonry walls are permitted in high-seismic regions. Unreinforced-masonry walls are permitted only in Seismic Design Categories A and B.

In buildings with rigid diaphragms or along a line of resistance in flexible diaphragms, horizontal forces due to lateral loads will be distributed to individual shear walls based on the relative stiffness of the walls. A stiff wall will require a proportionally larger force to cause the same deflection as a more flexible wall. Therefore, a larger percentage of the lateral loads will be resisted by the stiffer walls. Wall stiffness typically is measured as the wall rigidity, which is the shear required to cause a unit deflection of the wall and is defined as

$$R = \frac{V}{\Delta} \tag{5.94}$$

where V is the shear being resisted by the wall, and D is the deflection of the wall. The inverse of the wall rigidity is wall flexibility $\overline{\Delta}$, which is defined as the wall deflection due to a unit load such that

$$R = \frac{1}{\overline{\Delta}} \tag{5.95}$$

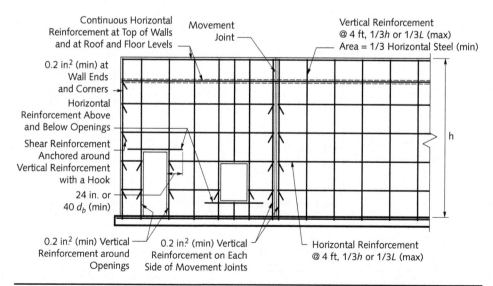

Figure 5.37 Minimum reinforcement requirements for special reinforced-masonry shear walls.

Seismic Design Category	Ordinary Reinforced-Masonry Shear Walls	Intermediate Reinforced-Masonry Shear Walls	Special Reinforced-Masonry Shear Walls
A	Permitted	Permitted	Permitted
B	Permitted	Permitted	Permitted
C	Permitted	Permitted	Permitted
D	Not permitted	Not permitted	Permitted
E	Not permitted	Not permitted	Permitted
F	Not permitted	Not permitted	Permitted

TABLE 5.2 Reinforced-Masonry Shear Walls Permitted in Various Seismic Design Categories

A shear wall can be characterized as a deep beam with a span that is, at most, several times the depth. The wall deflection thus can be determined by beam theory so that a deflection of the wall as shown in Fig. 5.38 is equal to

$$\Delta = \frac{VH^3}{3E_m I} + \frac{1.2VH}{AE_v} + \Delta_F \frac{2H}{L} \tag{5.96}$$

where H is the height of the wall, L is the wall length, and Δ_F is the vertical displacement of the foundation due to rocking. The first term represents bending or flexural deformation, the second term is the shear deformation, and the third term represents the effects of foundation rotation or uplift. The bending term considers the wall as a simple vertical cantilever beam with a moment of inertia I that includes returns or pilasters at the ends of the wall. The cross-sectional area A of the wall is the area of the web and omits the flange areas. E_m and E_v are Young's modulus and shear modulus, respectively, which, as described in Chap. 2, are given by

$$E_m = 900 f'_m \qquad \text{for concrete masonry} \tag{5.97a}$$

$$E_m = 700 f'_m \qquad \text{for clay masonry} \tag{5.97b}$$

$$E_v = 0.4 E_m \tag{5.98}$$

where f'_m is the masonry compressive strength. Although its contribution may be significant, the foundation rotation term typically is ignored in most design situations because difficulties in evaluating the stiffness of various foundation systems restrict the designer's ability to include the foundation rotation term. Foundation distortions and the resulting wall deflection can be as large as the shear or bending effects because some foundations are relatively flexible when compared with rigid walls.

Ignoring the effect of foundation rotation and noting that the wall flexibility $\bar{\Delta}$ is equal to the wall deflection due to a unit load, Eqs. (5.96) and (5.97) can be used to obtain

$$Em\bar{\Delta} = \left(\frac{H^3}{3I}\right) + \left(\frac{3H}{A}\right) \tag{5.99}$$

For walls with rectangular cross sections (no flanges), the cross-sectional area and moment of inertia are equal to

$$A = tL \tag{5.100}$$

$$I = \frac{tL^3}{12} \tag{5.101}$$

so that Eq. (5.96) can be further simplified to

$$tE_m\bar{\Delta} = 4\left(\frac{H}{L}\right)^3 + 3\left(\frac{H}{L}\right) \tag{5.102}$$

When determining the distribution of earthquake loads, it is the relative rigidity of the walls that is required and not necessarily the absolute value of the wall stiffness. Therefore, for walls with the same thickness, the leading terms in Eq. (5.102) can be ignored, and the relative rigidity of each wall is equal to

$$R_i = \frac{1}{\bar{\Delta}_i} = \frac{1}{\left[4\left(\frac{H}{L}\right)^3 + 3\left(\frac{H}{L}\right)\right]} \tag{5.103}$$

The force resisted by each wall along a line of resistance then is given by

$$V_i = \frac{R_i}{\sum\limits_{i=1}^{n} R_i} V \tag{5.104}$$

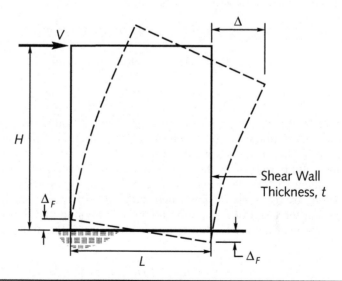

FIGURE 5.38   Shear wall deformation.

It should be noted the procedure just described for obtaining the distribution of earthquake loads to shear walls is an approximate method at best. As noted earlier, the contribution of the foundation rotation to the wall deflection is often neglected. In addition, calculations are based on elastic, uncracked masonry wall cross sections. While it may be justifiable to use uncracked section properties for extremely low levels of loading, the actual response of buildings during design-level earthquakes will be extremely nonlinear and result in cracking of the masonry shear walls. Determining the stiffness of cracked reinforced-masonry shear walls can be quite complex, and the stiffness of a cracked masonry wall varies significantly depending on the degree of cracking. Thus it is important to emphasize that the relative rigidity provides only an estimate of the distribution of earthquake loads and that the true distribution will not be completely predictable during a major earthquake. Nevertheless, a good design that uses walls of similar rigidity that are placed in a symmetric pattern will have a more predictable seismic response behavior, even with the significant cracking expected during large earthquake ground motion.

The discussion on wall stiffness so far has been focused on the behavior of cantilever masonry walls. Sometimes, the tops of masonry shear walls or piers are restrained by deep beams so that the deformation occurs with no rotation at the top of the wall. Figure 5.39 shows the differences in the deformed shape and moment demands on a cantilever shear wall compared with a wall that is prevented from rotating at both the bottom and top.

It should be recognized that in order to obtain the deformed shape shown in Fig. 5.39b, the restraining beam must possess both the stiffness and strength to resist the moment that develops at the top of the wall. From beam theory, the deflection of a wall or a pier that is prevented from rotating at the top is equal to

$$\Delta = \Delta_f + \Delta_s = \frac{VH^3}{12E_m I} + \frac{1.2VH}{AE_v} \tag{5.105}$$

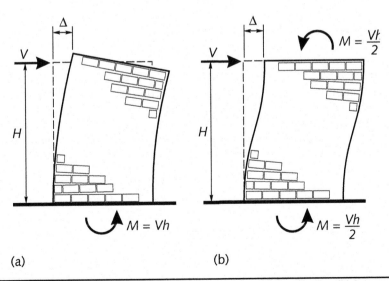

(a)   (b)

Figure 5.39 Shear wall deformation: (a) cantilever wall/pier; (b) fixed-fixed wall/pier.

where Δ_f and Δ_s are the wall deflections due to flexural and shear deformations, respectively. As with cantilever shear walls, the effect of foundation rotation is neglected to simplify the calculations. Noting that $E_v = 0.4E_m$, and substituting Eqs. (5.100) and (5.101) for the wall area and moment of inertia, that is,

$$\Delta = \left[\left(\frac{H}{L} \right)^3 + 3 \left(\frac{H}{L} \right) \right] \left(\frac{V}{tE_m} \right) \tag{5.106}$$

and neglecting the common terms, the relative rigidity of fixed-fixed walls or piers is given by

$$R = \frac{1}{\Delta} = \frac{1}{\left[\left(\frac{H}{L} \right)^3 + 3 \left(\frac{H}{L} \right) \right]} \tag{5.107}$$

Masonry shear walls that are constructed as part of buildings usually are constructed with openings for doors and windows. These openings will increase the wall deflection and consequently decrease rigidity. This section presents an approximate method for incorporating the effect of openings on wall rigidity. The method should not be applied to walls with large openings or walls with a configuration that requires the wall assemblage to be analyzed as a frame rather than as an individual wall. The method is based on determining the deflection of the solid wall and increasing this deflection due to the effect of the openings. The wall rigidity then is calculated from the total wall deflection.

Figure 5.40 illustrates the process of incorporating the effect of openings on the deflection of the wall. After the deflection of a solid cantilever wall is obtained, the deflection of a solid strip of wall equal to the height of the openings is subtracted and replaced by the deflection due to the piers around the openings. Thus the total wall deflection is given by

$$\Delta_{\text{wall}} = \Delta_{\text{solid}} - \Delta_{\text{solid strip}} + \Delta_{\text{piers}} \tag{5.108}$$

And the relative rigidity of the wall is equal to

$$R = 1/\Delta_{\text{wall}} \tag{5.109}$$

The deflections of the solid strip and pier typically are obtained assuming a fixed-fixed condition because there is usually a sufficient amount of wall above the openings to restrain rotation at the tops of the piers. Typically, the piers of a wall have the same thickness and masonry compressive strength. This means that the deflection of each pier due to a unit load can be calculated using Eq. (5.106), and the relative deflection of a pier with rotation fixed at the top and bottom can be calculated as

$$\overline{\Delta}_i = \left(\frac{H_i}{L_i} \right)^3 + 3 \left(\frac{H_i}{L_i} \right) \tag{5.110}$$

The deflection of the piers then is given by

$$\Delta_{\text{piers}} = \frac{1}{\sum_{i=1}^{n} \frac{1}{\overline{\Delta}_i}} \tag{5.111}$$

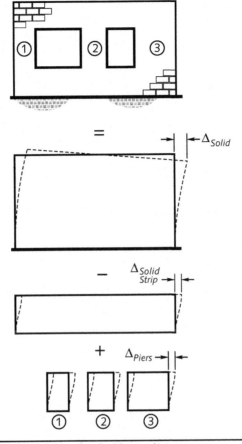

FIGURE 5.40   *Calculation of deflection of wall with openings.*

The lateral force to the wall will be distributed to individual piers based on the relative rigidity of each pier, which can be calculated from the deflection of the pier:

$$V_i = \frac{R_i}{\sum\limits_{i=1}^{n} R_i} V \qquad (5.112)$$

If the openings are at different elevations, this method can become significantly more complex.

Example 5.5.1
Determine the distribution of a 40-kip lateral to the two walls shown in Fig. 5.41, assuming that there is no foundation rotation.

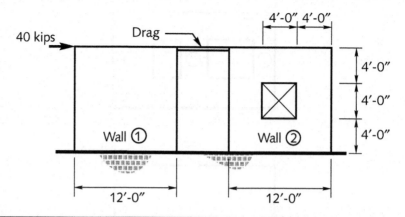

FIGURE 5.41 Distribution of lateral load to walls.

Solution

The relative rigidity of wall 1 is given by

$$\Delta_1 = 4\left(\frac{H}{12}\right)^3 + 3\left(\frac{H}{12}\right) = 4\left(\frac{12}{12}\right)^3 + 3\left(\frac{12}{12}\right) = 7.0$$

$$R_1 = \frac{1}{\Delta} = \frac{1}{7.0} = 0.143$$

For wall 2,

$$\Delta_{\text{solid wall}} = 4\left(\frac{12}{12}\right)^3 3\left(\frac{12}{12}\right) = 7.0$$

Assuming that the 4-ft-tall piers are fixed at the tops and bottoms by the portion of the wall above and below,

$$\Delta_{\text{solid strip}} = \left(\frac{H}{12}\right)^3 + 3\left(\frac{H}{12}\right) = \left(\frac{4}{12}\right)^3 + 3\left(\frac{4}{12}\right) = 1.037$$

$$\Delta_{\text{pier}} = \left(\frac{4}{4}\right)^3 + 3\left(\frac{4}{4}\right) = 4.0$$

$$\Delta_{\text{piers}} = \frac{1}{\dfrac{1}{\Delta_{\text{pier}}} + \dfrac{1}{\Delta_{\text{pier}}}} = \frac{1}{\dfrac{1}{4.0} + \dfrac{1}{4.0}} = 2.0$$

The deflection and rigidity of the entire wall 2 then are given by

$$\Delta_2 = \Delta_{\text{solid wall}} - \Delta_{\text{solid strip}} + \Delta_{\text{piers}}$$

$$= 7.0 - 1.037 + 2.0 = 7.963$$

$$R_2 = \frac{1}{\Delta_2} = \frac{1}{7.963} = 0.126$$

The force in each wall then is determined by the relative rigidity of the two walls:

$$F_1 = \frac{0.143}{0.143 + 0.126} 40 = 21.26 \text{ kips}$$

$$F_2 = \frac{0.126}{0.143 + 0.126} 40 = 18.74 \text{ kips}$$

5.5.1 Strength Design of Shear Walls

The strength design of shear walls follows the procedures described in Sec. 5.4 for members subjected to a combination of axial and flexural loads. Interaction diagrams need to be developed to determine whether walls can resist the axial and flexural loads for the applicable load combinations. Since shear walls typically contain distributed steel to satisfy minimum reinforcement requirements, several layers of reinforcement need to be considered in calculations for the interaction diagrams. The shear strength of walls is calculated using Eqs. (5.71) to (5.74), as described in Sec. 5.4.

In addition to possessing sufficient strength to resist axial, flexural, and shear demands, concrete masonry shear walls also must satisfy specific detailing requirements outlined in the code, which are required to ensure that walls possess adequate ductility. Since curvature ductility is related to displacement ductility, the curvature on the critical cross section can be used as a measure of wall ductility. At the yield displacement, the curvature on the cross section is given by the slope of the strain diagram, which is equal to

$$\varphi_y = \frac{\varepsilon_y}{d - c} \tag{5.113}$$

where c is the depth of the neutral axis, d is the distance to the extreme tension reinforcement, and ε_y is the yield strain of the reinforcing steel. With increasing wall deformation, the curvature on the cross section increases, and the curvature demand at the maximum displacement is given by

$$\varphi_u = \frac{\alpha \varepsilon_y}{d - c} \tag{5.114}$$

where $\alpha \varepsilon_y$ is the maximum tensile strain in the reinforcing steel at the maximum displacement. If we assume that the depth of the neutral axis does not change significantly after yield, Eqs. (5.113) and (5.114) can be used to determine the curvature ductility as follows:

$$\mu_\varphi = \frac{\varphi_u}{\varphi_y} \simeq \frac{\alpha \varepsilon_y}{\varepsilon_y} = \alpha \tag{5.115}$$

From Eq. (5.115) it can be seen that ductility demand on a wall can be evaluated with the tension reinforcement strain factor α. Table 5.3 summarizes the TMS 402-08/ ACI 530-08/ASCE 5-08 requirements for ductility capacity of various members using the tension reinforcement strain factor. Special reinforced and intermediate reinforced

Basic Seismic Force–Resisting System	R	$\dfrac{M_u}{V_u d_v}$	α
Special reinforced-masonry shear walls	5	≥1.0	4
		< 1.0	1.5
Intermediate reinforced- masonry shear walls	3.5	≥1.0	3.0
		<1.0	1.5
Walls loaded out-of plane	—	—	1.5
All others	All	≥1.0	1.5
	≥1.5	< 1.0	1.5
	<1.5	<1.0	None

TABLE 5.3   Minimum Tension Reinforcement Strain Factor α at Ultimate Limit State for Masonry Members

masonry walls, which are designed with higher values of R and are likely to exhibit significant nonlinear response during design-level earthquakes, must be capable of achieving higher values of α at the ultimate limit state. Members designed with lower values of R have less stringent requirements. Members with low aspect ratios, which will deflect primarily as a result of shear deformation, are not likely to develop flexural plastic hinges that require significant flexural ductility. Therefore, the code requirements for ductility are also less stringent for such members.

The ductility capacity of a cross section is determined primarily by two factors:

1. The amount of flexural reinforcement in the cross section
2. The amount of axial load on the cross section

The compressive strength of the masonry and the yield stress of the steel also have an effect on ductility, but their effect is less significant than the preceding two factors.

At the ultimate limit state, the depth of the compression block is determined by the area of masonry required to balance the tensile force in the reinforcing steel. Therefore, a larger steel area will result in a larger compression block, which corresponds to lower steel strains at the ultimate limit state, when the compressive strain in the masonry is equal to 0.0025 for concrete masonry or 0.0035 for clay masonry. Similarly, a larger axial compression load results in lower tensile strains because a larger compression block is required to resist the axial loads.

Figure 5.42 illustrates the calculations that need to be performed to determine whether the reinforcement ratio is acceptable. The neutral axis is selected so that the masonry wall cross section is in equilibrium with the axial load, which is determined using the load combination $D + 0.75L + 0.525Q_E$, where Q_E is the effect of the horizontal component of the earthquake load. Reinforcement in compression may be used in evaluating the equilibrium of the cross section even if it is not laterally supported by ties. Once the location of the neutral axis has been determined, the strain in the extreme tension reinforcement can be obtained and compared with the acceptable value of $\alpha\varepsilon_y$.

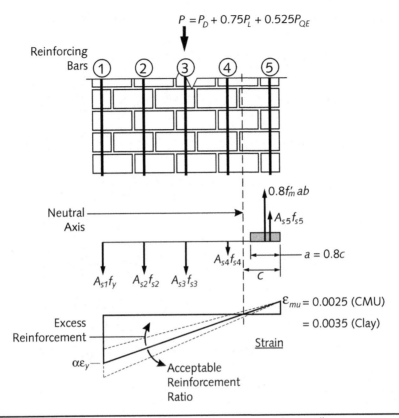

FIGURE 5.42 Verification of maximum reinforcement in masonry shear walls.

If the maximum tensile strain in the steel is less than $\alpha\varepsilon_y$, the wall is overly reinforced, and the amount of steel must be reduced or the wall redesigned.

An alternative approach may be used to simplify the calculations if the wall reinforcement is uniformly and evenly distributed. Figure 5.43 shows the free-body diagram of a shear wall with the maximum permissible amount of reinforcement.

From Fig. 5.43, the compression force in the masonry is equal to

$$C_m = 0.8(0.8)f_m'bc \tag{5.116}$$

The tensile force in the reinforcement is given by

$$T_s = f_y A_{s,\max}\left(\frac{L_w - c}{L_w}\right) - \frac{1}{2}f_y A_{s,\max}\frac{\varepsilon_y c}{L_w \varepsilon_{mu}} \tag{5.117}$$

and the compressive force in the steel reinforcement is equal to

$$C_s = f_y A_{s,\max}\left(\frac{c}{L_w}\right) - \frac{1}{2}f_y A_{s,\max}\frac{\varepsilon_y c}{L_w \varepsilon_{mu}} \tag{5.118}$$

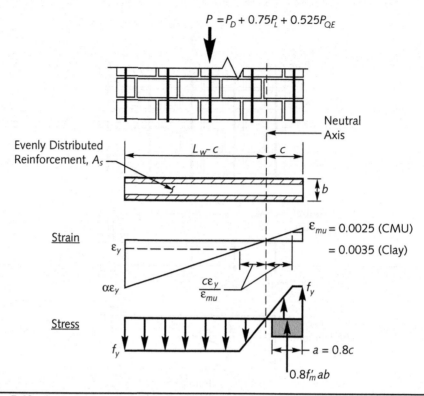

FIGURE 5.43 Approximate calculation of maximum reinforcement.

For equilibrium of forces on the cross section,

$$C_m + C_s = T_s + P \tag{5.119}$$

Substituting Eqs. (5.116), (5.117), and (5.118) into Eq. (5.119), we obtain

$$A_{s,max} = \frac{0.64 f'_m bc - P}{f_y \left(1 - \dfrac{2c}{L_w} \right)} \tag{5.120}$$

And since, from Fig. 5.43, the depth of the neutral axis is given by

$$c = \frac{\varepsilon_{mu} L_w}{\alpha \varepsilon_y + \varepsilon_{mu}} \tag{5.121}$$

the maximum permissible amount of reinforcement is given by

$$A_{s,max} = \frac{0.64 f'_m b \left(\dfrac{\varepsilon_{mu} L_w}{\alpha \varepsilon_y + \varepsilon_{mu}} \right) - P}{f_y \left(\dfrac{\alpha \varepsilon_y - \varepsilon_{mu}}{\alpha \varepsilon_y + \varepsilon_{mu}} \right)} \tag{5.122}$$

which means that the maximum steel per unit length of wall is equal to

$$
\frac{A_{s,max}}{L_w} = \frac{0.64 f'_m b \left(\dfrac{\varepsilon_{mu}}{\alpha \varepsilon_y + \varepsilon_{mu}} \right) - \dfrac{p}{L_w}}{f_y \left(\dfrac{\alpha \varepsilon_y - \varepsilon_{mu}}{\alpha \varepsilon_y + \varepsilon_{mu}} \right)}
\tag{5.123}
$$

or in terms of the reinforcement ratio on the gross cross section,

$$
\rho_{max} = \frac{A_{s,max}}{b L_w} = \frac{0.64 \left(\dfrac{\varepsilon_{mu}}{\alpha \varepsilon_y + \varepsilon_{mu}} \right) - \dfrac{P}{L_w b f'_m}}{\dfrac{f_y}{f'_m} \left(\dfrac{\alpha \varepsilon_y - \varepsilon_{mu}}{\alpha \varepsilon_y + \varepsilon_{mu}} \right)}
\tag{5.124}
$$

Note that Eqs. (5.122) to (5.124) are valid only for shear walls with relatively closely spaced vertical reinforcement that is evenly distributed along the length of the wall. However, they do provide a tool for obtaining a quick estimate of the maximum reinforcement for walls.

5.5.2 Allowable Stress Design of Shear Walls

The design of shear walls using allowable-stress design procedures is similar to the procedures described in Sec. 5.4.2 for members subjected to axial and flexural loads. Interaction diagrams can be developed using all the vertical reinforcing steel in the wall, as is typically done for strength design but with the assumptions of Sec. 5.4.2. However, the most common approach is to use only the jamb reinforcement at the ends of the wall to resist flexural loads and ignore the distributed steel between the jambs. This simplifies the design because the wall then can be evaluated as a singly reinforced member using Eqs. (5.88) through (5.91).

The TMS 402-08/ACI 530-08/ASCE 5-08 stipulates that for special reinforced masonry walls with a shear-span ratio M/Vd greater than 1.0 and an axial load P greater than $0.05 f'_m A_n$, the maximum reinforcement ratio, which is calculated using all the reinforcement in tension, should not exceed the following:

$$
\rho_{max} = \frac{n f'_m}{2 f_y \left(n + \dfrac{f_y}{f'_m} \right)}
\tag{5.125}
$$

When shear walls are not subjected to flexural tension, the allowable shear is determined as for flexural elements without flexural tension, as described in Sec. 5.3, or the entire shear is resisted by shear reinforcement, as described below. When shear walls are subjected to flexural tension, the average shear stress in the masonry is given by

$$
f_v = \frac{V}{bd}
\tag{5.126}
$$

The allowable shear is first determined assuming that the masonry alone resists the entire shear. For relatively squat walls with M/Vd less than 1.0,

$$F_v = \frac{1}{3}\left(4 - \frac{M}{Vd}\right)\sqrt{f'_m} < 80 - 45\frac{M}{Vd} \text{ psi} \qquad (5.127)$$

whereas for walls with M/Vd of 1.0 or greater,

$$F_v = \sqrt{f'_m} < 35 \text{ psi} \qquad (5.128)$$

where M is the moment corresponding to the wall shear demand, and d is the distance from the extreme compression fiber to the reinforcing steel. The ratio M/Vd always must be a positive number. As with flexural elements, if the calculated shear stress in the masonry exceeds the allowable shear stress F_v, reinforcing steel must be provided to resist the entire shear. The required area of shear reinforcement is given by

$$A_v = \frac{sV}{F_s d} \qquad (5.129)$$

where s is the spacing, and F_s is the allowable stress in the shear reinforcement. Since shear reinforcement must be placed parallel to the direction of the applied shear, shear reinforcement in shear walls is placed horizontally.

Even when shear reinforcement is used to resist the entire shear, the demand on shear walls is limited to the following values:

For $M/Vd < 1.0$,

$$F_v = \frac{1}{2}\left(4 - \frac{M}{Vd}\right)\sqrt{f'_m} < 120 - 45\frac{M}{Vd} \text{ psi} \qquad (5.130)$$

For walls with $M/Vd \geq 1.0$,

$$F_v = 1.5\sqrt{f'_m} < 75 \text{ psi} \qquad (5.131)$$

5.5.3 Examples

Example 5.5.2

Determine the horizontal and vertical reinforcing steel required at the first story of the special reinforced-masonry shear wall to resist the in-plane earthquake loads shown in Fig. 5.44. The wall is constructed with 8-in. medium-weight concrete masonry units. The specified masonry compressive strength is 1,500 psi, and Grade 60 steel is used as reinforcement. $S_{DS} = 0.9g$.

The weight of an 8-in.-thick solid grouted masonry with medium-weight units is 78 psf. Thus the loads at the base of the wall are as follows:

$$P_D = \frac{[2,905 + (3,450)2]22}{1,000} + \frac{78(22)(34.5)}{1,000} = 275 \text{ kips}$$

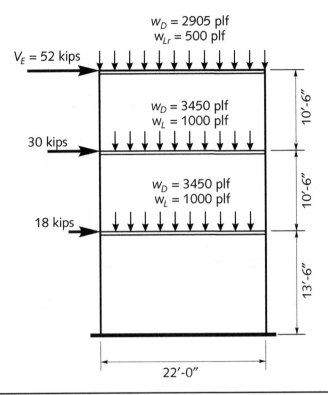

$w_D = 2905$ plf
$w_{Lr} = 500$ plf

$V_E = 52$ kips

$w_D = 3450$ plf
$w_L = 1000$ plf

10'-6"

30 kips

$w_D = 3450$ plf
$w_L = 1000$ plf

10'-6"

18 kips

13'-6"

22'-0"

Figure 5.44   Masonry shear wall.

$$P_L = \frac{2(1,000)22}{1,000} = 44 \text{ kips}$$

$$P_{Lr} = \frac{(500)22}{1,000} = 11 \text{ kips}$$

$$V_E = 52 + 30 + 18 = 100 \text{ kips}$$

$$M_E = 52(34.5) + 30(24) + 18(13.5) = 2757 \text{ kip-ft}$$

Strength Design Solution
The factored loads for the applicable load combinations are given by the following:

For $1.2D + 0.5L + 1.0E$,

$$P_u = [1.2 + 0.2(0.9)](275) + 0.5(44) = 402 \text{ kips}$$
$$V_u = 100 \text{ kips}$$
$$M_u = 2,757 \text{ kip-ft}$$

For 0.9D + 1.0E,

$$P_u = [0.9 - 0.2(0.9)](275) = 198 \text{ kips}$$
$$V_u = 100 \text{ kips}$$
$$M_u = 2{,}757 \text{ kip-ft}$$

An estimate for the amount of evenly distributed vertical steel required to resist the moment demand can be obtained from the following equation, which is based on typical strain distributions in concrete masonry walls:

$$A_{s,req} = \frac{2.5M_u}{\phi f_y d} - \frac{P_u}{f_y}$$

$$= \frac{2.5(2{,}757 \times 12)}{0.9(60)(22 \times 12)} - \frac{198}{60} = 2.5 \text{ in.}^2$$

We can try #4 bars spaced at 24 in. on center, with a total steel area of 2.4 in.². First, we calculate the axial strength of the wall as the first point on the interaction diagram:

$$A_n = 7.63(22 \times 12) = 2{,}014 \text{ in.}^2$$

$$r = \frac{t}{\sqrt{12}} = \frac{7.63}{\sqrt{12}} = 2.2 \text{ in.}$$

$$\frac{H}{r} = \frac{13.5(12)}{2.2} = 73.6 < 99$$

Therefore, noting that the reinforcing steel is not laterally confined by ties and cannot be used to resist compression loads, we get

$$P_n = 0.8\left[0.8f_m'(A_n - A_s) + f_y A_s\right]\left[1 - \left(\frac{H}{140r}\right)^2\right]$$

$$= 0.8\left[0.8(1.5)(2{,}014 - 2.4) + 60(0)\right]\left[1 - \left(\frac{73.6}{140}\right)^2\right]$$

$$= 1{,}931(0.724) = 1{,}398 \text{ kips}$$

$$\phi P_n = 0.9(1{,}398) = 1{,}258 \text{ kips}$$

The next point on the interaction diagram is the moment strength with no axial load. Table 5.4 provides the calculations for determination of the location of the neutral axis and flexural strength with no axial load.

From this table,

$$\phi M_n = 0.9\left(\frac{17{,}962}{12}\right) = 1{,}347 \text{ kip-ft}$$

	A (in.²)	d (in.)	ε_s (in./in.)	f_s (ksi)	P (kips)	M (kip-in.)
Masonry		7.024			128.6	16074.6
Bar 1	0.20	4.00	0.0019	—	—	—
Bar 2	0.20	28.00	−0.0015	−43.1	−8.6	−896.6
Bar 3	0.20	52.00	−0.0049	−60.0	−12.0	−960.0
Bar 4	0.20	76.00	−0.0083	−60.0	−12.0	−672.0
Bar 5	0.20	100.00	−0.0117	−60.0	−12.0	−384.0
Bar 6	0.20	124.00	−0.0152	−60.0	−12.0	−96.0
Bar 7	0.20	140.00	−0.0174	−60.0	−12.0	96.0
Bar 8	0.20	164.00	−0.0208	−60.0	−12.0	384.0
Bar 9	0.20	188.00	−0.0243	−60.0	−12.0	672.0
Bar 10	0.20	212.00	−0.0277	−60.0	−12.0	960.0
Bar 11	0.20	236.00	−0.0311	−60.0	−12.0	1,248.0
Bar 12	0.20	260.00	−0.0345	−60.0	−12.0	1,536.0
	2.40				0.0	17,962.0

TABLE 5.4 Equilibrium Calculations for No Axial Load (c = 17.56 in.)

For the balanced condition, the depth of the neutral axis is given by

$$c_b = \frac{d}{\left(\dfrac{f_y}{E_s\varepsilon_{mu}}+1\right)} = \frac{d}{\left[\dfrac{60}{29{,}000(0.0025)}+1\right]} = 0.547d$$

$$= 0.547(260) = 142.22 \text{ in.}$$

Table 5.5 provides the calculations for determination of the axial load and flexural strength at the balanced condition.
From this table,

$$\phi P_b = 0.9(1{,}006) = 906 \text{ kips}$$

$$\phi M_b = 0.9\left(\frac{81{,}678}{12}\right) = 6{,}126 \text{ kip-ft}$$

We can obtain an additional point on the interaction diagram and determine the axial and flexural strength when the neutral axis is equal to 80 in., as shown in Table 5.6
From this table,

$$\phi P_n = 0.9(503.5) = 453 \text{ kips}$$

$$\phi M_n = 0.9\left(\frac{63{,}306}{12}\right) = 4{,}748 \text{ kip-ft}$$

	A (in.²)	d (in.)	ε_s (in./in.)	f_s (ksi)	P (kips)	M (kip-in.)
Masonry		56.89			1041.7	78,246.7
Bar 1	0.20	4.00	0.0024	—	—	—
Bar 2	0.20	28.00	0.0020	—	—	—
Bar 3	0.20	52.00	0.0016	—	—	—
Bar 4	0.20	76.00	0.0012	—	—	—
Bar 5	0.20	100.00	0.0007	—	—	—
Bar 6	0.20	124.00	0.0003	—	—	—
Bar 7	0.20	140.00	0.0000	—	—	—
Bar 8	0.20	164.00	−0.0004	−11.1	−2.2	71.1
Bar 9	0.20	188.00	−0.0008	−23.3	−4.7	261.4
Bar 10	0.20	212.00	−0.0012	−35.6	−7.1	569.2
Bar 11	0.20	236.00	−0.0016	−47.8	−9.6	994.4
Bar 12	0.20	260.00	−0.0021	−60.0	−12.0	1,536.0
	2.40				1,006.2	81,678.6

TABLE 5.5 Equilibrium Calculations for Balanced Condition (c = 142.22 in.)

	A (in.²)	d (in.)	ε_s (in./in.)	f_s (ksi)	P (kips)	M (kip-in.)
Masonry		32.00			586.0	58,598.4
Bar 1	0.20	4.00	0.0024	—	—	—
Bar 2	0.20	28.00	0.0016	—	—	—
Bar 3	0.20	52.00	0.0009	—	—	—
Bar 4	0.20	76.00	0.0001	—	—	—
Bar 5	0.20	100.00	−0.0006	−18.1	−3.6	−116.0
Bar 6	0.20	124.00	−0.0014	−39.9	−8.0	−63.8
Bar 7	0.20	140.00	−0.0019	−54.4	−10.9	87.0
Bar 8	0.20	164.00	−0.0026	−60.0	−12.0	384.0
Bar 9	0.20	188.00	−0.0034	−60.0	−12.0	672.0
Bar 10	0.20	212.00	−0.0041	−60.0	−12.0	960.0
Bar 11	0.20	236.00	−0.0049	−60.0	−12.0	1,248.0
Bar 12	0.20	260.00	−0.0056	−60.0	−12.0	1,536.0
	2.40				503.5	63,305.6

TABLE 5.6 Equilibrium Calculations for c = 80 in.

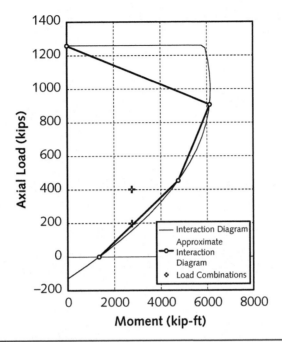

FIGURE 5.45 Interaction diagram for shear wall.

Figure 5.45 shows the interaction diagram and applied loads, indicating that the wall has sufficient flexural strength.

To design the wall to resist the shear demands, we must first calculate the shear corresponding to the nominal flexural strength. From the interaction diagram, the flexural strength at the maximum factored load of 402 kips is approximately equal to

$$\phi M_{n(P=402\text{kips})} \approx 1,347 + 402\left(\frac{4,748-1,347}{453-0}\right) = 4,365 \text{ kip-ft}$$

$$\rightarrow M_{n(P=402\text{kips})} = \frac{\phi M}{\phi} = \frac{4,365}{0.9} = 4,850 \text{ kip-ft}$$

Since a shear demand of 100 kips corresponds to a bending moment of 2,757 kip-ft, the shear demand corresponding to 1.25 times the wall nominal flexural strength of 4,850 kip-ft is equal to

$$\phi V_n \geq 1.25(100)\frac{4,850}{2,757} = 220 \text{ kips}$$

The shear also may be checked by verifying that the nominal shear strength exceeds 2.5 times the calculated shear demands:

$$V_n \geq 2.5V_v$$

$$\phi V_n \geq \phi 2.5V_v = 0.8(2.5)(100) = 200 \text{ kips} \leftarrow \text{Governs}$$

The shear span ratio for the wall is equal to

$$\frac{M}{Vd} = \frac{2757}{100(260/12)} = 1.27 > 1.0; \text{ Therefore, use } 1.0.$$

The maximum shear that the wall can be designed to resist thus is given by

$$V_n \leq 4A_n\sqrt{f'_m} = \frac{4(7.63 \times 264)\sqrt{1,500}}{1,000}$$

$$= 312 \text{ kips} > 200 \text{ kips} \qquad \qquad Okay!$$

We can try #4 bars ($A_v = 0.31$ in.²) spaced at 24 in. on center for the horizontal reinforcement. The shear strengths contributed by the masonry and steel are as follows:

$$\phi V_{mn} = \phi\left(2.25A_n\sqrt{f'_m} + 0.25P_u\right)$$

$$= 0.8\left[\frac{2.25(7.63 \times 264)\sqrt{1,500}}{1,000} + 0.25(198)\right]$$

$$= 180 \text{ kips}$$

$$\phi V_{ns} = \phi 0.5A_v f_y \frac{d}{s} = 0.8(0.5)(0.20)(60)\frac{260}{24} = 52 \text{ kips}$$

Then the total shear strength of the wall is equal to

$$\phi V_n = \phi V_{nm} + \phi V_{ns} = 180 + 52 = 232 \text{ kips} > 200 \text{ kips} \qquad \qquad Okay!$$

Therefore, #4 bars spaced at 24 in. on center are sufficient to resist the shear demands on the wall.

Verify that the horizontal and vertical reinforcements satisfy the minimum reinforcing and maximum spacing requirements:

$$\left(\frac{A_s}{st}\right)_{hor} = \frac{0.2}{24(7.63)} = 0.0011 > 0.0007 \qquad \qquad Okay!$$

$$\left(\frac{A_s}{st}\right)_{ver} = \frac{0.20}{24(7.63)} = 0.0011 > 0.0007 \qquad \qquad Okay!$$

$$\left(\frac{A_s}{st}\right)_{hor} + \left(\frac{A_s}{st}\right)_{ver} = 0.0011 + 0.0011 = 0.0022 > 0.002$$

As required by TMS 402-08/ACI 530-08/ASCE 5-08, the vertical reinforcement must be at least one-third the horizontal reinforcement:

$$\left(\frac{A_s}{st}\right)_{ver} = 0.0011 > \frac{1}{3}\left(\frac{A_s}{st}\right)_{hor} = \frac{0.0011}{3} = 0.0004 \qquad \qquad Okay!$$

Finally, the maximum reinforcement ratio needs to be checked using the load combination $D + 0.75L + 0.525Q_E$:

$$P = P_D + 0.75P_L + 0.525P_{Q_E}$$
$$= 235 + 0.75(44) + 0.525(0) = 268 \text{ kips}$$

The calculations for determining the neutral axis and maximum strain in the reinforcing steel are presented in Table 5.7. The neutral axis depth of 48.48 in. is obtained by trial and error.

The maximum tensile strain in the reinforcing steel (Bar 12) is equal to

$$\varepsilon_{s,\max} = 0.0109 = 5.27\varepsilon_y > 4\varepsilon_y \qquad\qquad\qquad Okay!$$

The maximum reinforcement ratio also may be checked using the approximate Eq. (5.124):

$$\rho_{\max} = \frac{A_{s,\max}}{bL_w} = \frac{0.64\left(\dfrac{\varepsilon_{mu}}{\alpha\varepsilon_y + \varepsilon_{mu}}\right) - \dfrac{P}{L_w bf'_m}}{\dfrac{f_y}{f'_m}\left(\dfrac{\alpha\varepsilon_y - \varepsilon_{mu}}{\alpha\varepsilon_y + \varepsilon_{mu}}\right)}$$

$$\Delta = \left(\frac{H^3}{3E_m I}\right) + \left(\frac{1.2H}{AG_m}\right) + \left(\Delta_F \frac{2H}{L}\right)$$

Flexure   Shear   Foundation rotation

	A (in.²)	d (in.)	ε_s (in./in.)	f_s (ksi)	P (kips)	M (kip-in.)
Masonry		19.39			355.2	39,992.1
Bar 1	0.20	4.00	0.0023	60.0	12.0	1,536.0
Bar 2	0.20	28.00	0.0011	30.6	6.1	637.2
Bar 3	0.20	52.00	−0.0002	−5.3	−1.1	−84.1
Bar 4	0.20	76.00	−0.0014	−41.1	−8.2	−460.8
Bar 5	0.20	100.00	−0.0027	−60.0	−12.0	−384.0
Bar 6	0.20	124.00	−0.0039	−60.0	−12.0	−96.0
Bar 7	0.20	140.00	−0.0047	−60.0	−12.0	96.0
Bar 8	0.20	164.00	−0.0060	−60.0	−12.0	384.0
Bar 9	0.20	188.00	−0.0072	−60.0	−12.0	672.0
Bar 10	0.20	212.00	−0.0084	−60.0	−12.0	960.0
Bar 11	0.20	236.00	−0.0097	−60.0	−12.0	1,248.0
Bar 12	0.20	260.00	−0.0109	−60.0	−12.0	1,536.0
	2.40				268.0	46,036.4

TABLE 5.7   Equilibrium Calculations for Maximum Reinforcement Ratio

$$\rho = \frac{A_s}{bL_w} + \frac{2.4}{7.63(22 \times 12)} = 0.0012 < \rho_{\max} \qquad \textit{Okay!}$$

Allowable-Stress Design Solution
Using the alternate basic load combinations of the 2009 IBC, which permit a one-third increase in the allowable stresses, the moment at the base of the wall is given by

$$M = \frac{M_E}{1.4} = \frac{2,757}{1.4} = 1,969 \text{ kip-ft}$$

The applied shear must be multiplied by 1.5 for allowable-stress design:

$$V = \frac{1.5V_E}{1.4} = \frac{1.5(100)}{1.4} = 107.1 \text{ kips}$$

Taking into account the vertical component of earthquake loads, the axial loads for the applicable load combinations are given by the following:

For $D + L + E/1.4$,

$$P = \left[1 + \frac{0.2(0.9)}{1.4}\right](275) + (44) = 354 \text{ kips}$$
$$V = 107.1 \text{ kips}$$
$$M = 1,969 \text{ kip-ft}$$

For $0.9D + 1.0E/1.4$,

$$P = \left[0.9 - \frac{0.2(0.9)}{1.4}\right](275) = 212 \text{ kips}$$
$$V = 107.1 \text{ kips}$$
$$M = 1,969 \text{ kip-ft}$$

By inspection, the lower axial load will govern the design of the wall when using allowable-stress design procedures. Assuming uncracked properties, check for flexural tension:

$$\frac{P}{A_n} - \frac{M}{S_n} = \frac{(212)1,000}{(7.63)(22 \times 12)} - \frac{(1,969 \times 12)(1,000)6}{(7.63)(22 \times 12)^2}$$
$$= -162 \text{ psi}$$

Therefore, flexural tension exists. Assuming that only jamb reinforcement at the end of the wall resists the bending moment, we can try two #4 bars at each end of the wall with $d = 260$ in.

$$n = \frac{29,000,000}{750(1,500)} = 21.5$$

$$\rho = \frac{2 \times 0.2}{7.63 \times 260} = 0.0002$$

$$n\rho = 21.5\left(\frac{2 \times 0.2}{7.63 \times 260}\right) = 0.0043$$

$$\frac{P}{bdF_b} = \frac{212(1,000)}{7.63(260)(667)} = 0.160$$

$$\frac{nP}{F_s bd} = \frac{21.5(212)(1,000)}{32,000(7.63)(260)} = 0.072$$

For the allowable moment based on the masonry compressive stress,

$$k = \sqrt{\left(n\rho - \frac{P}{bdF_b}\right)^2 + 2n\rho} - \left(n\rho - \frac{P}{bdF_b}\right) = 0.338$$

$$j = \left(1 - \frac{k}{3}\right) = 0.887$$

From Eq. (5.90), the allowable moment is given by

$$M = \frac{1}{2}F_b jkbd^2 - P\left(d - \frac{h}{2}\right) = 2{,}031 \text{ kip-ft} > 1{,}969 \text{ kip-ft} \qquad Okay!$$

If the cross section is governed by the steel tensile stress,

$$k = \sqrt{\left(n\rho + \frac{nP}{F_s bd}\right)^2 + 2\left(n\rho + \frac{nP}{F_s bd}\right)} - \left(n\rho + \frac{nP}{F_s bd}\right) = 0.321$$

$$j = \left(1 - \frac{k}{3}\right) = 0.893$$

$$M = A_s F_s jd + P\left(\frac{h}{2} - \frac{kd}{3}\right) = 2{,}089 \text{ kip-ft} > 1{,}969 \text{ kip-ft} \qquad Okay!$$

Check the maximum reinforcement using Eq. (5.125):

$$\rho_{max} = \frac{nf'_m}{2f_y\left(n + \frac{f_y}{f'_m}\right)} = \frac{21.5(1.5)}{2(60)\left(21.5 + \frac{60}{1.5}\right)} = 0.0044 > 0.0002 \qquad Okay!$$

Since flexural tension exists, the shear stress in the wall is given by Eq. (5.126):

$$f_v = \frac{V}{bd} = \frac{(107.1)(1,000)}{(7.63)(260)} = 54 \text{ psi}$$

$$\frac{M}{Vd} = \frac{2,757}{100(21.67)} = 1.27 > 1.0$$

Therefore, the allowable shear considering the masonry alone is obtained from Eq. (5.128), including the one-third increase in allowable stresses:

$$F_v = 1.33\left(1.0\sqrt{f'_m}\right) = 1.33\sqrt{1,500} = 51.5 \text{ psi} \leftarrow \text{Governs.}$$
$$= 1.33(35) = 46.6 \text{ psi}$$

The allowable shear stress using masonry alone is less than the demand; therefore, shear reinforcement is required. Checking with Eq. (5.131) to see if the wall needs to be made thicker, we see that

$$F_v = \left(1.5\sqrt{f'_m}\right)1.33 = 77.2 \text{ psi} \qquad \leftarrow \text{Governs} > f_v \qquad\qquad Okay!$$
$$= 75(1.33) = 99.8 \text{ psi}$$

Assuming #5 bars, the required spacing is obtained from Eq. (5.129):

$$s = \frac{A_v F_s d}{V} = \frac{0.31(32,000)(21.67 \times 12)}{107.1 \times 1,000} = 24.1 \text{ in.}$$

We can use #5 bars spaced at 24 in. on center. Then

$$\left(\frac{A_s}{st}\right)_{hor} = \frac{0.31}{24(7.63)} = 0.0017 > 0.0007 \qquad\qquad Okay!$$

The vertical reinforcement must be at least one-third the horizontal reinforcement. We can try #4 bars at 32 in. on center as the minimum vertical reinforcement between the jamb steel:

$$A_s = 0.20 \text{ in.}^2 \geq A_{s,min} = 0.2 \text{ in.}^2 \qquad\qquad Okay!$$

$$s = 32 \text{ in.} < 48 \text{ in.} \qquad\qquad Okay!$$

$$s << \frac{H}{3}; \frac{L_w}{3} \qquad\qquad Okay!$$

$$\left(\frac{A_s}{st}\right)_{ver} = \frac{0.20}{32(7.63)} = 0.0008 > 0.0007 \qquad\qquad Okay!$$

$$\left(\frac{A_s}{st}\right)_{hor} + \left(\frac{A_s}{st}\right)_{ver} = 0.0017 + 0.0008 = 0.0025 > 0.002 \qquad\qquad Okay!$$

$$\left(\frac{A_s}{st}\right)_{ver} > \frac{1}{3}\left(\frac{A_s}{st}\right)_{hor} = \frac{0.0017}{3} = 0.00056 \qquad\qquad Okay!$$

5.6 Design of Walls Loaded Out of Plane (Slender Walls)

Section 5.2 provided procedures for calculating out-of-plane loads on walls due to earthquake and wind loads. This section will discuss the strength design of masonry walls that are loaded out of plane. A major consideration is the fact that the lateral displacement often is comparable with the wall thickness. This means that secondary effects (also known as P-δ effects) that exist as a result of the application of loads on the wall's deformed shape can be a significant part of the wall demand and must be considered in the analysis.

The TMS 402-08/ACI 530-08/ASCE 5-08 requirements for out-of-plane design are valid only for walls that satisfy one of the following requirements:

1. $P_u \leq 0.05 f'_m A_g$

2. $0.05 f'_m A_g < P_u \leq 0.20 f'_m A_g;$ $\dfrac{H'}{t} \leq 30$

    where P_u is the factored axial load at the location of maximum moment, A_g is the gross cross-sectional area, and f'_m is the masonry compressive strength. H' is the effective height, and t is the wall thickness. There are no design requirements for walls with axial loads greater than $0.2 f'_m A_g$.

Derivation of the equations for evaluation of walls loaded out of plane that include P-δ effects is based on a wall span with pinned-pinned boundary conditions. When such boundary conditions do not exist, alternate equations should be used, such as the method described in Sec. 5.2. Figure 5.46 shows the loads on a deformed wall spanning a height H. In addition to the wall moment resulting from the out-of plane lateral wind

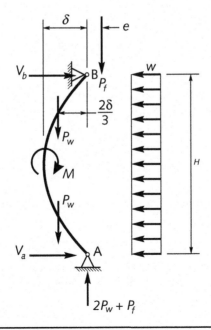

FIGURE 5.46   Free-body diagram of slender wall loaded out of plane.

or earthquake load w, there is also moment due to the eccentricity e of the floor (or roof) load P_f. We can assume that the wall is deformed in the shape of a parabola such that the weight of each half of the wall P_w is located at a distance of $\frac{2}{3}\delta$ from the supports, where δ is the displacement at wall mid-height. Then, by taking moments about the bottom support, we obtain

$$\frac{wH^2}{2} + 2P_w\left(\frac{2}{3}\delta\right) - P_f e - V_b H = 0 \tag{5.132}$$

This means that the horizontal reaction at the top of the wall is given by

$$V_b = \frac{wH}{2} + \frac{4P_w\delta}{3H} - \frac{P_f e}{H} \tag{5.133}$$

Then, by taking moment about the mid-height for loads on the upper part of the wall,

$$M = \frac{V_b H}{2} + P_f(e+\delta) + \frac{P_w\delta}{3} - \frac{wH^2}{8} \tag{5.134}$$

and substituting Eq. (5.133) into Eq. (5.134), we get

$$M = \frac{wH^2}{8} + \frac{P_f e}{2} + \left(P_f + P_w\right)\delta \tag{5.135}$$

From Eq. (5.135) it can be seen that the maximum moment, which occurs at mid-height, depends on the wall displacement, which, in turn, depends on the wall moment. Figure 5.47 shows the variation in stiffness used for calculation of the wall deflections. When the moment demand is less than the cracking moment M_{cr}, the wall deflection δ is calculated as follows:

$$\delta = \frac{5MH^2}{48E_m I_g} \tag{5.136}$$

where E_m is the masonry modulus of elasticity, and I_g is the gross, uncracked moment of inertia. If the maximum moment is greater than M_{cr}, the wall deflection is calculated from

$$\delta = \frac{5M_{cr}H^2}{48E_m I_g} + \frac{5(M - M_{cr})H^2}{48E_m I_{cr}} \tag{5.137}$$

where I_{cr} is the cracked moment of inertia. The cracking moment of the wall is given by

$$M_{cr} = S_n f_r \tag{5.138}$$

where S_n is the section modulus, and the modulus of rupture f_r can be found in Table 3.1.8.2 of TMS 402-08/ACI 530-08/ASCE 5-08. The cracked moment of inertia can be determined using the equation determined by the Structural Engineers Association of California (SEAOC)[5]:

$$I_{cr} = nA_{se}(d-c)^2 + \frac{bc^3}{3}$$ (5.139)

where

n = modular ratio
d = effective depth of reinforcement
c = depth of neutral axis
b = width of compression block

A_{se} is the effective area of reinforcement that takes into account the effect of axial load P on the cross section and is given by

$$A_{se} = A_s + \frac{P_u}{f_y}\frac{h}{2d}$$ (5.140)

where A_s and f_y are the area and yield stress of tension reinforcement, respectively.

Since the moment depends on the deflection and the deflection depends on the moment, it is apparent that a solution to Eq. (5.135) can be reached by a trial-and-error process in which a displacement at mid-height is assumed, and the maximum moment is calculated using Eq. (5.135). The displacement then is calculated using either Eq. (5.136) or (5.137) and compared with the original guess. If the calculated displacement is close to the assumed displacement, the calculated moment can be used to design the wall. Otherwise, a new guess is made, and the process is continued until convergence is achieved. Figure 5.48 shows a flowchart describing the basic procedure for performing analysis of walls subjected to out-of-plane loads using the preceding trial-and-error procedure. The approach is not particularly cumbersome and usually converges to a solution after a few trials.

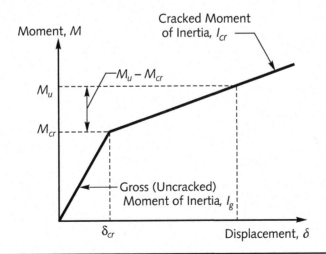

FIGURE 5.47 Out-of-plane stiffness of masonry walls.

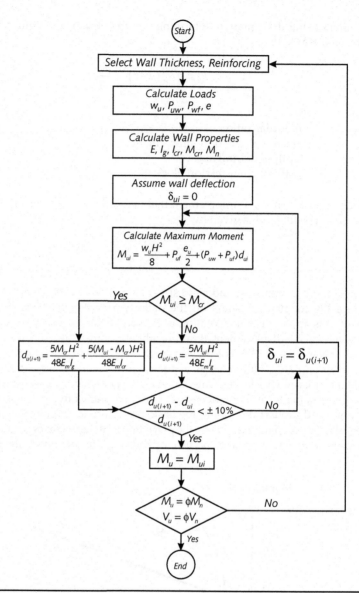

Figure 5.48 Flowchart for design of masonry walls loaded out of plane using an iterative procedure.

In lieu of performing the iterations described in Fig. 5.48, a closed-form solution can be developed by substituting Eq. (5.135) into Eq. (5.137) to obtain the following equation, when $M > M_{cr}$[5]:

$$\delta_u = \frac{\left(\dfrac{wH^2}{8} + \dfrac{P_f e}{2}\right) - M_{cr}\left(1 - \dfrac{I_{cr}}{I_g}\right)}{\dfrac{48E_m I_{cr}}{5H^2} - \left(P_w + P_f\right)} \tag{5.141}$$

And similarly, if $M < M_{cr}$,

$$\delta_u = \frac{\dfrac{wH^2}{8} + \dfrac{P_f e}{2}}{\dfrac{48 E_m I_g}{5H^2} - \left(P_w + P_f\right)} \tag{5.142}$$

After the wall displacement and corresponding out-of-plane moment on a wall are determined, the cross section needs to be designed to resist the demands. The design process essentially involves satisfying the following two goals: First, the out-of-plane moment strength of the wall must be no less than the factored out-of-plane moment demand, including the effects of axial load and P - δ effects. That is,

$$\phi M_n \geq M_u \tag{5.143}$$

Second, the horizontal deflection at mid-height under service loads (without load factors) δ_s must be no greater than $0.007H$.

Moments and displacements for checking the strength of the cross section are determined using the procedure described earlier with factored loads for all applicable load combinations. On the other hand, when checking the displacement limit, service loads are used. In addition, although shear demands on walls loaded out of plane are not usually significant, the out-of-plane shear strength of walls should be checked against the factored shear demand.

As described in Sec. 5.4, the nominal flexural strength for a singly reinforced cross section is given by

$$M_n = A_s f_y \left(d - \frac{a}{2}\right) + \frac{P_u}{\phi}\left(\frac{h}{2} - \frac{a}{2}\right) \tag{5.144}$$

where A_s, f_y, and d are the area, yield stress, and effective depth of the reinforcement; P_u is the factored axial load; h is the wall thickness; and a is the depth of the equivalent compressive block, which is given by

$$a = \frac{A_s f_y + P_u / \phi}{0.8 b f'_m} \tag{5.145}$$

where b is the width of the compression area. The TMS 402-08/ACI 530-08/ASCE 5-08 stipulates that when concrete masonry units are placed in running bond, or if fully grouted bond beams are used at a spacing of no more than 48 in., the effective width of the compression area for each reinforcing bar should not exceed the smallest of the following:

1. The center-to-center bar spacing

2. Six times the nominal wall thickness

3. 72 in.

For masonry not placed in running bond and with bond beams spaced further apart than 48 in., the width of the compression area used to calculate the moment capacity should not exceed the length of the masonry unit in which the reinforcing bar is placed.

When the masonry is partially grouted, Eqs. (5.144) and (5.145) should be modified to account for the voids in the cross section.

Walls loaded out of plane also must satisfy the minimum and maximum requirements outlined in the previous sections. The TMS 402-08/ACI 530-08/ASCE 5-08 stipulates that for walls loaded out of plane, the strain in the reinforcing steel must be at least 1.5 times the yield strain when the masonry attains the maximum usable strain. Using a procedure similar to that used in Sec. 5.5.1 for shear walls, we obtain the following equation for walls with only tension steel:

$$\rho_{max} = \frac{A_{s,max}}{bd} = \frac{0.64 f_m' \left(\dfrac{\varepsilon_{mu}}{\alpha \varepsilon_y + \varepsilon_{mu}} \right) - \dfrac{P}{bd}}{f_y} \tag{5.146}$$

where ε_y and f_y are the steel yield strain and stress, respectively, and P is the axial load from the load combination $D + 0.75L + 0.525Q_E$. Since the steel yield reinforcement strain factor and maximum usable strain are given by

$$\alpha = 1.5$$

$$\varepsilon_{mu} = 0.0025$$

the maximum reinforcement ratio is given by

$$\rho_{max} = 0.64 \frac{f_m'}{f_y} \left(\frac{0.0025}{1.5 f_y / E_s + 0.0025} \right) - \frac{P}{bdf_y} \tag{5.147}$$

5.6.3 Examples

Example 5.6.1
Design the masonry wall shown in Fig. 5.49 to resist out-of-plane earthquake loading. The short-period design spectral acceleration $S_{DS} = 1.46g$. The masonry is constructed in running bond with 12-in. thick fully grouted medium-weight concrete masonry units (weight = 124 psf) and Type S mortar. Masonry compressive strength is 1,500 psi, and steel is Grade 60. The wall-to-roof connection is shown in Fig. 5.50.

Strength Design Solution
Since the joist girders are 24 ft away from the wall, the roof width tributary to wall is equal to 0.5(24) =12 ft. Then

$$\text{Roof dead load} = 16(12) = 192 \text{ lb/ft}$$

$$\text{Roof live load} = 20(12) = 240 \text{ lb/ft}$$

Self-weight of wall at mid-height:

$$\text{Parapet} = 124(2.5) = 310 \text{ lb/ft}$$

$$\text{Wall} = 124(0.5)(29) = 1{,}798 \text{ lb/ft}$$

$$\text{Total Weight of Wall at Mid-height} = 2{,}108 \text{ lb/ft}$$

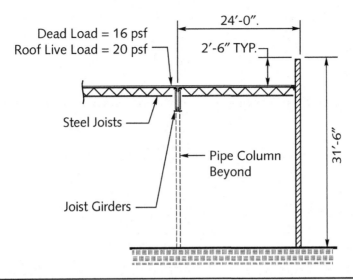

FIGURE 5.49   Masonry wall.

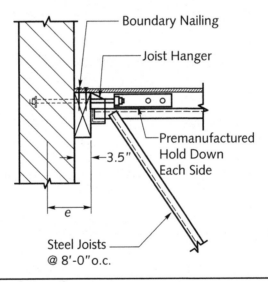

FIGURE 5.50   Wall-to-roof connection.

The axial loads at the critical wall cross section, which occurs at wall mid-height, are given by

$$P_{fD} = 192 \text{ lb/ft}$$

$$P_{fLr} = 240 \text{ lb/ft}$$

$$P_{w} = 2,108 \text{ lb/ft}$$

From Fig. 5.50, the eccentricity of the roof reaction, which occurs at the edge of the 3.5-in. wide ledger is given by

$$e = \frac{t}{2} + 3.5 = \frac{11.63}{2} + 3.5 = 9.3 \text{ in.}$$

From Eq. (5.2), the out-of-plane earthquake load on the structural wall is equal to

$$F_p = 0.4 S_{DS} I W_p$$
$$= 0.4(1.46)(1.0)(124) = 72.4 \text{ psf}$$

The load combinations that need to be considered for earthquake design are as follows:

$$1.2D + E$$

$$0.9D + E$$

Taking into account the vertical component of earthquake load, the load combinations become

$$(1.2 + 0.2 \times 1.46)D + E_h = 1.49D + E_h$$

$$(0.9 - 0.2 \times 1.46)D + E_h = 0.61D + E_h$$

For the first load combination, the axial loads on the wall are given by

$$P_{uf} = 1.49(192) = 268 \text{ lb/ft}$$

$$P_{uw} = 1.49(2,108) = 3,141 \text{ lb/ft}$$

$$P_u = P_{uf} + P_{uw} = 268 + 3,141 = 3,409 \text{ lb/ft}$$

$$\frac{P_u}{A_g} = \frac{3,409}{12(11.63)} = 24.4 \text{ psi} \leq 0.05 f_m' \qquad\qquad Okay!$$

Try two layers of #5 bars spaced at 16 in. on center ($A_s = 0.23$ in.^{2}/ft, $d = 9.2$ in.), and determine the wall properties.

$$I_g = \frac{b(t)^3}{12} = \frac{12(11.63)^3}{12} = 1,573 \text{ in.}^4/\text{ft}$$

From the modulus of rupture, f_r for solid-grouted masonry with Type S mortar with flexural stresses normal to bed joints is 163 psi. The cracking moment thus is given by

$$M_{cr} = \left(f_r + \frac{P_u}{A_g} \right) S_n = \left(163 + 24.4 \right) \frac{12(11.63)^2}{6}$$

$$= 50,694 \text{ lb-in./ft} = 4,225 \text{ lb-ft /ft}$$

$$a = \frac{P_u/\phi + A_s f_y}{0.8 f_m' b} = \frac{3,409/0.9 + 0.23(60,000)}{0.8(1,500)12} = 1.22 \text{ in.}$$

$$c = \frac{1.22}{0.8} = 1.53 \text{ in.}$$

$$A_{se} = A_s + \frac{P_u t}{2df_y} = 0.23 + \frac{3{,}409(11.63)}{2(9.2)(60{,}000)} = 0.27 \text{ in.}^2/\text{ft}$$

The cracked moment of inertia then is calculated from Eq. (5.139):

$$n = \frac{E_s}{E_m} = \frac{29{,}000{,}000}{900(1{,}500)} = \frac{29{,}000{,}000}{1{,}350{,}000} = 21.5$$

$$I_{cr} = nA_{se}(d-c)^2 + \frac{bc^3}{3}$$

$$= 21.5(0.27)(9.2 - 1.53)^2 + \frac{12(1.53)^3}{3} = 356 \text{ in.}^4/\text{ft}$$

The wall deflection can be calculated directly using Eq. (5.141), assuming that the wall is cracked:

$$\delta_u = \frac{\left(\dfrac{w_u H^2}{8} + \dfrac{P_{uf} e}{2}\right) - M_{cr}\left(1 - \dfrac{I_{cr}}{I_g}\right)}{\dfrac{48E_m I_{cr}}{5h^2} - \left(P_{uw} + P_{uf}\right)}$$

$$= \frac{\left[\dfrac{72.4(29 \times 12)^2}{8(12)} + \dfrac{268(9.3)}{2}\right] - 4{,}225(12)\left(1 - \dfrac{356}{1{,}573}\right)}{\dfrac{48(1{,}350{,}000)(356)}{5(29 \times 12)^2} - 3{,}409}$$

$$= 1.54 \text{ in.}$$

Therefore, the moment demand, including P-δ effects, is equal to

$$M_u = \frac{w_u H^2}{8} + \frac{P_{uf} e}{2} + \left(P_{uf} + P_{uw}\right)\delta_u$$

$$= \frac{72.4(29)^2}{8} + \frac{268}{2}\left(\frac{9.3}{12}\right) + 3{,}409\left(\frac{1.54}{12}\right)$$

$$= 8{,}152 \text{ lb-ft/ft}$$

M_u is greater than M_{cr}, so the assumption of a cracked wall is correct. The flexural strength when the reinforcement is not at the center of the wall is given by

$$M_n = A_s f_y\left(d - \frac{a}{2}\right) + \frac{P_u}{\phi}\left(\frac{h}{1} - \frac{a}{2}\right) \tag{5.148}$$

$$\phi M_n = \phi \left[A_s f_y \left(d - \frac{a}{2} \right) + \frac{P_u}{\phi} \left(\frac{h}{2} - \frac{a}{2} \right) \right]$$

$$= 0.9 \left[(0.23 \times 60,000) \left(9.2 - \frac{1.22}{2} \right) + \frac{3,409}{0.9} \left(\frac{11.63}{2} - \frac{1.22}{2} \right) \right] \frac{1}{12}$$

$$= 10,369 \text{ lb-ft /ft} > M_u = 8,150 \text{ lb-ft /ft} \qquad\qquad\qquad Okay!$$

For the second load combination, the axial loads on the wall are given by

$$P_{uf} = 0.61(192) = 117 \text{ lb/ft}$$

$$P_{uw} = 0.61(2,108) = 1,286 \text{ lb/ft}$$

$$P_u = P_{uf} + P_{uw} = 117 + 1,286 = 1,403 \text{ lb/ft}$$

$$\frac{P_u}{A_g} = \frac{1,403}{12(11.63)} = 10.1 \text{ psi} \le 0.05 f_m' \qquad\qquad\qquad Okay!$$

With two layers of #5 bars at 16 in. on center (A_s = 0.23 in.^{2}/ft, d = 9.3 in.):

$$M_{cr} = \left(f_r + \frac{P_u}{A_g} \right) S_n = (163 + 10.1) \frac{12(11.63)^2}{6}$$

$$= 46,826 \text{ lb-in./ft} = 3,902 \text{ lb-ft/ft}$$

$$a = \frac{P_u/\phi + A_s f_y}{0.8 f_m' b} = \frac{1,403/0.9 + 0.23(60,000)}{0.8(1,500)12} = 1.07 \text{ in.}$$

$$c = \frac{1.07}{0.8} = 1.34 \text{ in.}$$

$$A_{se} = A_s + \frac{P_u t}{2d f_y} = 0.23 + \frac{1,403(11.63)}{2(9.3)(60,000)} = 0.24 \text{ in.}^2/\text{ft}$$

$$I_{cr} = n A_{se} (d - c)^2 + \frac{bc^3}{3}$$

$$= 21.5(0.24)(9.2 - 1.34)^2 + \frac{12(1.34)^3}{3} = 328 \text{ in.}^4/\text{ft}$$

Assuming that the wall is cracked, the wall deflection can be calculated directly using Eq. (5.141):

$$\delta_u = \frac{\left(\dfrac{w_u H^2}{8} + \dfrac{P_{uf} e}{2} \right) - M_{cr} \left(1 - \dfrac{I_{cr}}{I_g} \right)}{\dfrac{48 E_m I_{cr}}{5h^2} - \left(P_{uw} + P_{uf} \right)}$$

$$
= \frac{\left[\dfrac{72.4(29 \times 12)^2}{8(12)} + \dfrac{117(9.3)}{2}\right] - 3{,}902(12)\left(1 - \dfrac{328}{1{,}573}\right)}{\dfrac{48(1{,}350{,}000)(328)}{5(29 \times 12)^2} - 1{,}403}
$$

$$
= 1.63 \text{ in.}
$$

Therefore, the moment demand, including P - δ effects, is equal to

$$
\begin{aligned}
M_u &= \frac{w_u H^2}{8} + \frac{P_{uf}e}{2} + \left(P_{uf} + P_{uw}\right)\delta_u \\
&= \frac{72.4(29)^2}{8} + \frac{117}{2}\left(\frac{9.3}{12}\right) + 1{,}403\left(\frac{1.63}{12}\right) \\
&= 7{,}847 \text{ lb-ft /ft}
\end{aligned}
$$

M_u is greater than M_{cr}, so the assumption of a cracked wall is correct. The flexural strength is given by

$$
\begin{aligned}
\phi M_n &= \phi\left[\left(A_s f_y\right)\left(d - \frac{a}{2}\right) + \frac{P_u}{\phi}\left(\frac{h}{2} - \frac{a}{2}\right)\right] \\
&= 0.9\left[(0.23 \times 60{,}000)\left(9.2 - \frac{1.07}{2}\right) + \frac{1{,}403}{0.9}\left(\frac{11.62}{2} - \frac{1.07}{2}\right)\right]\frac{1}{12}
\end{aligned}
$$

$$
= 9{,}586 \text{ lb-ft/ft} > M_u = 7{,}845 \text{ lb-ft/ft} \qquad\qquad \textit{Okay!}
$$

The deflection at mid-height also must be checked against the limiting value of $0.007H$ under service load conditions. The service-level axial loads are equal to

$$
P_f = 192 + 240 = 432 \text{ lb/ft}
$$

$$
P_w = 2{,}108 \text{ lb/ft}
$$

$$
P = P_f + P_w = 432 + 2{,}108 = 2{,}540 \text{ lb/ft}
$$

$$
\begin{aligned}
M_{cr} &= \left(f_r + \frac{P}{A_g}\right)S_n = \left(163 + \frac{2{,}540}{11.63 \times 12}\right)\frac{12(11.63)^2}{6} \\
&= 49{,}017 \text{ lb-in./ft} = 4{,}085 \text{ lb-ft/ft}
\end{aligned}
$$

$$
a = \frac{P + A_s f_y}{0.8 f'_m b} = \frac{2{,}540 + 0.23(60{,}000)}{0.8(1{,}500)12} = 1.13 \text{ in.}
$$

$$
c = \frac{1.13}{0.8} = 1.41 \text{ in.}
$$

$$
A_{se} = A_s + \frac{P_u t}{2 d f_y} = 0.23 + \frac{2{,}540(11.63)}{2(9.2)(60{,}000)} = 0.26 \text{ in.}^2/\text{ft}
$$

$$I_{cr} = nA_{se}(d-c)^2 + \frac{bc^3}{3}$$

$$= 21.5(0.26)(9.3 - 1.41)^2 + \frac{12(1.41)^3}{3} = 359 \text{ in.}^4/\text{ft}$$

Assuming that the wall is cracked under service loads, the wall deflection can be calculated from Eq. (5.141):

$$\delta_s = \frac{\left(\dfrac{wH^2}{8} + \dfrac{P_f e}{2}\right) - M_{cr}\left(1 - \dfrac{I_{cr}}{I_g}\right)}{\dfrac{48E_m I_{cr}}{5h^2} - \left(P_w + P_f\right)}$$

$$= \frac{\left[\dfrac{72.4(29 \times 12)^2}{1.4(8 \times 12)} + \dfrac{432(9.3)}{2}\right] - 4{,}085(12)\left(1 - \dfrac{359}{1{,}573}\right)}{\dfrac{48(1{,}350{,}000)(359)}{5(29 \times 12)^2} - 2{,}540}$$

$$= 0.82 \text{ in.} < 0.007H = 0.007(29)(12) = 2.4 \text{ in.} \qquad \textit{Okay!}$$

The assumption that the wall is cracked under service loads needs to be verified:

$$M_s = \frac{wH^2}{8} + \frac{P_f e}{2} + \left(P_f + P_w\right)\delta_s$$

$$= \frac{72.4(29)^2}{1.4(8)} + \frac{432}{2}\left(\frac{9.3}{12}\right) + 2{,}540\left(\frac{0.82}{12}\right)$$

$$= 5{,}777 \text{ lb-ft/ft} > M_{cr} = 4{,}085 \text{ lb-ft/ft} \qquad \textit{Okay!}$$

Reinforcement must be checked against maximum reinforcement limits using the load combination $D + 0.75L + 0.525Q_E$. The axial at the wall mid-height is given by

$$P = 2{,}300 + 0.75(0) + 0.525(0) = 2{,}300 \text{ lb/ft}$$

Then, using Eq. (5.146),

$$\rho_{max} = 0.64\frac{f'_m}{f_y}\left(\frac{0.0025}{1.5f_y/E_s + 0.0025}\right) - \frac{P}{bdf_y}$$

$$= 0.64\frac{1.5}{60}\left(\frac{0.0025}{1.5(0.00207) + 0.0025}\right) - \frac{2.3}{12(9.2)(60)}$$

$$= 0.0068$$

And the steel provided is equal to

$$\rho = \frac{A_s}{bd} = \frac{0.23}{12(9.2)} = 0.0021 < \rho_{max} \qquad \textit{Okay!}$$

Allowable-Stress Design Solution

The wall can be designed using the alternate load combinations for working stress design given in Sec. 3.2.1. The appropriate load combinations thus are

$$D+L+\frac{E}{1.4}$$

$$0.9D+\frac{E}{1.4}$$

where D, L, and E represent the dead, floor live, and earthquake loads, respectively. Note that $E = \rho E_h + E_v$. E_h is the horizontal component of the earthquake load, and E_v is the vertical component of the earthquake load. The redundancy factor ρ is equal to 1.0. By including the vertical component of the earthquake load, which is equal to $0.2S_{DS}D$ ($S_{DS} = 1.46$g), the load combinations for allowable-stress design become

Load combination 1:

$$\left(1+\frac{0.2\times1.46}{1.4}\right)D+L+\frac{E_h}{1.4}=1.21D+\frac{E_h}{1.4}$$

Load combination 2:

$$\left(1-\frac{0.2\times1.46}{1.4}\right)D+L+\frac{E}{1.4}=0.79D+\frac{E}{1.4}$$

Note that load combination 1 is used to represent the condition with the maximum axial load, whereas load combination 2 is used to represent the condition with the minimum axial load. For load combination 1, the axial load on the wall is given by

$$P=1.21\left(P_f+P_w\right)$$
$$=1.21(192+2,108)=2,783 \text{ lb/ft}$$

And the maximum moment on the wall is

$$M=\frac{F_pH^2}{8(1.4)}+\frac{P_fe}{2}$$
$$=\frac{72.4(29)^2}{8(1.4)}+\frac{1.21(192)(9.3)}{2(12)}=5,526 \text{ lb-ft}$$

For load combination 2,

$$P=0.79(192+2108)=1,817 \text{ lb/ft}$$

$$M=5,495 \text{ lb-ft}$$

First, the wall must be checked for the effects of only axial loads. The radius of gyration is equal to

$$r=\sqrt{\frac{I}{A}}=\frac{t}{\sqrt{12}}=\frac{11.63}{\sqrt{12}}=3.36 \text{ in.}$$

And the slenderness ratio is given by

$$\frac{H}{r} = \frac{29(12)}{3.36} = 103.5 > 99$$

Therefore, from Eq. (5.76), the allowable compressive force due to axial load alone is given by (including the one-third increase in allowable stresses)

$$P_a = \left(0.25 f'_m A_n + 0.65 A_{st} F_s\right)\left(\frac{70r}{H}\right)^2\left(\frac{4}{3}\right)$$

$$= \left[0.25(1,500)(11.63 \times 12) + 0.65(0)\right]\left(\frac{70}{103.5}\right)^2\left(\frac{4}{3}\right)$$

$$= 31,919 \text{ lb/ft} > 2,856 \text{ lb/ft} \qquad\qquad\qquad Okay!$$

The compressive stress due to the combination of axial and flexural loads should not exceed

$$F_b = \frac{1}{3} f'_m\left(\frac{4}{3}\right) = \frac{1,500}{3}\left(\frac{4}{3}\right) = 667 \text{ psi}$$

And the allowable steel stress is

$$F_s = 24,000\left(\frac{4}{3}\right) = 32,000 \text{ psi}$$

Assuming that the effective depth d is 9 in. for two layers of vertical reinforcement, an estimate of the steel required is

$$A_{s,\text{req}} \approx \frac{M}{0.85 F_s d} = \frac{5,526 \times 12}{32,000(0.85)(9)} = 0.27 \text{ in.}^2/\text{ft}$$

We can try #5 bars spaced at 16 in. on center:

$$d = 9.2 \text{ in.}$$

$$A_s = 0.31\left(\frac{12}{16}\right) = 0.23 \text{ in.}^2/\text{ft}$$

$$\rho = \frac{A_s}{bd} = \frac{0.23}{12(9.2)} = 0.0021$$

$$n = \frac{E_s}{E_m} = \frac{29,000,000}{900(1,500)} = 21.5$$

$$n\rho = 21.5(0.0021) = 0.045$$

For the masonry, using the axial load from load combination 1,

$$\frac{P}{bd F_b} = \frac{2783}{12(9.2)(667)} = 0.038$$

$$k = \sqrt{\left(n\rho - \frac{P}{bdF_b}\right)^2 + 2n\rho} - \left(n\rho - \frac{P}{bdF_b}\right)$$

$$= \sqrt{(0.045 - 0.038)^2 + 2(0.045)} - (0.045 - 0.038)$$

$$= 0.293$$

$$j = 1 - \frac{k}{3} = 0.902$$

And from Eq. (5.90),

$$M = \frac{1}{2}F_b jkbd^2 - P\left(d - \frac{h}{2}\right)$$

$$= \frac{667}{2}(0.902)(0.293)(12)(9.2)^2 - 2,856\left(9.2 - \frac{11.63}{2}\right)$$

$$= 80,101 \text{ lb-in./ft}$$

$$= 6,675 \text{ lb-ft/ft}$$

For the reinforcing steel,

$$\frac{P}{bd}\left(\frac{n}{F_s}\right) = \frac{2,783}{12(9.3)}\left(\frac{21.5}{32,000}\right) = 0.017$$

$$k = \sqrt{\left(n\rho + \frac{nP}{F_s bd}\right)^2 + 2\left(n\rho + \frac{nP}{F_s bd}\right)} - \left(n\rho + \frac{nP}{F_s bd}\right)$$

$$= 0.295$$

$$j = 1 - \frac{k}{3} = 0.901$$

$$M = A_s F_s jd + P\left(\frac{h}{2} - \frac{kd}{3}\right) = 6,223 \text{ lb-ft/ft} \qquad \leftarrow \text{Governs}$$

$$> 5,526 \text{ lb-ft/ft} \qquad\qquad\qquad\qquad\qquad\qquad \textit{Okay!}$$

For load combination 2,

$$\frac{P}{bdF_b} = \frac{1817}{12(9.2)(667)} = 0.025$$

$$k = \sqrt{\left(n\rho - \frac{P}{bdF_b}\right)^2 + 2n\rho} - \left(n\rho - \frac{P}{bdF_b}\right) = 0.280$$

$$j = 1 - \frac{k}{3} = 0.907$$

$$M = \frac{1}{2}F_b jkbd^2 - P\left(d - \frac{h}{2}\right) = 6{,}656 \text{ lb-ft/ft}$$

$$\frac{P}{bd}\left(\frac{n}{F_s}\right) = \frac{1{,}817}{12(9.2)}\left(\frac{21.5}{32{,}000}\right) = 0.011$$

$$k = \sqrt{\left(n\rho + \frac{nP}{F_s bd}\right)^2 + 2\left(n\rho + \frac{nP}{F_s bd}\right)} - \left(n\rho + \frac{nP}{F_s bd}\right)$$

$$= 0.283$$

$$j = 1 - \frac{k}{3} = 0.906$$

$$M = A_s F_s jd + P\left(\frac{h}{2} - \frac{kd}{3}\right) = 5{,}868 \text{ lb-ft/ft} \qquad \leftarrow \text{Governs}$$

$$> 5{,}495 \text{ lb-ft/ft} \qquad\qquad\qquad\qquad\qquad \textit{Okay!}$$

5.7 Connections in Masonry

Connections in masonry usually are achieved by the use of anchor bolts or by reinforcement development and splices. Anchor bolts may be used to connect structural elements constructed with other elements such as steel or wood to masonry walls and for attaching nonstructural components. Reinforcement development and splices often are relied on for attachment of masonry to concrete elements such as foundations or floors.

5.7.1 Anchor Bolts

Anchor bolts are embedded in the masonry and transfer loads by resisting tension, shear, or a combination of both, as shown in Fig. 5.51. Anchor bolts may be bent-bar anchor bolts, headed anchor bolts, or plate anchor bolts. Other types of anchor bolts, such as postinstalled expansion and epoxy anchors, are also used in concrete masonry, but they are not specifically addressed by TMS 402-08/ACI 530-08/ASCE 5-08. Bolts not addressed by the code may be tested in accordance with the procedures outlined in American Society for Testing and Materials (ASTM) *Standard Test Methods for Strength of Anchors in Concrete Elements* (ASTM E 488)[6] under conditions similar to those expected during service. A minimum of five tests must be performed, and the allowable loads should not exceed 20 percent of the average tested strength.

All anchor bolts must be solidly grouted in masonry. The TMS 402-08/ACI 530-08/ASCE 5-08 strength design requirements stipulate that all bolts must be grouted in place with at least ½ in. of grout between the bolt and the masonry, as shown in Fig. 5.51. Also, ¼-in. anchor bolts are permitted to be placed in bed joints, but this is not common practice.

The allowable loads on anchor bolts depend on the effective embedment depth, the bolt diameter, and the strength of the masonry and bolt material. Figure 5.51 illustrates the effective embedment depth for the various types of anchor bolts. For bent-bar anchor bolts, the effective embedment depth is measured as the distance perpendicular from the surface of the masonry to the bearing surface of the bent end minus one bolt diameter. The effective embedment depth for headed and plate anchor bolts is equal to the length of embedment perpendicular from the surface of the masonry to the bearing surface of the anchor head. The TMS 402-08/ACI 530-08/ASCE 5-08 requirements specify that the minimum effective embedment depth of anchor bolts l_b is four bolt diameters or 2 in., whichever is greater.

Anchor bolts loaded in tension fail by one of three modes:

1. Breakout of a conically shaped section of masonry, as shown in Fig. 5.52
2. Failure of the anchor bolt in tension by yield and fracture of the steel
3. For bent-bar anchor bolts, by anchor pullout as the bolt straightens and pulls out of the masonry

When using strength design procedures, the nominal axial tensile strength of anchor bolts is given by the smaller of the following equations:

Masonry tensile breakout:

$$B_{anb} = 4A_{pt}\sqrt{f_m'}$$ (5.149)

Steel tensile yield:

$$B_{ans} = A_b f_y$$ (5.150)

Anchor pullout (bent-bar bolts out):

$$B_{anp} = 1.5 f_m' e_b d_b + 300\pi \left(l_b + e_b + d_b \right) d_b$$ (5.151)

where d_b and A_b are the bolt diameter and cross-sectional area, f_m' is the compressive strength of the masonry, f_y is the yield stress of the anchor bolt, and e_b is the projected leg extension of a bent-bar anchor, as shown in Fig. 5.51.

As shown in Fig. 5.52, the term A_{pt} in Eq. (5.149) represents the projected tension area of a right circular cone on the masonry surface. The projected area of the right circular breakout cone is given by

$$A_{pt} = \pi l_b^2$$ (5.152)

where l_b is the effective embedment length of the bolt. The projected area should be reduced by any portion that lies outside the masonry or overlaps an open cell or open head joint. In addition, when the projected areas of adjacent bolts overlap, the value of A_{pt} should be adjusted so that no area is included more than once. Figure 5.53 shows some examples of adjustments to the projected tension area.

When the projected area of the breakout cones of adjacent bolts overlap, as shown in Fig. 5.53, the projected area used to calculate the allowable load on each bolt should be reduced by one-half the overlapping area. Then the effective area is equal to

$$A_{pt}' = \pi l_b^2 - \frac{1}{2} l_b^2 (\theta - \sin\theta)$$ (5.153)

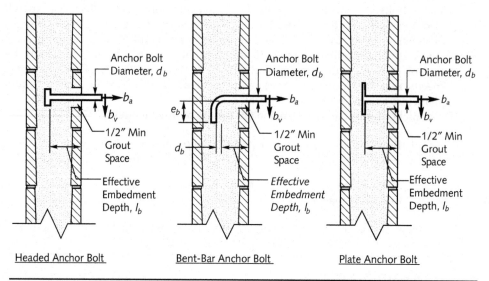

FIGURE 5.51   Anchor bolts in masonry.

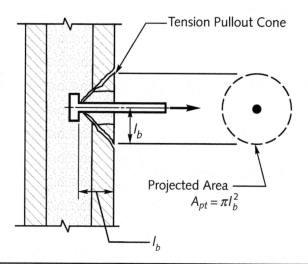

FIGURE 5.52   Tension failure of anchor bolt by masonry breakout.

where θ is in radians and is given by

$$\theta = 2\cos^{-1}\left(\frac{s}{2l_b}\right) \tag{5.154}$$

and s is the spacing between the bolts.

The capacity reduction factor used for the design of anchor bolts depends on the failure mode being considered. The capacity reduction factors for anchor bolts

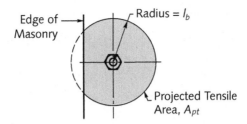

Reduction Bolt Due to Edge Distance

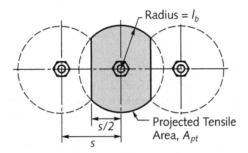

Reduction Bolt Due to Close Bolt Spacing

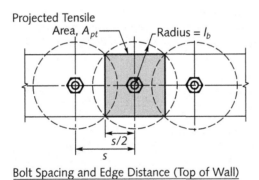

Bolt Spacing and Edge Distance (Top of Wall)

FIGURE 5.53   Reduction in projected tensile area due to edge distance and spacing.

controlled by masonry breakout, bolt steel, and anchor pullout are 0.5, 0.9, and 0.65, respectively.

For strength design, the capacity of anchor bolts loaded in shear is determined by one of four failure modes, which are represented by the following equations:

Masonry shear breakout:

$$B_{vnb} = 4A_{pv}\sqrt{f'_m}$$ (5.155)

Masonry shear crushing:

$$B_{vnc} = 1,050\sqrt[4]{f'_m A_b}$$ (5.156)

Anchor shear pryout:

$$B_{vpry} = 2.0B_{anb} = 8A_{pv}\sqrt{f'_m}$$ (5.157)

Steel shear yielding:

$$B_{vns} = 0.6A_b f_y$$ (5.158)

where A_{pv} is one-half the projected area of the right circular cone on the masonry surface, which is used for calculating the shear capacity of anchor bolts, and is given by

$$A_{pv} = \frac{\pi l_{be}^2}{2}$$ (5.159)

where l_{be} is the distance to the edge of the masonry measured in the direction of the applied shear load. Figure 5.54 illustrates the projected shear area. As with the tension pullout cone, A_{pv} must be reduced to account for overlapping areas of closely spaced bolts or for portions of the area that lie outside the masonry.

For Eqs. (5.155) to (5.157), when the shear strength is controlled by the masonry, a capacity reduction factor of 0.5 should be used. For Eq. (5.158), which determines the shear strength based on failure of the steel anchor bolt, the capacity reduction factor is 0.9.

Anchor bolts that are loaded in shear usually are offset from the face of the masonry by the thickness of a ledger or other connecting element. This means that in-plane shear loads often induce tension on anchor bolts that must be considered, as illustrated in Fig. 5.55. The eccentricity in the connection is resisted by a couple formed by the tension in

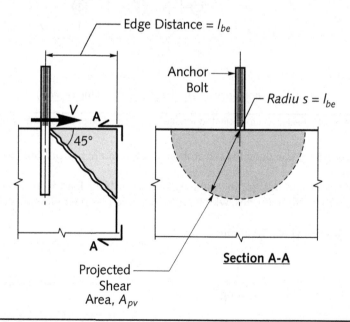

Figure 5.54 Projected shear area for anchor bolts in masonry.

the anchor bolt and bearing of the ledger on the masonry. If the moment arm for the couple is assumed to be five to six times the distance from the bolt centerline to the edge of the ledger, the tension due to offset shear loads is given by

$$T = \frac{6Vt_L}{5d} \tag{5.160}$$

where t_L is the thickness of the ledger, and d is the distance from the bolt centerline to the edge of the ledger.

A linear interaction relationship provides a conservative estimate of the allowable loads on bolts subjected to both shear and tension. Thus, the allowable loads are determined with the following equation:

$$\frac{b_{af}}{\phi B_{an}} + \frac{b_{vf}}{\phi B_{vn}} \le 1 \tag{5.161}$$

where b_{af} and b_{vf} are the factored tension and shear loads, respectively.

When using allowable-stress design procedures, the allowable load on bolts in tension is given by lowest value determined from the following equations:

Masonry tensile breakout:

$$B_{ab} = 1.25 A_{pt} \sqrt{f_m'} \tag{5.162}$$

Steel tensile yield:

$$B_{as} = 0.6 A_b f_y \tag{5.163}$$

Anchor pullout (bent-bar bolts):

$$B_{ap} = 0.6 f_m' e_b d_b + \left[120\pi \left(l_b + e_b + d_b\right)d_b\right] \tag{5.164}$$

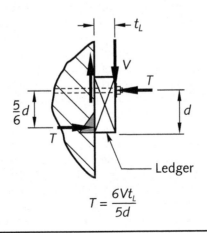

FIGURE 5.55 Eccentricity of shear loads due to ledger.

And the allowable shear on an anchor bolt B_v is the smallest of the values obtained from the following equations:

Masonry shear breakout:

$$B_{vb} = 1.25 A_{pv} \sqrt{f_m'}$$

(5.165)

Masonry shear crushing:

$$B_{vc} = 350 \sqrt[4]{f_m' A_b}$$

(5.166)

Anchor shear pryout:

$$B_{vpry} = 2.0 B_{ab} = 2.5 A_{pv} \sqrt{f_m'}$$

(5.167)

Steel shear yielding:

$$B_{vs} = 0.36 A_b f_y$$

(5.168)

And for combined tension and shear loads on an anchor bolt:

$$\frac{b_a}{B_a} + \frac{b_v}{B_v} \le 1$$

(5.169)

5.7.2 Reinforcement Development and Lap Splices

Reinforced-concrete masonry is designed to behave as a composite material made up of concrete masonry units, grout, and steel reinforcement. For this to occur, the bond between the reinforcement and the grout must be capable of transferring stresses between the two materials. Figure 5.56 shows the stresses in a straight reinforcing bar embedded in masonry. As the figure shows, it can be assumed that the bond stress between the grout and the reinforcement is uniform along its length. Therefore, the force that can be developed is proportional to the depth of embedment or development length of the reinforcing bar. In hooked bars, the development length also determines the force that can be transferred. However, the presence of a hook means that more force can be transferred over a shorter length because bearing stresses also contribute to the resistance of load, as shown in Fig. 5.57. If the development length is not sufficient to transfer the applied forces, reinforcing bars will pull out of the masonry, and this could result in failure of the structure. To prevent this from occurring, an adequate development length, which is the distance over which the steel stresses are transferred to the masonry, must be provided.

Limits on the length of a reinforcing steel bar that can be stored, transported, or constructed efficiently means that reinforcement is rarely placed in structures without splicing. For example, the vertical reinforcement in concrete masonry walls typically does not extend continuously from the foundation to the roof and needs to be spliced at some point along the wall height. Splicing typically is achieved by lapping the reinforcing bars as shown in Fig. 5.58. While welded or mechanical splices also may be used, construction constraints make lap splices the most common form of splice used in concrete masonry. In a lap splice, the force is transferred from one bar to the concrete and then from the concrete to the other bar by using a mechanism similar to that shown

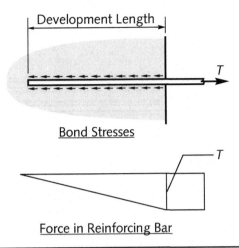

Bond Stresses

Force in Reinforcing Bar

FIGURE 5.56 Bond stresses and development length in straight reinforcing bars.

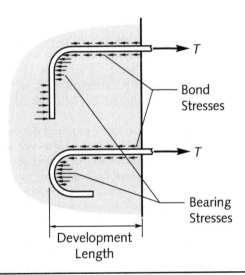

FIGURE 5.57 Development of hooked bars.

in Fig. 5.56. To ensure that a concrete masonry structure performs as designed, adequate lap splice lengths over which steel forces can be transferred from one reinforcing bar to the next must be provided.

The 2009 IBC requirements for lap splice lengths in masonry structures are different depending on whether allowable-stress design or strength design procedures are used. This can be confusing because the TMS 402-08/ACI 530-08/ASCE 5-08 requirements provide identical requirements for the minimum development length for reinforcing bars irrespective of which method is used for design (allowable-stress or strength design). The TMS 402-08/ACI 530-08/ASCE 5-08 stipulates that all straight reinforcing

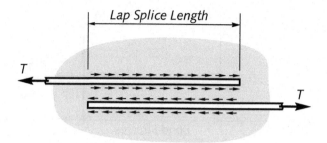

<figure>FIGURE 5.58 Reinforcement lap splice.</figure>

bars must extend from the point of maximum moment in either direction for a distance no less than the development length l_d, which is given by the following equation:

$$l_d = \frac{0.13d_b^2 f_y \gamma}{K\sqrt{f_m'}} \geq 12 \text{ in.} \tag{5.170}$$

where

d_b = diameter of reinforcing bar
f_y = yield stress of reinforcing bar
γ = reinforcement size factor
f_m' = masonry compressive strength
K = least of cover or clear spacing between adjacent bars

Note that there is no upper limit on the required development length based on TMS 402-08/ACI 530-08/ASCE 5-08 requirements. However, for strength design, 2009 IBC Sec. 2108.2 indicates that the computed value of l_d need not exceed $72d_b$. Bar development failures usually occur due to cracks on the surface of the masonry between or adjacent to reinforcement. Thus the factor K is introduced to account for the fact that development length depends on the spacing between bars or the masonry cover. The value of K shall not exceed the smaller of the masonry cover, the clear spacing between bars, or 5 bar diameters.

The reinforcement factor γ takes into consideration the fact that larger bars require relatively longer development lengths. They are outlined in the code as follows:

γ = 1.0 for #3 through #5 bars
γ = 1.3 for #6 through #7 bars
γ = 1.5 for #8 through #11 bars

For epoxy-coated bars, development lengths shall be increased by 50 percent.

The TMS 402-08/ACI 530-08/ASCE 5-08 also requires that the reinforcement extends beyond the point at which it is no longer required a distance equal to the effective depth of the member or $12d_b$, whichever is greater. However, this requirement does not apply at the supports of simple spans and at the free ends of cantilevers.

The lengths of lap splices typically are calculated in a manner similar to the development length. However, the 2009 IBC, which is the basis of design for most jurisdictions, overwrites the TMS 402-08/ACI 530-08/ASCE 5-08 lap-splice provisions and employs the traditional approach in which different formulas are used to calculate the required development lengths for allowable-stress design and strength design. For allowable-stress design, the required splice length is calculated as follows:

$$l_d = 0.002d_b f_s \geq 40d_b \qquad (5.171)$$

where f_s is the tensile or compressive stress in the reinforcing bars. When the tensile stress in the reinforcement is greater than 80 percent of the allowable tensile stress, the lap-splice length must be increased by 50 percent.

The TMS 402-08/ACI 530-08/ASCE 5-08 also stipulates that welded and mechanical splices must be capable of developing 125 percent of the yield strength f_y of the reinforcing bar in tension or compression. Noncontact lap splices should not be spliced further apart that one-fifth the required lap length or 8 in. This means that in a typical concrete masonry wall, bars may be placed in adjacent cells. For strength design, 2009 IBC Sec. 2108.3 further stipulates that the welded splices are not permitted in the plastic hinge zones in special and intermediate reinforced-masonry walls. The welded reinforcement must comply with ASTM A706 requirements. Per the 2009 IBC Sec. 2108.3, the mechanical splices must be classified as Type 1 and Type 2 as defined by ACI 318-08. Type 1 splices may not be used in the zones of plastic hinging in intermediate and special reinforced-masonry walls.

The dimensions for standard 180- and 90-degree hooks are shown in Fig. 5.59. Standard stirrup and tie anchorage hooks require a 90- or 135-degree bend plus an extension of at least 6 bar diameters. For standard hooks in tension, different values are

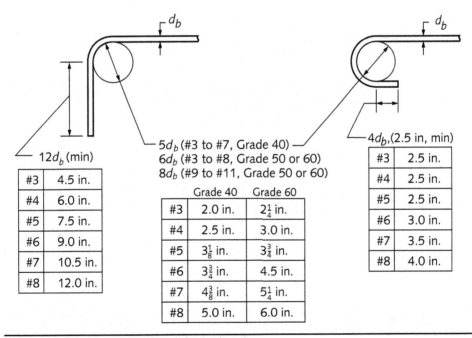

$12d_b$ (min)	
#3	4.5 in.
#4	6.0 in.
#5	7.5 in.
#6	9.0 in.
#7	10.5 in.
#8	12.0 in.

$5d_b$ (#3 to #7, Grade 40)
$6d_b$ (#3 to #8, Grade 50 or 60)
$8d_b$ (#9 to #11, Grade 50 or 60)

	Grade 40	Grade 60
#3	2.0 in.	$2\frac{1}{4}$ in.
#4	2.5 in.	3.0 in.
#5	$3\frac{1}{8}$ in.	$3\frac{3}{4}$ in.
#6	$3\frac{3}{4}$ in.	4.5 in.
#7	$4\frac{3}{8}$ in.	$5\frac{1}{4}$ in.
#8	5.0 in.	6.0 in.

$4d_b$, (2.5 in, min)	
#3	2.5 in.
#4	2.5 in.
#5	2.5 in.
#6	3.0 in.
#7	3.5 in.
#8	4.0 in.

FIGURE 5.59 Standard hooks for reinforcing bars in masonry.

provided for allowable-stress and strength design. With strength design, a standard hook is equivalent to an embedment length l_e of 13 bar diameters. This means that the required development can be reduced by the preceding amount if there is a hook at the end of a bar. When using allowable-stress design, a standard hook is equivalent to an embedment length l_e of 11.25 bar diameters. This means that the required development can be reduced by the preceding amount if there is a hook at the end of a bar. The effect of hooks must be neglected for bars in compression.

At the end of special reinforced-masonry shear walls, horizontal reinforcement used to resist shear must be bent around vertical reinforcement with a 180-degree hook. At wall intersections, horizontal shear reinforcement must be bent around the edge vertical reinforcing with a 90-degree standard hook and extend horizontally into the intersecting wall for a length equal to or greater than the development length l_d. Figure 5.60 illustrates the required anchorage of horizontal reinforcement at wall ends and intersections.

5.7.3 Examples

Example 5.7.1

Determine whether the ¾-in. diameter anchor bolts can resist the combined tension and shear loads on the wall connected to a flexible diaphragm, as shown in Fig. 5.61. The bolts have an effective yield stress of 36 ksi, an effective embedment length of 6 in., and are spaced 7 in. on each side of the roof framing. The projected extension of the bent bar is 4 in., and the project's quality-assurance plan stipulates that the shanks of bent-bar bolts are free of debris, oil, and grease when the bolts are installed. The masonry compressive strength is 1,500 psi, and the weight of the wall is 78 psf. $S_{DS} = 1.2g$.

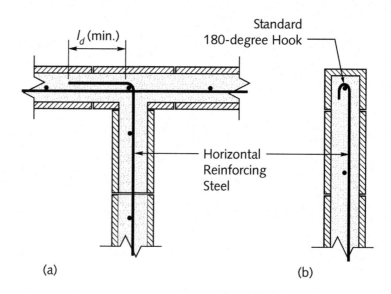

Figure 5.60 Anchorage of wall reinforcement at (a) wall intersections; (b) wall ends.

Strength Design Solution
Consider only the load combination $1.2D + 1.0E$, which governs by observation. The out-of-plane anchorage force for a wall connected to a flexible diaphragm is given by

$$F_p = 0.8S_{DS}I_E W_p$$

Assuming that $I_E = 1.0$,

$$F_p = 0.8S_{DS}I_E W_p = 0.8(1.2)(1.0)W_p = 0.96W_p$$

The tributary wall weight is equal to

$$W_p = 78\left(\frac{20}{2} + 3.5\right) = 1{,}053 \text{ lb/ft}$$

Therefore, the tension on the connection is given by

$$F_p = 0.96(1{,}053) = 1{,}011 \text{ lb/ft}$$

Note that the anchorage force must be greater than the following:

$$F_p > 400S_{DS}I_E = 400(1.2)(1.0) = 480 \text{ lb/ft} \qquad \textit{Okay!}$$

$$F_p > 280 \text{ lb/ft} \qquad \textit{Okay!}$$

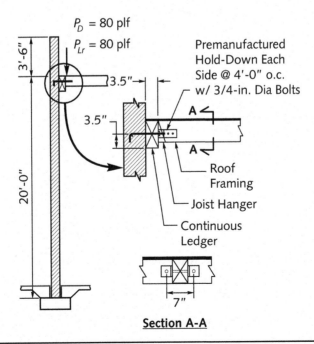

Section A-A

FIGURE 5.61 Anchor bolts in masonry wall.

The shear on the connection is equal to

$$(1.2 + 0.2S_{DS})P_D = (1.2 + 0.24)80 = 115 \text{ lb/ft}$$

And since the shear is offset from the face of the masonry, the tension due to the shear is given by

$$T_D = T_{Lr} = \frac{6Vt_L}{5d} = \frac{6(115)3.5}{5(3.5)} = 138 \text{ lb/ft}$$

Thus the total shear and tension at the connection are given by

$$b_{af} = F_p + T_D = 1{,}011 + 138 = 1{,}149 \text{ lb/ft}$$
$$b_{vf} = 115 \text{ lb/ft}$$

Calculate the tensile strength of the bolts using the smaller of Eqs. (5.149) to (5.151). For each bolt,

$$A_{pt} = \pi l_b^2 = \pi(6)^2 = 113 \text{ in.}^2$$

However, since $s < l_b$, there is overlap of the projected cone area of adjacent bolts. From Eq. (5.154),

$$\theta = 2\cos^{-1}\left(\frac{s}{2l_b}\right) = 2\cos^{-1}\left(\frac{7}{2 \times 6}\right) = 1.9 \text{ radians}$$

And the modified projected area is obtained from Eq. (5.153):

$$A_{pt}' = \pi l_b^2 - \frac{1}{2}l_b^2(\theta - \sin\theta)$$
$$= 113 - \frac{1}{2}(6)^2(1.9 - \sin 1.9) = 96 \text{ in.}^2$$

The allowable tension on a bolt due to masonry breakout thus is given by Eq. (5.149):

$$\phi B_{anb} = \phi 4 A_{pt}' \sqrt{f_m'} \qquad \leftarrow \text{Governs}$$

From Eq. (5.150), the allowable load due to yielding of each anchor bolt is equal to

$$\phi B_{ans} = \phi A_b f_y$$
$$= 0.9(0.44)36{,}000 = 14{,}256 \text{ lb}$$

And from Eq. (5.151), the allowable tension on a bolt due to anchor pullout is given by

$$\phi B_{an} = \phi\left[1.5 f_m' e_b d_b + 300\pi(l_b + e_b + d_b)d_b\right]$$
$$= 0.65\{1.5(1{,}500)(4)(0.75) + \left[300\pi(6 + 4 + 0.75)0.75\right]\}$$
$$= 9{,}327 \text{ lb}$$

For the shear capacity, we use Eqs. (5.156), (5.157), and (5.158), ignoring the masonry shear breakout failure mode because the edge distance is large:

Masonry shear crushing:

$$\phi B_{vnc} = \phi(1,050)\sqrt[4]{f'_m A_b}$$
$$= 0.5(1,050)\sqrt[4]{(1,500)(0.44)}$$

$$= 2,661 \text{ lb} \qquad \leftarrow \text{Governs}$$

Anchor shear pryout:

$$\phi B_{vpry} = \phi 8 A_{pt} \sqrt{f'_m}$$
$$= 0.5(8)(96)\sqrt{1,500} = 14,872 \text{ lb}$$

Steel shear yielding:

$$\phi B_{vns} = \phi 0.6 A_b f_y$$
$$= 0.9(0.6)(0.44)(36,000) = 8,554 \text{ lb}$$

For two bolts spaced at 48 in. on the center and using the smallest of the values for tension and shear, the interaction equation for the load combination $1.2D + E$ is given by

$$\frac{b_{af}}{\phi B_{an}} + \frac{b_{vf}}{\phi B_{vn}} = \frac{1,026(4)}{2(7,436)} + \frac{115(4)}{2(2,661)} = 0.36 \leq 1.0 \qquad\qquad Okay!$$

Allowable-Stress Design Solution

For the load combination $D + L_r$, the axial and shear loads are

$$b_a = T_D + T_{Lr} = 96 + 96 = 192 \text{ lb/ft}$$

$$b_v = P_D + P_{Lr} = 80 + 80 = 160 \text{ lb/ft}$$

For the load combination $D + E/1.4$, the vertical component of the gravity load is given by

$$\frac{0.2 S_{DS}}{1.4} D = \frac{0.2(1.2)}{1.4} D = 0.17D$$

Then, the axial and shear loads are as follows:

$$b_a = \frac{F_p}{1.4} + 1.17 T_D = \frac{1,011}{1.4} + 1.17(96) = 1,125 \text{ lb/ft}$$

$$b_v = 1.17 P_D = 1.17(80) = 94 \text{ lb/ft}$$

From the strength design solution,

$$A_{pt}' = \pi l_b^2 - \frac{1}{2} l_b^2 (\theta - \sin\theta)$$

$$= 113 - \frac{1}{2}(6)^2(1.9 - \sin 1.9) = 96 \text{ in.}^2$$

The allowable tension in each bolt is given by the following equations:

Masonry tensile breakout:

$$B_{ab} = 1.25A_{pt}\sqrt{f_m'}$$

$$= 1.25(96)\sqrt{1,500} = 4,648 \text{ lb} \qquad \leftarrow \text{Governs}$$

Steel tensile yield:

$$B_{as} = 0.6A_b f_y$$

$$= 0.6(0.44)36,000 = 9,504 \text{ lb}$$

Anchor pullout (bent-bar pullout):

$$B_{ap} = 0.6f_m' e_b d_b + \left[120\pi \left(l_b + e_b + d_b\right)d_b\right]$$

$$= 0.6(1,500)(4)(0.75) + \left[120\pi (6+4+0.75)0.75\right] = 6,753 \text{ lb}$$

For the shear capacity,

Masonry shear crushing:

$$B_{vc} = 350\sqrt[4]{f_m' A_b}$$

$$= 350\sqrt[4]{(1,500)(0.44)} = 1,774 \text{ lb} \qquad \leftarrow \text{Governs}$$

Anchor shear pryout:

$$B_{vpry} = 2.5A_{pt}\sqrt{f_m'}$$

$$= 2.5(96)\sqrt{1,500} = 9,295 \text{ lb}$$

Steel shear yielding:

$$B_{vs} = 0.36A_b f_y$$

$$= 0.36(0.44)(36,000) = 5,702 \text{ lb}$$

Using the smallest of the values in tension and shear, the interaction equation for the load combination $D + L_r$ is given by

$$\frac{b_a}{B_a} + \frac{b_v}{B_v} = \frac{4(192)}{2(4,648)} + \frac{4(160)}{2(1,774)} = 0.26 \leq 1.0 \qquad Okay!$$

And for the load combination $D + E/1.4$,

$$\frac{b_a}{B_a} + \frac{b_v}{B_v} = \frac{4(1,125)}{2(4,648)} + \frac{4(94)}{2(1,774)} = 0.59 \leq 1.0 \qquad Okay!$$

References

1. International Code Council (ICC). 2009. *International Building Code* (2009 IBC). ICC, Washington, DC.

2. Masonry Standards Joint Committee (MSJC). 2008. *Building Code Requirements for Masonry Structures* (TMS 402-08/ACI 530-08/ASCE 5-08). TMS, Boulder, CO; ACI, Farmington Hills, MI; and ASCE, Reston, VA.

3. Structural Engineering Institute (SEI) of the American Society of Civil Engineers (ASCE). 2006. *Minimum Design Loads for Buildings and Other Structures, Including Supplements Nos. 1 and 2* (ASCE/SEI 7-05). ASCE, Reston, VA.

4. American Concrete Institute (ACI). *Building Code Requirements for Structural Concrete* (ACI 318-08) *and Commentary*. ACI, Farmington Hills, MI.

5. Structural Engineers of Southern California (SEAOSC). 1982. Test Report on Slender Walls. SCCACI-SEAOSC Task Committee on Slender Walls, Los Angeles, CA.

6. American Society for Testing and Materials (ASTM). 2010. *Standard Test Methods for Strength of Anchors in Concrete Elements* (ASTM E488). ASTM, Philadelphia, PA.

CHAPTER 6

Design of Structural Timber

6.1 Introduction

Timber has been one of the principal building materials used by humans since we began to build primitive structures. Timber continues to be one of the main materials used for residential and commercial buildings in many parts of the world today. This chapter covers many of the design considerations that a practicing designer will have to face during the design of typical light-frame timber buildings. The chapter does not attempt to address the issues of glued-laminated construction (glulam), nor does it try to rehash the detailed basic design concepts of how to design a simple beam or column. These concepts can be found in many of the typical textbooks used for introductory classes in timber design.[6] Rather, this chapter tries to address many of the more advanced issues of design of light-frame timber buildings that designers are faced with every day. Some of the issues do not have consensus as to how they should be addressed by the designer and therefore require engineering judgment on the designer's part, whereas other issues have been discussed in sufficient forums to form a consensus as to how the design should be handled. These issues will be covered with options for the designer to consider when addressing the solution. Finally, some of the less frequent details or restrictions imposed by the 2005 American Forest and Paper Association (AF&PA) *National Design Specification for Engineered Wood Construction with 2005 Supplement* (ANSI/AF&PA NDS-2005)[1] are not included in this chapter, and the designer is referred to either the commentary of the AF&PA, *National Design Specification for Engineered Wood Construction—Commentary,*[1a] or other reputable texts[6] for discussion.

The basis of design for this chapter will be the ANSI/AF&PA NDS-2005 and the 2008 AF&PA *Special Design Provisions for Wind and Seismic* (ANSI/AF&PA SDPWS-08).[2] Additional references to various trade publications (design aides), research reports, and student graduate theses will be made to provide clarification or justification for adjustments that influence the design process. In general, promotion of proprietary products will not be done, other than to state that classes of products are available to the designer as a potential solution should traditional solutions fail to address the design issue adequately.

Timber is still one of the traditional construction materials with high specific strength (strength/weight). It is also a material that is classified as a "green" material because it requires a relatively low amount of energy to produce, is renewable, and actually sequesters carbon. These characteristics continue to make timber design as one that is attractive to many designers.

As with all construction materials, timber has its own weaknesses, and one of the design objectives is to use the strengths of the material while minimizing the effects of the weaknesses. In this chapter, the word *timber* is used to describe the material instead of the term *wood*. The difference is that *timber* is the form of the material used in structures and includes the growth characteristics common to the material (i.e., knots, wane, slope of grain, and general orthotropic mechanical properties). *Wood* is the term used to describe the form where everything is perfect, with straight grain, no knots, and so on.

While timber does have an anisotropic nature, the design process has been simplified by developing all the mechanical properties and strengths with the assumption that the material is homogeneous or at least having only two principle axes for the properties (i.e., parallel to grain and perpendicular to grain). By developing the design properties with this assumption, if the design process also makes the same assumptions, the errors are self-correcting. Therefore, designers do not have to worry about how the material is formed; they only have to worry about whether the stresses being designed for are oriented parallel or perpendicular to the grain direction of the member being designed. Usually this translates into whether the stresses are oriented parallel or perpendicular to the principal axis of the member being designed.

6.2 Basic Design Philosophy

Timber design has adopted a general philosophy through various codes and standards committees' actions. Though some may interpret the decisions differently, an interpretation is that for all applied loads (i.e., wind, snow, live, etc.), the structural system should remain completely elastic, and for seismic loading, the system will be forced into the inelastic range, but the system should not collapse.

Since timber is for the most part a brittle material, in that when it fails in tension, bending, and so on, it fails in a rather catastrophic manner, the designer must achieve any ductility from the system through conscientious design of the connections. An analogy might be that the design is using glass members connected with ductile connections. Therefore, where appropriate, the design process will differentiate between design objectives for seismic design versus all other loadings.

Timber buildings have a history of good performance in high-wind events and earthquakes. This is due to the highly redundant load path available in light-frame construction, high specific strength, and high energy dissipation associated with the distributed yielding that goes on in the connections. Most structures are assumed to have damping coefficients on the order of 5 to 10 percent of critical damping. Wood light-frame construction has damping coefficients on the order of 30 to 40 percent of critical damping, which is attributable to the yielding of nails and friction between the various layers of the structural system.

For wind loading, the main objective is to ensure that the load path is complete and strong enough to resist the applied loading. For seismic design, the objective is a bit more complex, but basically it is to ensure that the objective for wind is achieved as well as providing high ductility. A better way of looking at the additional design objective might be to provide sufficient displacement capacity to the system by using connections that will yield the fastener and not the timber member itself.

Since timber and wood are viscoelastic materials, designers need to consider the duration that the load is applied and be aware that the material will relax and/or creep with time. When timber is loaded, its strength depends on the length of time the load is imposed. The longer the duration of the load, the weaker the material is. Therefore, ANSI/AF&PA NDS-2005 has included a duration-of-load factor [time-effect factor for

Type of Shortest Load	Assumed Cumulative Duration over 50 Years	Value of C_D
Dead load	Permanent	0.9
Occupancy live loads	10 years	1.0
Snow load	2 months	1.15
Construction load	7 days	1.25
Wind or seismic load	10 minutes	1.6
Impact load	Impact	2.0

TABLE 6.1   Load-Duration Factor Values for ASD

load and resistance factor design (LRFD)] that is based on tests that were conducted on small, clear wood specimens during the 1940s by Wood[27] and later by Barrett and Foschi[5] on full-size lumber. Table 6.1 presents the values used for the load duration factor C_D for all allowable stress design (ASD) applications. The duration used is the duration of the shortest duration of any of the loads included in the load combination used. Values of the time effect factor λ for the LRFD are given in Appendix N of the ANSI/AF&PA NDS-2005. The C_D and λ are not applied to either the modulus of elasticity or compression perpendicular to grain design values.

A final design philosophy issue that has been imposed by committee decisions made for the ANSI/AF&PA NDS-2005 is that when the LRFD methodology was introduced into the ANSI/AF&PA NDS-2005, the committees involved made a conscientious decision to impose a restriction on the method. The restriction was that the design of any member or connection using the LRFD method could not result in more than a 10 percent change in the final design. This decision essentially has made the two design methodologies equivalent, and therefore, this chapter does not distinguish between them. In some instances, one or the other design methodology can provide some advantage in span or similar aspect, but the differences are not large. The ANSI/AF&PA NDS-2005 has the two methodologies side by side in the recent editions and distinguishes between the two methods with a format conversion factor and having different bases for the time effect of load-duration factor (10-minute duration for LRFD and 10-year duration for ASD).

6.2.1   Basic Framing and Load Path

Timber light-frame construction has a redundant gravity load path that is illustrated in Fig. 6.1. The figure shows that the gravity loads are resisted by a load path where the load is applied to either a truss or beam that transfers the load to the framing of the walls (i.e., top plate, studs, and bottom plate). Then the load is transferred through a platform (floor system) and on into the wall framing of the next story below until it reaches the foundation. While this looks like a simple load path for which simple statics are used to determine the loading in any given member, it is in fact quite a complicated structural system that performs more like a system of beams on elastic foundations than the rigid-body monolithic members often used in the design of heavier, less redundant structural systems such as concrete or masonry.

If one considers the gravity load path in more detail, it becomes obvious that there are significant system effects; however, it becomes more difficult to move loads

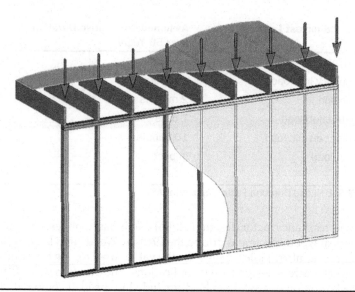

Figure 6.1   Gravity load path in light-framed wood buildings.

horizontally over great distances. Consider a simple point load applied to the second-story floor in Fig. 6.1. The floor beam will act as a simple beam that is part of a series of parallel beams that are sheathed with wood structural panels. This results in a system effect in that the beams on either side of the loaded beam will assist in carrying the load, but the number of beams that will share the load is limited, and research by Li,[18] Stiess,[24] and Stark,[23] has shown that with typical sheathing thicknesses, the number of beams sharing the load is about two on each side, with the amount of sharing decreasing quickly as the distance from the load increases. However, even if this load sharing is effective, the beam to which the load is applied carries most of the load.

The load then is transferred directly to the wall framing through bearing on the top plate of the wall or is transferred to a rim joist through a joist hanger, and the rim joist then transfers the load to the top plate of the wall. The top plate (or rim-joist/top-plate combination) then acts like a beam on an elastic foundation, where the beam spreads the load over a few studs, and the studs act as the foundation for the beam. This is the same action as found in railroad tracks or grade beams and is classified as a *beam on elastic foundation*. Hetényi[10] derived the governing equations for these types of problems in the 1940s, and when applied to timber light-frame construction with 2×12-in. nominal rim joists and 2×4-in. nominal studs, the decay length for a point load (i.e., a floor joist) is approximately two stud spacings. Since in reality all the floor joists would be loaded, the combined loading results in the load from each floor joist effectively being transferred to the nearest stud, which then transfers the load down to the next-lower level. These loads accumulate as the number of floors contributing increases.

For lateral loads (wind and seismic), the load path becomes a bit more complicated and is simplified for illustration in Fig. 6.2. First, the out-of-plane (horizontal) forces on the roof and walls (Fig. 6.3) are transferred to the horizontal diaphragms through simple beam behavior. This means that one-half of the load on the wall is transferred up to the diaphragm above and one-half of the load is transferred down to the diaphragm below. This load then is transferred through deep beam behavior of the diaphragm to the shear

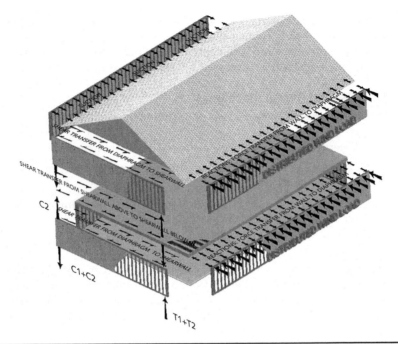

FIGURE 6.2   Lateral load path in timber light-framed construction.

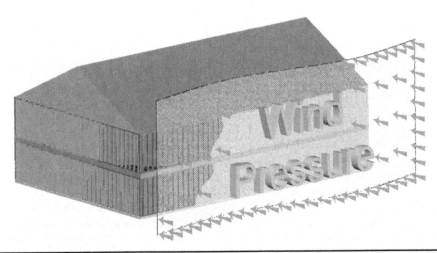

FIGURE 6.3   Lateral load on building illustrating wall studs acting as simple beams to distribute the load to the upper as well as to the lower diaphragms.

walls that are oriented parallel to the loading direction and are supporting the diaphragm. The shear walls then transfer the load to the diaphragm below through racking and overturning action.

Gravity loads to some extent can resist the overturning forces induced by the lateral loads. However, how effective the gravity loads are in countering the uplift forces is a controversial subject, and a consensus has not been reached as to how much gravity load can be activated to counter the uplift force on the end stud of a wall segment, and the designer must use some judgment as to how this will be accounted for in the design. This topic of design will be discussed further in the section on shear-wall design.

6.2.2   General Design Mechanics

Design of timber structures consists of the following elements:

1. Determining the loads applied to the member or connection of interest

2. Determining the characteristic strength parameter for the species, grade, and action being checked

3. Determining the environmental and end-use adjustment factors appropriate for the action being considered

4. Determining the minimum size required to satisfy the design, $F_{applied} \leq F_{resistance}$.

Since wood is a natural composite, it is affected by the moisture content and temperature of the surrounding environment. For this reason, the characteristic design values need to be factored in, to account for the effects of the environment. The general format used by the ANSI/AF&PA NDS-2005 to accomplish this takes on the form Eq. (6.1) for ASD or Eq. (6.2) for LRFD:

$$F' = F^* C_D C_m C_T \ldots \tag{6.1}$$

$$F' = \varphi^* F^* K_F^* \lambda^* C_m C_T \ldots \tag{6.2}$$

where

F' = adjusted design value

F = characteristic or referenced design value of the property

C_i = adjustment factors to adjust the characteristic value to the allowable design value to account for environmental conditions other than the standard conditions

The adjustments are outlined in the appropriate sections of the ANSI/AF&PA NDS-2005 and AF&PA *National Design Specification Supplement – Design Values for Wood Construction* (NDS Supplement),[1b] but for design of most structures, most adjustment factors have a value of 1.0. The exceptions are C_D for ASD (or λ for LRFD), C_M, and C_t. Other end-use adjustment factors are discussed in the sections to which they pertain.

The C_D is the ASD load-duration factor. Timber is a natural composite that is viscoelastic. Therefore, two things occur. Wood is weaker the longer it is required to resist a given load, and creep occurs. C_D is the adjustment used to account for this behavior when the design is being completed in ASD format. It has a value of 1.0 for "normal" duration loads such as floor live load. For loads in load combinations that include loads of shorter duration (such as snow, construction, wind, and seismic loads), the parameter has a value greater than 1.0. For loads with expected accumulated duration over a 50-year window greater than 10 years (normal live-load duration), the parameter has a value of less than 1. For instance, C_D has a value of 0.9 for permanent loads such as dead loads. It is important to note that the shortest-duration load in any given load combination dictates the value of C_D.

C_M is the moisture content adjustment factor. If the structure is to be constructed with the lumber having a moisture content of less than 19 percent (typical for kiln-dried lumber), and the structure will remain below 19 percent moisture content (typical for most buildings that are protected from the weather by the outside finish envelop), C_M has a value of 1.0. If the lumber has a moisture content greater than 19 percent at any time during construction or use, the value of C_M is less than 1.0. This is due to the fact that timber is weaker when it is moist.

C_t is the temperature adjustment factor and is used to adjust the referenced design value to account for the fact that wood is weaker as it is exposed to prolonged elevated temperatures. Normal daily fluctuations in temperature usually are not considered to affect the performance, but extended exposure above 100°F (38°C) will reduce the strength and stiffness of timber. The values to be used for elevated temperatures are given in the ANSI/AF&PA NDS-2005 for each type of design (i.e., timber members or connections).

The following are LRFD-specific adjustment factors:

K_F = the LRDF format adjustment factor, which converts the referenced design value from ASD format to LRFD format. The values change depending on which design property is being used. The values are presented in ANSI/AF&PA NDS-2005, App. N, Table N1.

ϕ = the resistance factor. It changes the value for each property being checked. The values for ϕ are presented in ANSI/AF&PA NDS-2005, App. N, Table N2.

λ = the time-effect factor. It accomplishes the same adjustment as C_D, only for when the design is using the LRFD format. λ has a different basis of time than C_D in that LRFD is calibrated to a 10-minute duration (wind and seismic loading), whereas ASD is calibrated to a 10-year duration (normal live loads). Therefore, λ has a value of 1.0 for wind and seismic loads, and all other load conditions other than impact loads have values less than 1.0. It should be noted that ANSI/AF&PA NDS-2005, App. N.3.3, Table N3, provides the values for λ that are associated with each of the American Society of Civil Engineers (ASCE) *Minimum Design Loads for Buildings and Other Structures, Including Supplements Nos. 1 and 2* (ASCE/SEI 7-05)[21] load combinations.

6.3 Flexural Members

Beams are the main member for transferring loads horizontally in the structure. In timber, the design of rectangular members, whether they are sawn lumber or glued-laminated beams (glulam) in straight or slightly curved (such as cambered beams) sections are designed essentially the same. The slight differences in glulam design will be identified when appropriate.

In essence, beam design entails four checks on performance:

1. Bending strength

2. Shear strength

3. Deflection

4. Bearing (both on the beam of interest and on the material on which the beam is bearing)

Due to the typical spans over which timber beams are used, deflection is most likely to govern the design, followed by the bending strength and bearing strength. Shear failures are rare, and unless the application is either a short, heavily loaded beam or a heavily loaded beam with multiple spans, shear strength will not govern the design.

6.3.1 Deflection

Since deflection governs many of the designs of timber beams, it is often efficient to check this behavior first. Determination of the expected deflection of a beam is based on standard engineering mechanics analysis. Since most beams are designed for a uniformly distributed load, the deflection can be determined easily using the Eq. (6.3):

$$\Delta = \frac{5wL^4}{384E'I} \leq \Delta_{allowable} \tag{6.3}$$

where

Δ = calculated maximum deflection

$\Delta_{allowable}$ = allowable deflection criteria for the design

w = uniformly distributed load

L = length of the beam

E' = adjusted modulus of elasticity

I = moment of inertia

6.3.2 Bending Strength

For simple bending problems, Eq. (6.4) governs the design:

$$f_b = \frac{Mc}{I} = \frac{M}{S} \leq F_b' \tag{6.4}$$

where

f_b = actual bending stress

M = maximum applied moment on the beam

c = distance from the neutral axis to the extreme fiber position on the beam

I = moment of inertia for the beam

S = section modulus of the beam (I/c)

F_b' = adjusted bending strength of the beam for the species, grade, and end use conditions of the application

Adjustment Factors

Since lumber is graded with several assumptions about the application and end-use conditions, the nominal strength or other mechanical properties need to be adjusted to account for differences in the grading assumptions and final application. Therefore, it is in the best interest if the designer is aware of the assumptions that have a direct effect on the end design values. These assumptions and the associated adjustments include

- The adjustments for duration of load, moisture content, temperature, and LRFD format changes were discussed earlier.

- When the member is placed in bending, it will be oriented such that the bending will be about the strong axis. This says that if the application were to use the board on flat, then an adjustment of $C_{fu} = 1.0$ to 1.2 can be made to increase the bending strength. This is so because the location of knots in the member with respect to the neutral axis is one of the grading criteria, and if the bending is changed to the other axis, the knot located near the edge of the board has less effect on strength.

- In most applications, timber beams are placed such that simple beam assumptions are applicable, the tops of the beams are all on one plane, and the loads are applied such that the tops of the beams are placed in compression. This allows sheathing to be attached to all the beams and acts as bracing for lateral torsional response. However, when the beams are not placed such that sheathing can provide continuous bracing for the compression flange, the beam stability factor C_L is used to reduce the bending strength accordingly. Section 3.3.3 of ANSI/AF&PA NDS-2005 provides the equation necessary to calculate the C_L adjustment factor.

- Lumber grading rules are based on a member that has a cross section of 2 × 12 in. nominal (1.5 × 11.5 in. actual). In addition, the allowable knot size for any given cross section is based on the relative size of the knot compared with the overall cross section. However, what really causes the weakening effect of a knot is not so much the knot itself, but the fact that the grain of the wood curves around the knot. This results in the bending stresses having a component oriented perpendicular to grain, which is the weakest direction for lumber to be stressed. The larger the knot, the larger is this effect. Therefore, smaller cross sections of the same species and grade are stronger than the larger pieces because the allowable knot size is smaller, and therefore, the amount of grain deviation is smaller. Therefore, when the beam cross section is smaller than 12 in. nominal, an adjustment of $C_F = 0.9$ to 1.5 may be used.

- If the beam is preservative-treated to prevent rot or insect attack, some species require incising in order to gain penetration of the chemicals used. The process of incising is essentially punching holes into the sides of the piece of lumber before the pressure treatment is applied. This incising process damages the physical structure of the wood and results in an adjustment of $C_i = 0.8$ on the bending strength.

- When joists, studs, and other repetitive members with 2 to 4 in. in nominal thickness are used with at least three members in parallel, with a spacing of no more than 24 in., and have a load-distribution element (i.e., sheathing or mechanical fasteners) that forces such members to deflect together, then the bending strength can be adjusted to account for the system effects and gain additional design strength. The adjustment $C_r = 1.15$.

Glulam beams have three adjustment factor changes from lumber to account for small differences in the products. To account for a volume effect due to finger joints used to produce long laminations and other growth and manufacturing characteristics that affect strength, the adjustment C_V is used. In essence, the adjustment accounts for the fact that the larger the beam, the more strength-affecting characteristics are present. Therefore, the adjustment factor C_V is determined using the Eq. (6.5):

$$C_V = \left(\frac{21}{L}\right)^{1/x} \left(\frac{12}{d}\right)^{1/x} \left(\frac{5.125}{b}\right)^{1/x} \leq 1.0 \qquad (6.5)$$

where

L = length of the bending member between inflection points (ft)

D = depth of bending member (in.)

b = width of the widest piece used in the layup (usually the width of the beam) (in.)

x = 20 for southern pine and 10 for all other species

Second, when adjusting the design values for glulam, the designer should be aware of the fact that the moisture content threshold at which the material is considered wet is 16 percent rather than 19 percent. Finally, curvature factor C_c, characteristic of glulam members, takes into account the stresses developed in individual laminations due to curving that remains after gluing such laminations together into a composite member. The adjustment factor C_c is computed using Eq. (6.5a):

$$C_c = 1 - 200(t/R)^2 \qquad (6.5a)$$

where

t = thickness of lamination (in.)

R = radius of curvature of inside lamination face (in.)

$t/R \leq 1/100$ for hardwoods and southern pine

$t/R \leq 1/125$ for other softwoods

6.3.3 Shear Strength

Wood is relatively weak in shear parallel to grain, but it cannot fail in shear perpendicular to grain because it will fail in compression perpendicular to grain before the shear failure can occur in this direction. Therefore, designers need to concern themselves only with checking the shear parallel to grain when designing beams. Since most timber beams are rectangular in cross section, the equation used for this check simplifies to the form:

$$f_v = \frac{VQ}{Ib} = \frac{3V}{2A} \leq F_v' \qquad (6.6)$$

where

f_v = calculated shear stress

F_v' = adjusted design shear stress

V = maximum shear force on the beam

Q = calculated statical moment of area

I = moment of inertia for the cross section

B = width of the beam

A = cross-sectional area

The maximum bending shear stress occurs at the neutral axis of the beam at the cross section with the highest shear force, and the only adjustments that affect the design shear stress are the load duration, wet service factor, and temperature factors.

6.3.4 Bearing

The last check that a designer needs to do is to make sure that the beam being designed and the member supporting the beam can resist contact stresses without crushing. This is accomplished with Eq. (6.7):

$$f_{c\perp} = \frac{P}{A} \leq F'_{c\perp} \qquad (6.7)$$

where

$f_{c\perp}$ = calculated compression stress on the contact area between the beam and the supporting member

$f_{c\perp}$ = adjusted design value for compression perpendicular to the grain for either the beam of interest or the supporting member, if applicable

P = reaction force of the beam

A = contact area of the beam at the reaction

There are three adjustment factors that can be used with the bearing stress: the wet service, the temperature, and the bearing-area factors. The *bearing-area factor* adjusts for the actual bearing area being affected by the bending of the wood fibers. This effectively increases the bearing length by ⅜ in., and since the bearing-area factor is always greater than 1.0, it is conservative to ignore it in design. Equation (6.8) is used to determine the bearing-area factor C_b:

$$C_b = \frac{l_b + 0.375}{l_b} \qquad (6.8)$$

where C_b = bearing-area factor

l_b = actual contact length of bearing

When a design has critical deflection criteria, the designer might consider limiting the compression perpendicular to the grain strength to one-half the reference design deformation value. When the deformation is limited to 0.02 in., the compression perpendicular to the grain strength is predicted by Eq. (6.9):

$$F_{c\perp0.02} = 0.73 \, F_{c\perp} \qquad (6.9)$$

6.4 Tension Members

Tension members occur in several locations in a light-frame building and are the easiest member to design. Tension members occur in timber structures in such places as trusses, end posts in shear walls resisting uplift, and possibly permanent bracing for truss systems. The design process essentially is to ensure that the net section of the member is sufficient to resist the stresses imposed. Equation (6.10) is used for tension member design:

$$f_t = \frac{P}{A} \leq F'_t \qquad (6.10)$$

where

f_t = calculated tension stress due to the loads

F'_t = adjusted design value for tension parallel to the grain

P = applied load

A = net area of the cross section of the member (i.e., the gross area minus holes or notches for connections, etc.)

6.5 Compression Members

Compression members include columns and members with combined flexure and compression (beam-columns). Historically, the design of columns followed a process of classifying the column as short, intermediate, or slender and then using one of three equations to size the column. Short columns were designed to prevent crushing parallel to the grain, and long columns were designed to prevent Euler buckling from occurring. This changed to directly designing the column with an adjustment factor C_p that consists of a single equation that covers the entire range of columns.

The initial step is to determine the column effective length l_e, which is the length adjusted to account for end fixity. Since wood connections are difficult to make "rigid" due to over drilling of the bolt holes and the crushing of the wood around the connection, ANSI/AF&PA NDS-2005 provides design guidance on adjusting the length of the column for end conditions using an adjustment factor K_e in App. G. Therefore, the effective length is defined as $l_e = K_e l$, where l and l_e are the actual and effective lengths for the column, and K_e is the length adjustment factor. The designer also has the option to determine the effective length through alternative methods that are based on engineering mechanics.

The slenderness ratio for the column now can be determined as the larger of l_e/d_1 and l_e/d_2, where d_1 and d_2 are the actual cross-sectional dimensions of the column. The slenderness ratio of the column cannot exceed 50 for the application, with an exception that it does not exceed 75 during construction. The column stability factor then can be determined using Eq. (6.11):

$$C_p = \frac{1 + \left(\dfrac{F_{cE}}{F^*_c}\right)}{2c} - \sqrt{\left[\frac{1 + \left(\dfrac{F_{cE}}{F^*_c}\right)}{2c}\right]^2 - \frac{\dfrac{F_{cE}}{F^*_c}}{c}} \qquad (6.11)$$

where

F^*_c = reference compression design value parallel to the grain adjusted for all the end-use conditions except for the column stability factor C_p

$$F_{cE} = \frac{0.0822 E'_{min}}{\left(l_e/d\right)^2}$$

$c = 0.8$ for sawn lumber, 0.85 for round timber poles and piles, and 0.9 for structural glulam and structural composite lumber

The column then can be designed following Eq. (6.12):

$$f_c = \frac{P}{A} \le F_c^* C_P = F_c' \tag{6.12}$$

where f_c is the actual compression stress due to the loads, and F_c' is the adjusted design value.

For example, consider a column that is 30 ft long that supports a warehouse roof, where the loading consists of an 18,000-lb roof live load and a 14,000-lb dead load. The controlling allowable stress design (ASD) load combination is $D + L_r$, or the axial load for design would be 32 kips. The applied stress would be $f_c = 32,000/A$. The unbraced length of the column is 30 ft. Referencing Table 5B of the NDS Supplement[1b] stressed primarily in compression for an identification number 3 (Douglas Fir), the following mechanical properties are identified:

$E = 1.9 \times 10^6$ psi

$E_{min} = 0.98 \times 10^6$ psi

$F_c = 2,300$ psi for four or more laminations (i.e., ≥ 6.0 cross-sectional dimensions)

$F_c = 1,850$ psi for less than four laminations (i.e., ≤ 6.0 cross-sectional dimensions)

Since roof live load is considered a construction load, use the load-duration factor $C_D = 1.25$. The column would have been manufactured with a moisture content of less than 16 percent, and it will be assumed that it will be kept at less than 16 percent during service; therefore, the moisture adjustment factor $C_M = 1.0$. A warehouse would not have elevated temperatures, so the temperature adjustment factor $C_t = 1.0$. Finally, the column stability factor must be determined:

$$F_c^* = 2,300(1.25) = 2,875 \text{ psi}$$

$l_e = 360(1.0) = 360$ in. (See App. G of ANSI/AF&PA NDS-2005 for effective length factors for pinned base and pinned top. It is difficult to provide fixity in timber construction.)

$$F_{CE} = \frac{0.0822 E'_{min}}{\left(l_e/d\right)^2} = \frac{(0.822)(980000)}{\left(360/d\right)^2}$$

Assuming that a square cross section will be used because it is located in a general interior location for the warehouse, a trial-and-error solution provides the answer that a column with a cross section of 8.75 × 9 in. will work with a capacity of almost 37 kips. The calculations are as follows:

$F_{CE} = 28.32$ psi

$$C_P = \frac{1 + \dfrac{475.9}{2,875}}{2(0.9)} - \sqrt{\left[\frac{1 + \left(\dfrac{475.9}{2,875}\right)}{2(0.9)}\right]^2 - \frac{\dfrac{475.9}{2,875}}{0.9}} = 0.162$$

$$F_c' = F_c^* C_P = 2{,}875(0.162) = 465.8 \text{ psi}$$

$$f_c = 32{,}000/78.75 = 406.3 \text{ psi} < 465.8 \text{ psi} \qquad \textit{Okay!}$$

A small subgroup of columns includes columns with a tapered cross section. The ANSI/AF&PA NDS-2005 provides a method to estimate the effective cross-sectional dimension if the cross section is rectangular and has a constant taper. The effective dimension d can be determined using Eq. (6.13). In this equation, a adjusts for the fixity effects at the ends of the columns.

$$d = d_{min} + (d_{max} - d_{min})[a - 0.15(1 - d_{min}/d_{max})] \qquad (6.13)$$

where

d_{min} = minimum dimension for the face of the column of interest

d_{max} = maximum dimension for the face of the column of interest

$a = 0.70$ for large end fixed, small end unsupported, or simply supported

$a = 0.30$ for small end fixed, large end unsupported, or simply supported

$a = 0.50$ for both ends simply supported and the tape in one direction

$a = 0.70$ for both ends simply supported and the tape toward both ends of the column

$d = d_{min} + (d_{max} - d_{min})/3$ for tapered columns with any other type of end supports

In addition, the compressive stress at any cross section within the length of the column shall not exceed F_c'.

For columns, the end bearing stress f_c shall be less than F_c^*, and if $f_c > 0.75\ F_c^*$, a 20-gauge or thicker metal or other appropriate material bearing plate shall be used. The bearing plate shall have sufficient stiffness to distribute the bearing forces evenly across the reaction surface. If the column is bearing on another member such that stresses perpendicular to the grain are induced, the stresses perpendicular to the grain must satisfy the inequality $f_{c\perp} \leq f_{c\perp}'$ The nonuniform distribution of stresses due to bending stresses need not be accounted for when making these checks on stress.

6.5.1 Combined Bending and Tension

The design checks for simple bending plus tension consist of two simple interaction equations that essentially check how much of the capacity is used up by either of the actions. These two equations are shown as Eqs. (6.14) and (6.15):

$$\frac{f_t}{F_t'} + \frac{f_b}{F_b^*} \leq 1.0 \qquad (6.14)$$

and

$$\frac{f_b - f_t}{F_b^{**}} \leq 1.0 \qquad (6.15)$$

where

F_b^* = adjusted design valued, which is adjusted using all appropriate factors except for C_L

F_b^{**} = adjusted design valued, which is adjusted using all appropriate factors except for C_V

An example of this condition might be the bottom chord of a truss that has ½-in. gypsum wallboard attached. Thus, if the bottom chord of a truss system that was spaced at 24 in. on center were to have a 4.5-kip tension due to dead and snow load, and the element had a span of 10 ft between panel points, one might consider that there would be a moment

$$M = (2.5 \text{ psf})(2 \text{ ft})(10 \text{ ft})^2 / 8 = 62.5 \text{ ft-lb}$$

For a 2×4-in. nominal southern pine graded as No. 2 and better, the maximum bending stress would be

$$f_b = M/S = (62.5 \text{ ft-lb})(12 \text{ in.-ft}) / (3.063 \text{ in.}^3) \approx 245 \text{ psi}$$

(See Table 1B in NDS Supplement for cross sectional area, A, and the section modulus, S, corresponding to the dressed size of the section.) The tension stress would be

$$f_t = P/A = 4,500/5.25 \approx 857 \text{ psi}$$

The allowable design stresses for tension would be

$$F_t' = F_t C_D C_M C_t C_F C_i = (825 \text{ psi})(1.15)(1.0)(1.0)(1.0)(1.0) \approx 948 \text{ psi}$$

(*Note:* C_F is included in the tabulated values for southern pine in Table 4B of the NDS Supplement.)

$$F_b' = F_b C_D C_M C_t C_L C_F C_{fu} C_i C_r = (1,500 \text{ psi})(1.15)(1.0)(1.0)(1.0)(1.0)(1.0)(1.0)(1.0) = 1,725 \text{ psi}$$

It should be noted that C_F is included in the tabulated values for southern pine in Table 4B of the NDS Supplement, $C_L = 1.0$ due to it being a tension member and braced by gypsum, and $C_{fu} = 1.0$ due to bending about the strong axis. If the member had been a parallel-chord truss with a 4×2-in. nominal chord, then C_{fu} equals 1.1. C_r would equal 1.0 because the system is a truss, and trusses do not use C_r due to the frame effect rather than a series of beams.

Now the checks can be made:

$$\frac{f_t}{F_t'} + \frac{f_b}{F_b'} = \frac{875}{947} + \frac{245}{1,725} = 0.92 + 0.14 = 1.07 > 1.0$$

By this check, the member failed and is overstressed by 7 percent. The second check is

$$\frac{f_b - f_t}{F_b^{**}} = \frac{245 - 875}{1,725} = -0.37 < 1.0$$

However, this check is to check the net bending stress, and because the member is primarily a tension member, the check in not really necessary. Since 92 percent of the capacity is used in tension and only 14 percent in bending, the member can be brought into compliance using either thicker or wider lumber. If the member had been principally a bending member, it would have been beneficial to focus on using wider lumber to increase the section modulus the most.

6.5.2 Biaxial Bending

For applications that experience biaxial bending or bending plus axial forces, the ANSI/AF&PA NDS-2005 requires that an interaction equation be used to check the sufficiency of the member. The ANSI/AF&PA NDS-2005 differentiates between bending plus tension and bending plus compression because compression requires the inclusion of buckling in the check. The interaction equation that is used for combined compression and bending is shown as Eq. (6.16):

$$\left(\frac{f_c}{F_c'}\right)^2 + \frac{f_{b1}}{F_{b1}'\left[1-\left(\frac{f_c}{F_{cE1}}\right)\right]} + \frac{f_{b2}}{F_{b2}'\left[1-\left(\frac{f_c}{F_{cE2}}\right)-\left(\frac{f_{b1}}{F_{bE}}\right)^2\right]} \le 1.0 \tag{6.16}$$

where

$$f_c < F_{cE1} = \frac{0.822E_{min}^2}{\left(\dfrac{l_{e1}}{d_1}\right)^2}$$ for either uniaxial edgewise bending or biaxial bending

$$f_c < F_{cB2} = \frac{0.822E_{min}'}{\left(\dfrac{l_{e2}}{d_2}\right)^2}$$ for uniaxial flatwise bending or biaxial bending

$$f_{b1} < F_{bE} = \frac{1.20E_{min}'}{(R_B)^2}$$ for biaxial bending

f_{b1} = actual edgewise bending stress (i.e., bending about the strong axis)

f_{b2} = actual flatwise bending stress (i.e., bending about the weak axis)

d_1 and d_2 = cross-sectional dimensions of the member for wide and narrow faces, respectively.

$$R_b = \sqrt{\frac{l_e d}{b^2}}$$

Thus one might think of an application for biaxial bending to be something like a corner column in a post-frame building, where the wall purlins are anchored to the column only at 36-in. centers. The column then would be braced for buckling and would have an effective length of 36 in. in both directions. The mechanical properties for a column of southern pine graded as No 2 and better with a 6×6-in. nominal cross section are found in Table 4D of the NDS Supplement as

F_b = 850 psi
  (*Note:* Since $F_b C_F < 1,150$ psi, $C_m = 1.0$ per footnotes in Table 4E of NDS Supplement.)

F_c = 525 psi

$$E = 1,200,000 \text{ psi}$$

$$E_{min} = 440,000 \text{ psi}$$

$C_D = 1.6$ if bending is caused by either wind or seismic lateral forces; all other adjustments $= 1.0$

$$F'_b = 525(1.6) = 840 \text{ psi}$$

$$F'_{b1} = F'_{b2} = 850(1.6) = 1,360 \text{ psi}$$

If the loads are caused by dead load plus wind load, and the factored loads were 6 kips axial, and the maximum moments were 8.1 kip-in. in each direction, then the actual stresses would be

Axial: $f_c = P/A = 6,000 \text{ lb}/(5.5 \times 5.5 \text{ in.}^2) \approx 198 \text{ psi}$

Bending: $f_{b1} = f_{b2} = M/S = 8,100 \text{ in.-lb}/27.72 \text{ in.}^3 \approx 292 \text{ psi}$

$$F_{cE1} = F_{cE2} = \frac{0.822(440,000)}{\left(\dfrac{36}{5.5}\right)^3} \approx 8,442 \text{ psi}$$

$$F_{bE} = \frac{1.20(440,000)}{(36)\left(\dfrac{5.5}{5.5}\right)^2} \approx 80.667 \text{ psi}$$

$$\left(\frac{198}{840}\right)^2 + \frac{292}{1,360\left(1 - \dfrac{198}{8,442}\right)} + \frac{292}{\left(1 - \dfrac{198}{8,442} - \dfrac{292}{80,667}\right)^2} = 0.495 \leq 1.0 \qquad Okay!$$

A final consideration is when the loading causes the stresses to be oriented at an angle to the grain of the wood rather than parallel or perpendicular to the grain. This can happen when one member is cut to accommodate a member oriented at an angle other than 90 degrees, as shown in Fig. 6.3. When this type of stress orientation occurs, the allowable design stress is adjusted following what is called Hankinson's formula, which is given by Eq. (6.17):

$$F'_\theta = \frac{F_c^* F'_{c\perp}}{F_c^* \sin^2 \theta + F'_{c\perp} \cos^2 \theta} \tag{6.17}$$

where θ is the angle between the direction of the grain and the direction of the load in degrees, as illustrated in Fig. 6.4.

6.6 Connections

Connections are the locations where most structural failures occur. This is due to a number of things, such as connections being left to the end of a design project when there is more work than budget for the project. It is also due to some of the checks that should be completed when designing timber structures being left as nonmandatory

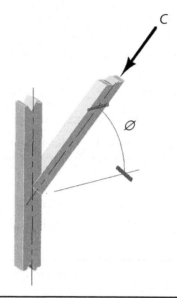

FIGURE 6.4 Relationship between load direction and orientation of grain.

design checks rather than making them mandatory. This latter item will be covered in the latter part of this section.

Timber connections can be classified into dowels, bearing enhancement, rivets, plates, and adhesives. This section will cover dowels, bearing enhancement, and rivets; plates and adhesives are considered to be proprietary as far as this discussion is concerned. This is so because the required design information for metal plate connectors (MPCs), such as used in MPC trusses is not available to the average designer, and use of adhesives in field-assembled applications requires special inspection and knowledge of the particular adhesive mechanical properties of the particular manufacturer before the connections can be designed and constructed properly.

Several design issues pertain to all mechanical connections in timber. The most obvious one is that the members being connected need to be checked to ensure that they have sufficient resistance for the loading at their net section (i.e., location of the connection). Another is that the connection must be detailed such that the shrinkage and expansion associated with changes in moisture content need to be provided for when configuring the connection to avoid splitting and checking of the members. The shrinkage and swelling of wood, along with the fact that the material is a viscoelastic composite, means that friction-transfer connections such as those used in steel design will not function in timber structures, and therefore, most connection design in wood is based on finger-tight concepts of bolt tightening rather than minimum torque on the fastener.

Additionally, the ANSI/AF&PA NDS-2005 does not address connection configurations that use multiple sizes or types of fasteners. The basis of the ANSI/AF&PA NDS-2005 is the performance of a single fastener, with the assumption that all the fasteners will be the same and that adjustments are made for performance issues such as multiple fasteners in a row. The metal portions of the connection, such as steel side plates, should be designed following the design specification for the material used, such as the American Iron and Steel Institute (AISC) *Specification for Structural Steel Buildings* (ANSI/AISC 360-05)[2a]. Appendix D of American Concrete Institute (ACI) *Building Code Requirements for Structural Concrete* (ACI 318-08)[29], is applied when evaluating the strength of anchorage via steel connector elements in concrete.

As with design of timber members, timber connection design values need to be adjusted to account for end-use conditions other than those assumed as the standard conditions for the particular fastener. The factors for most adjustments, such as load duration, are the same as those used for designing the members, with the exception that timber connections cannot be designed for impact loads. The highest value of C_D is 1.6 for designing connections. The values for each adjustment are provided in Chaps. 10 through 13 of the ANSI/AF&PA NDS-2005. For timber connections, the adjustment factors include the load-duration factor C_D, wet-service factor C_M, temperature-adjustment factor C_t, group action factor C_g, geometry factor C_Δ, penetration-depth factor C_d, end-grain factor C_{eg}, metal side-plate factor C_{st}, diaphragm factor C_{di}, and toe nail factor C_{tn}. With the exception of C_D and C_M, the adjustment factors typically have values of 1.0. The group action factor will be discussed in the section on large dowel connections.

The wet-service factor is an adjustment that has a significant effect on connections, especially on smooth dowels and/or when high temperatures are also present. The adjustment depends on the type of fastener used and the initial and in-service moisture contents of the wood. If the connection is constructed with dry wood (≤19 percent) and is kept dry, the moisture-content adjustment is 1.0. However, if the wood changes between moisture contents greater than 19 percent and less than 19 percent, the strength of the connection is reduced. The values for the adjustments are given in ANSI/AF&PA NDS-2005 in Table 10.3.3.

The temperature-adjustment factor C_t corrects for the fact that wood is weaker at elevated temperatures due to its viscoelastic nature. The maximum temperature at which timber can be used is 150°F, and the adjustments are broken into three ranges, as shown in Table 6.2. It should be noted that the elevated temperature must be a sustained elevated value, not a transient value. For instance, roof systems often can reach temperatures of over 200°F during a sunny afternoon, but the temperature decreases in the evening. Most of the time, this type of transient temperature change is not corrected for in the design of timber members.

Preservative and fire-retardant treatments of timber use chemicals that cause deterioration of the metal components of timber connections at an accelerated rate. It is recommended that either stainless steel or galvanized fasteners and plates be used with these types of treatments, and the designer is referred to the treatment manufacturer for recommendations on how to design connections to be corrosion-resistant.

The final adjustment that is common to the lateral design of dowel, split-ring, and shear-plate connections is the group action factor, C_g. This factor is intended to account for the shear-lag effect that occurs in multiple-fastener connections, where the efficiency of individual fasteners decreases with an increase in the number within a row. The

Moisture Content	$T \leq 100°F$	$100°F < T \leq 125°F$	$125°F < T \leq 150°F$
≤19%	1.0	0.8	0.7
>19%	1.0	0.7	0.5

TABLE 6.2 Temperature Factor C_t Values for Connections

phenomenon of shear lag is not unique to timber connections and was observed historically in riveted-steel and adhesive connections.

For wood connections, the group-action factor was derived following the elastic response of connections and is not associated directly with the capacity of the connections. Heine et al.[13] and Anderson[4] conducted numerical and experimental investigations of the group-action factor for capacity and showed that the reduction due to multiple bolts is far larger than the group-action factor indicates. In fact, the reduction on capacity can be as much as 40 percent. Therefore, it is important that additional checks on the design of multiple-bolt connections be completed.

The general equation governing the group-action factor is applicable to dowel-type connections with diameters of $0.25 \leq D \leq 1.0$ in., split-ring, and shear-plate connections. Dowel-type connections with fastener diameters of less than ¼ in. are not considered to have a group-action effect due to the assumption that all the fasteners will yield before the wood fails. The reference design value is multiplied by C_g and is determined from Eq. (6.18):

$$C_g = \left\{ \frac{m(1-m^{2n})}{n[(1+R_{EA}m^n)(1+m)-1+m^{2n}]} \right\} \left(\frac{1+R_{EA}}{1-m} \right) \tag{6.18}$$

where

n = number of fasteners in a row

R_{EA} = smaller value of $(E_s A_s)/(E_m A_m)$ or $(E_m A_m)/(E_s A_s)$

E_s and E_m = modulus of elasticity of the side and main members, respectively (psi)

A_s and A_m = cross-sectional area of the side and main members, respectively (in.)

$m = u - \sqrt{u^{2-1}}$

$u = 1 + \gamma \dfrac{s}{2}\left(\dfrac{1}{E_m A_m} + \dfrac{1}{E_s A_s} \right)$

s = center-to-center distance between individual fasteners in a row (in.)

γ = load-slip modulus of the connection (lb/in.)

= $180,000(D^{1.5})$ for dowel-type fasteners with wood main and side members

= $270,000(D^{1.5})$ for dowel-type fasteners with metal side members and wood main members

= 400,000 and 500,000 lb/in. for smaller and larger split rings or shear plate, respectively (The size differences are described in the section on shear-plate and split-ring connections.)

Tabulated values for the group-action factor for some of the more common timber connection configurations are presented in ANSI/AF&PA NDS-2005, Tables 10.3.6A–D.

Some common terminology for timber connections is necessary for clear communication of requirements for spacing and so on. Some of these terms are illustrated in Fig. 6.5. The following terms will be defined for all timber connections:

Row is defined as the direction aligned with the direction of the load applied.

Spacing is the center-to-center distance between fasteners in a row or a column.

End distance is the distance from the end of a piece of lumber to the center of the nearest bolt.

Edge distance is the distance from the side of a piece of lumber to the centerline of the nearest bolt hole.

There are two conditions to consider, one when the loading causes bearing stresses toward the edge and one when the loading causes tension stresses toward the edge. The first condition is called the *loaded edge*, and the second is called the *unloaded edge*.

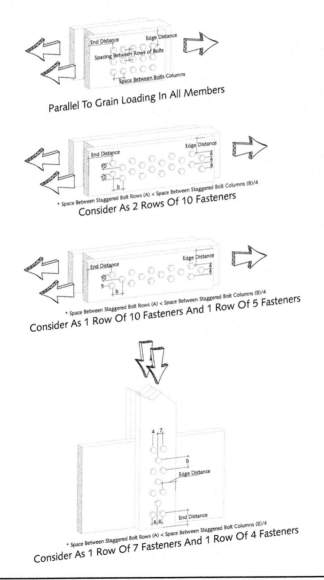

FIGURE 6.5   Basic geometry terms used in the design of timber connections.

6.6.1 Dowel Connections

Dowel connections consist of nails, screws, lag screws, bolts, and dowels and often require lead holes to prevent splitting of the lumber when used. In addition, the resistance is also affected by the number of fasteners in a row, the spacing of the fasteners, and the locations of the fasteners with respect to the edges and ends of the lumber. The required and allowable maximum dimensions of the required lead holes for various dowel-type fasteners are presented in Table 6.3 along with the minimum penetration requirements.

Terminology for driven fasteners is a bit different from that for bolted connections. For connections where the fastener fully penetrates only one member of the connection, the *side member* is defined as the member where the head of the fastener is located, and the *main member* is the member with the point of the fastener. For bolted connections, the *main member* is defined as the thickest member included in the connection, and all other members are defined as *side members*.

All screw-type threaded fasteners are to be screwed into the wood, not driven with a hammer or pneumatic tool. Driving with a hammer damages the wood that is supposed to work integrally with the threads to provide withdrawal resistance for the fastener. Often, soap or other lubricant is used on the threads to facilitate screwing the fastener into the wood. There are no reductions in performance for using these lubricants.

A special configuration of nails and spikes is the toenail. The dimensional requirements of toenails are illustrated in Fig. 6.6. In recent editions of the ANSI/AF&PA NDS-2005,

Fastener Type	Minimum Lead-Hole Size	Minimum Penetration into Main Member (Excluding Tapered Tip)
Lag screws Smooth-shank portion Threaded portion	= Diameter of shank = 60–75% D for $G > 0.6$ = 40–75% D for $G \leq 0.5$ = 0 for $D \leq \frac{3}{8}$ in. and $G \leq 0.5$	$4D$
Nails and spikes	$\leq 90\%$ D for $G > 0.6$ $\leq 75\%$ D for $G \leq 0.6$	$6D$
Bolts	($D + 1/32$ in.) $-$ ($D + 1/16$ in.)	
Wood screws	Withdrawal: = 90% D for $G > 0.6$ = 70% D for $0.5 < G \leq 0.6$ = 0 for $G \leq 0.5$ Lateral: Shank portion = D for $G > 0.6$ Thread portion = root D for $G > 0.6$ Shank portion = 7/8 D for $G \leq 0.6$ Thread portion = 7/8 D_{root} for $G \leq 0.6$	$6D$
Drift bolts or pins	0 to 1/32 in. smaller than D	

TABLE 6.3 Lead-Hole Requirements for Dowel-Type Fasteners

what used to be called *slant nails* have been combined with toenails in an effort to reduce confusion. Historically, toenails were oriented at the end of a piece of lumber, and slant nails were oriented along the edge of a piece of lumber. Due to the confusion between these two configurations, the conservative design values for orientations at the end of a member are now used for both configurations. The design of toenail configurations simply includes adding an adjustment factor to the normal withdrawal or lateral design value adjustments. The adjustment factor for withdrawal is 0.67, and no adjustment is made for moisture effects. The adjustment factor for lateral design is 0.83.

Withdrawal Design

Withdrawal design for dowel connections is limited to nails, screws, and lag screws. Bolts, drift pins, split rings, and shear plates are not allowed to be loaded with forces parallel to the axis of the dowel and therefore are not included in this section. All withdrawal reference design values for timber fasteners are determined from empirical equations. Further discussion of these equations and their development is provided in the *Wood Handbook*[26].

Withdrawal Strength of Lag Screws

The withdrawal resistance of lag screws depends on the density of the wood from which the screw is being withdrawn (i.e., the density of the main member), the form of the shank of the fastener (i.e., smooth or deformed), and the moisture content of the wood. The reference withdrawal design value for lag screws is given by Eq. (6.19):

$$W = 1,800G^{3/2}D^{3/4}$$

(6.19)

where

G = specific gravity for the species of wood being used

D = lag-screw diameter

The design withdrawal values are calculated and presented for various specific gravities and diameters of screw in Table 11.2A of ANSI/AF&PA NDS-2005. The

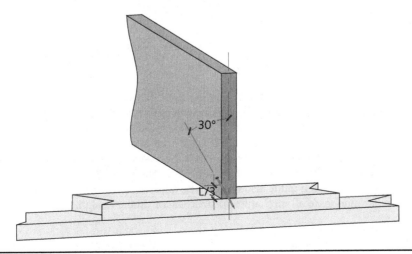

FIGURE 6.6 Configuration of a toenail connection.

reference design value also needs to be adjusted to account for end-use conditions in a similar manner to all mechanical properties for timber design.

Lag screws have geometric restrictions imposed on their design for withdrawal loads to prevent splitting and ensure that the fasteners function at their intended capacity. The restrictions are that the minimum spacing between lag screws is $4D$, the minimum end distance is $4D$, and the minimum edge distance is $1.5D$. For lag screw connections not adhering to these edge and spacing limitations, the design strength should be considered zero.

Withdrawal Strength of Wood Screws

The withdrawal resistance of wood screws also depends on the density of the wood from which the screw is being withdrawn (i.e., the density of the main member), the form of the shank of the fastener (i.e., smooth or deformed), and the moisture content of the wood. The reference withdrawal design value for lag screws is given by Eq. (6.20):

$$W = 2,850G^2D \tag{6.20}$$

where

G = specific gravity for the species of wood being used

D = diameter of the lag screw

The design withdrawal values are calculated and presented for various specific gravities and diameters of screw in Table 11.2B of ANSI/AF&PA NDS-2005. The reference design value also needs to be adjusted to account for end-use conditions in a similar manner to all mechanical properties for timber design.

Withdrawal Strength of Nails and Spikes

The withdrawal resistance of nails and spikes also depends on the density of the wood from which the nail or spike is being withdrawn (i.e., the density of the main member), the form of the shank of the fastener (i.e., smooth or deformed), and the moisture content of the wood. The reference withdrawal design value for nails and spikes is given by Eq. (6.21):

$$W = 1,380G^{5/2}D \tag{6.21}$$

where

G = specific gravity for the species of wood being used

D = diameter of the nail or spike

Nails and spikes are not to be loaded in withdrawal from the end grain of wood. In other words, the design value for withdrawal from end grain is zero. The design withdrawal values are calculated and presented for various specific gravities and diameters of nails in Table 11.2C of ANSI/AF&PA NDS-2005. There is a special column for threaded nails, which typically are manufactured from hardened steel, and some proprietary fasteners are not affected by changes in moisture content in the manner that smooth-shank nails are. The reference design value also needs to be adjusted to account for end-use conditions in a similar manner to all mechanical properties for timber design.

Lateral Design

Dowel connections can be separated into two major groups for lateral design—small- and large-diameter dowels. The division occurs at the fastener diameter of 0.25 in. and the division is made because fasteners smaller than 0.25 in. do not cause differences in resistance when the orientation of the load with respect to the grain direction changes, whereas fasteners larger than 0.25 in. experience differences when the load is oriented parallel or perpendicular to the grain of the wood. In addition to the grain-orientation issue, small-dowel connections do not experience shear-lag effects that are called group-action effects in ANSI/AF&PA NDS-2005. For larger-diameter fasteners, the effectiveness of the individual fastener decreases as the number of fasteners in a row increases due to shear-lag effects. Similar phenomena have been observed in other materials and adhesive joints. Therefore, the discussion of dowel connections will be divided into two subsections to cover these differences.

Historically, the lateral design of timber connections was governed by set configurations of fasteners. This limited the design of timber structures because unless the connections fit one of the approved configurations, extensive testing was required before new configurations could be used. In the early 1990s, a change was made to adopt a mechanics-based methodology developed by Johansen[16]. Currently, virtually all modern design codes, worldwide, have adopted this yield theory as the basis of the design of timber connections for lateral forces. Different countries use variations of the theory in that the base variables are defined slightly differently from country to country, but the basic theory is the same. In the United States, the method is called the yield theory. While the equations look complicated, the availability of spreadsheets makes using the equations relatively easy, and values for some of the more common connection configurations (i.e., lumber dimensions, species, dowel size, etc.) are presented in tabular form in ANSI/AF&PA NDS-2005, Tables 11A–R. Some of the basic assumptions of this theory are as follows:

1. The faces of the members being connected are in contact with each other, but the friction is not accounted for.
2. The load acts perpendicular to the axis of the fastener.
3. The edge and end distances, as well as the spacing between fasteners, are sufficient to not cause splitting (the required geometric spacing requirements are provided in Sec. 11.5 of ANSI/AF&PA NDS-2005).
4. The minimum penetration of the fastener into the member with the point of the fastener for driven fasteners is greater than the minimum defined penetration for the type of fastener being used.

The general yield theory assumes that the mechanics between the bending strength of the dowel and the crushing strength of the wood will determine whether the dowel will bend or the wood will crush. This results in the occurrence of several different yield configurations that are called *yield modes* and are illustrated in Fig. 6.7. In essence, the more slender the dowel, the more likely it is that the dowel will bend rather than the wood crushing significantly. *Slenderness* is the length of the dowel in a given member divided by the diameter of the dowel. The larger this ratio, the more likely it is that the dowel will bend. Therefore, in common dimensional lumber, the slenderness of nails results in Modes III and IV being the dominant modes of yield, whereas large-diameter bolts tend to have Modes I and II as the predominate yield mode. For thick glued-laminated timber, even large bolts may have sufficient slenderness to cause the fastener to bend and yield in Modes III and IV. For seismic design, the only source of ductility is

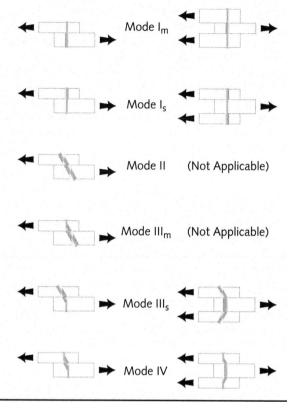

FIGURE 6.7 Illustration of yield modes of dowel connections in wood. (*American Wood Council.*)

from the connections, and designers are encouraged to use connectors that will yield in the higher modes to provide toughness to the structure. Yield Modes I and II are know to be brittle in failure, whereas Yield Modes III and IV are far more ductile.

The equations governing the yield theory are given in Table 6.4*a* for single-shear and Table 6.4*b* for double-shear connections. Yield Modes II and III$_m$ cannot occur in double-shear connections. The equation governing the design of a particular connection is the one that has the minimum value of all the equations.

where $k_1 = \dfrac{\sqrt{R_e + 2R_e^2(1+R_t+R_t^2)+R_t^2R_e^3} - R_e(1+R_t)}{(1+R_e)}$

$k_2 = -1 + \sqrt{2(1+R_e) + \dfrac{2F_{yb}(1+2R_e)D^2}{3F_{em}l_m^2}}$

$k_3 = -1 + \sqrt{\dfrac{2(1+R_e)}{R_e} + \dfrac{2F_{yb}(2+R_e)D^2}{3F_{em}l_s^2}}$

D = fastener diameter (in.)

Governing Equation	Yield Mode
$Z = \dfrac{Dl_m F_{em}}{R_d}$	I_m
$Z = \dfrac{Dl_s F_{es}}{R_d}$	I_s
$Z = \dfrac{k_1 Dl_s F_{es}}{R_d}$	II
$Z = \dfrac{k_2 Dl_m F_{em}}{\left(1 + 2R_e\right)R_d}$	III_m
$Z = \dfrac{k_3 Dl_s F_{em}}{\left(1 + 2R_e\right)R_d}$	III_s
$Z = \dfrac{D^2}{R_d}\sqrt{\dfrac{2F_{em}F_{yb}}{3\left(1 + R_e\right)}}$	IV

TABLE 6.4a   Yield Theory Equations for Single-Shear Connections

F_{yb} = dowel bending yield strength (psi)

R_d = for ¼ in. ≤ D ≤ 1.0 in., $4K_\theta$ for Yield Modes I_m and I_s, $3.6K_\theta$ for Yield Mode II, $3.2K_\theta$ for Yield Modes III_m, III_s, and IV.

= K_θ for all Yield Modes when $D < $ ¼ in., and $K_D K_\theta$ for threaded fasteners with $D \geq$ ¼ in.

$R_e = F_{em}/F_{es}$

$R_t = l_m/l_s$

l_m = length of dowel in the main member

l_s = length of dowel in the side member

F_{em} = dowel bearing strength in the main member (psi) (given in Table 11.3.2 of ANSI/AF&PA NDS-2005 for wood and Table 11.3.2B for wood structural panels; for fasteners with $D \geq$ ¼ in. and inserted into the end grain of the member, the perpendicular bearing strength shall be used)

F_{es} = dowel bearing strength in the side member (psi) (given in Table 11.3.2 of ANSI/AF&PA NDS-2005 for wood and Table 11.3.2B for wood structural panels)

$K_\theta = 1 + 0.25(\theta/90)$

K_D = 2.2 for $D \leq 0.17$ in. and $10D + 0.5$ for 0.17 in. < D < 0.25 in.

θ = maximum angle of load to the direction of the grain ($0 \leq \theta \leq 90$ degrees) for any member in the connection

Governing Equation	Yield Mode
$Z = \dfrac{Dl_m F_{em}}{R_d}$	I_m
$Z = \dfrac{2Dl_s F_{es}}{R_d}$	I_s
$Z = \dfrac{2k_3 Dl_s F_{em}}{(2 + R_e) R_d}$	III_s
$Z = \dfrac{2D^2}{R_d} \sqrt{\dfrac{2F_{em} F_{yb}}{3(1 + R_e)}}$	IV

TABLE 6.4b Yield-Theory Equations for Double-Shear Connections

When the connection includes loads that are oriented at an angle to the grain rather than either parallel or perpendicular to grain, *Hankinson's formula*, given at the end of the biaxial bending section as Eq. (6.17) must be used.

Small Dowels

As stated earlier, small dowels consist of fasteners that are less than ¼ in. in diameter. This class of fasteners typically consists of nails, spikes, screws, and so on. Connections using small dowels have a group-action factor of 1.0 due to the assumption that all the fasteners will yield before significant wood crushing or failure occurs. Observations of nailed connections for wood to wood and sheathing to wood have provided support for this assumption. Therefore, the lateral-connection design value is given in Eq. (6.22):

$$Z = nZ' \tag{6.22}$$

where

n = number of fasteners

Z' = reference design value that has been adjusted for all the end-use conditions

This equation is applicable to nails, staples, and wood screws. The spacing requirements are defined as sufficient to prevent splitting. The *Timber Construction Manual*[3] provides some guidance for spacing to prevent splitting.

When nails and spikes are loaded by a combination of withdrawal and lateral forces, the adjusted design value is determined using Eq. (6.23):

$$Z'_\alpha = \frac{(W'p)Z'}{(W'p)\cos\alpha + Z'\sin\alpha} \tag{6.23}$$

where

p = penetration depth of the threaded portion of the fastener into the main member

α = angle of the load with respect to the surface of the member

When wood screws and lag screws are loaded by a combination of withdrawal and lateral forces, the adjusted design value is determined using Eq. (6.24):

$$Z'_a = \frac{(W'p)Z'}{(W'p)\cos^2 \alpha + Z' \sin^2 \alpha} \tag{6.24}$$

where

p = penetration depth of the threaded portion of the fastener into the main member

α = angle of the load with respect to the surface of the member

Nails and spikes have two additional adjustments that should be included in a designer's consideration. The first is for when the fastener is driven into the end grain of a member. As stated earlier, the withdrawal value for this condition is zero, but the lateral adjustment factor is 0.67 for nails, spikes, wood screws, and lag screws. The withdrawal adjustment for end-grain effects is 0.75 for wood screws and lag screws.

The second adjustment for nails and spikes is the *diaphragm factor*. This factor is used only when diaphragms are being designed using engineering mechanics rather than tabulated values. The adjustment is intended to account for the effective increase in strength achieved in diaphragms due to the interaction of sheathing panels when the diaphragm is loaded. The diaphragm factor C_{di} is 1.1.

Large Dowels

Large dowels pose additional design concerns compared with small dowels. Other than the fact that a significant amount of the cross section of the member has to be removed to insert the large dowel into the member and therefore the member needs to be checked for net section strength, geometry and group-action effects need to be included in the checks.

The geometry effects include the spacing between dowels in a row, the spacing between rows, and the edge distance. The minimum distances for full design strength (C_Δ = 1.0) are presented in Table 6.5.

When steel plates or other components that do not shrink and swell in the same manner and direction as the member to which they are attached are used, the maximum spacing between the outer rows of fasteners in the connection cannot be more than 5 in. to prevent splitting in the wood member. When larger distances are required for the connection, the steel plates can be divided into multiple plates with spaces between them to allow the wood member to shrink and swell.

The group-action factor discussed earlier is only one of the checks for this effect that the designer is responsible for. The issue is that experimental tests by Anderson[4] have shown that splitting, block shear, row tear-out, and so on are all possible failure mechanisms for the connection. Therefore, the nonmandatory App. E is included in ANSI/AF&PA NDS-2005. This appendix was voted to be included in the LRFD provisions as mandatory design checks, but the ASD committee did not vote the same way, and therefore, it is not included in the main provisions of ANSI/AF&PA NDS-2005. This difference allows the designer to use other rationale for checking for these failure mechanisms, but it is important that checks be made. The one check that is not included in the appendix is for splitting because no agreed-on method of making this check has been developed.

Direction of Load	Minimum Distance and Associated C_Δ	Minimum Distance for Full Design Value	Adjustment Factor C_Δ for Distance Between Min and Min for Full Design
Spacing between fasteners in a row			
Parallel to grain	3D	4D	$C_\Delta = \dfrac{\text{Actual spacing}}{\text{Minimum spacing for } C_\Delta = 1.0}$
Perpendicular to grain	3D	Spacing requirements of attached member	
Spacing between rows			
Parallel to grain	1.5D		
Perpendicular to grain $1/D \leq 2.0$	2.5D		(*Note:* 1/D is smaller of main or side-member values.)
Perpendicular to grain $2 < 1/D < 6.0$	5*l* + 10D		
Perpendicular to grain $1/D \geq 6.0$	5D		
End-distance requirements			
Parallel to grain compression	2D ($C_\Delta = 0.5$)	4D	$C_\Delta = \dfrac{\text{Actual end distance}}{\text{Minimum end distance for } C_\Delta = 1.0}$
Parallel to grain tension	3.5D (softwood)	7D (softwood)	
	2.5D (hardwood)	5D (hardwood)	
Perpendicular to grain	2D	4D	
Edge distance			
Parallel to grain $1/D \leq 6.0$	1.5D		(*Note:* 1/D is the smaller of main or side-member values.)
Parallel to grain $1/D > 6.0$	Greater of 1.5D or half the spacing between rows		
Perpendicular to grain loaded edge	4D		
Perpendicular to grain loaded edge	1.5D		

TABLE 6.5 Geometric Requirements and Adjustments for Dowel Connections

6.6.2 Split-Ring and Shear-Plate Connectors

Split rings and shear plates are connectors that work in conjunction with a bolt or lag screw to form one of the highest capacity connectors for timber construction. The connectors are classified as bearing-enhancement connectors due to the fact that the connectors essentially increase the bearing capacity of the bolt over a larger diameter. An example of a split-ring connection is shown in Fig. 6.8*a*, and a shear-plate connection is shown in Fig. 6.8*b*. Shear-plate connections can be either between two shear plates (one in each of the main and side members) or between a shear plate and a steel side member. The bolt essentially works to keep the split ring engaged in the wood members and provides the shear capacity for the shear-plate connection.

Typical split rings are manufactured in either 2.5- or 4-in. diameters. Shear plates are manufactured in either 2⅝- or 4-in. diameters. Both types of connectors require special tools to cut the groove or daps in the wood members. Both types of fasteners require precise drilling because they both require full bearing contact with the sides of the dap or groove cut in the wood member. A concern with both types of connections is that it is difficult or impossible to verify proper installment after the fact because the size of the dap or hole cannot be checked unless the two members being joined are separated.

Reference design values are based on the specific gravity of the timber members being joined, and the species are broken into four groups—A through D. Group A is the

(a)

(b)

Fig. 6.8 Illustrations of (*a*) a split-ring connection configuration and (*b*) a shear-plate connection configuration.

highest density (≥0.60), followed by groups B (0.49 ≤ G < 0.6), C (0.42 ≤ G < 0.49), and D (G < 0.42). The reference design values are also for one shear plane (i.e., one split ring or one shear plate with either a shear plate in a wood member or a steel plate as the side member). The reference design values for split rings and shear plates used with bolts are given in the Tables 12.2A and 12.2B of ANSI/AF&PA NDS-2005, respectively. Design values for both types of connectors used with lag screws are given in Table 12.2.3 of ANSI/AF&PA NDS-2005.

Adjustments to the reference design values for these two connectors include an increase for steel side plates used with shear plates and potential decreases for the penetration effects of lag screws, loading at an angle to the grain, use of the connectors in the end grain, and the typical geometric corrections for configuration of the connectors. These adjustments are given in Secs. 12.2.3 through 12.3.7 of ANSI/AF&PA NDS-2005.

6.6.3 Timber Rivets

Timber rivets are unique connectors for timber construction. They were conceived originally as a connector for glued-laminated timber but have been accepted for use in dimensional lumber and timber in Canada. The ANSI/AF&PA NDS-2005 still restricts their use to only Douglas fir–larch and southern pine glued-laminated timber members.

The connection is truly a rivet in that the nail-type fastener is deformed as it is driven through a steel plate (ASTM A36 steel a minimum of ⅛-in. thick). The fastener is originally an oval that is deformed into a somewhat circular cross section as it is forced through the plate. The fasteners are not to be driven fully flush with the surface of the steel plate but rather only until they are fully seated into the hole in the steel plate. The head of the fastener will be a bit above the surface of the plate, but the process of driving the fastener provides a secure connection that does not have to have movement before the load starts to be resisted, as is the case with bolts. Finally, the long dimension of the oval cross section is to be oriented parallel to the grain of the wood for the rivet to function as planned.

The order in which the fasteners are driven is also important to optimal performance. The fasteners should be driven with the outside holes being filled first. Then the middle row should be filled, and the rows then are filled by splitting the space between the completed holes so that the wood is pinched between the rivets. In addition to acting similar to a nail, the rivet acts like a wedge in a wooden tool handle, where the wood is spread and pinched between the wedge and the head of the tool. The ANSI/AF&PA NDS-2005 specifies that the rivets should be placed around the perimeter of the connection first, and then the interior should be filled following a spiral pattern.

Timber rivet connection design is based on empirical methods where two failure mechanisms are allowed. Either the rivets will bend and withdraw from the wood, or the wood will fail in tension or shear. However, this connection also could be used easily to produce an energy dissipating mechanism for the structure. For instance, if the steel plate were shaped properly, the plate would yield in a predefined location and provide a ductile moment frame that could be repaired after an earthquake by cutting out the yielded plate and welding a new piece in its place after the frame was straightened.

The rivets are not to have any predrilled holes in the wood; the maximum penetration of the rivet is 70 percent of the thickness of the wood for single-sided connections, and the points should not overlap in double-sided connections. (If the connection is configured such that the points must overlap, then the spacing and location of the rivets with respect to the other side must follow the requirements of Secs. 13.3.1 and 13.3.2 of ANSI/AF&PA NDS-2005.

The design equations of the ANSI/AF&PA NDS-2005 are set up such that the design represents a one-sided connection, and the spacing of the fasteners at their points is represented by s_p and s_q for the dimensions parallel and perpendicular to grain, respectively. These dimensions are always measured parallel and perpendicular to the grain regardless of the orientation of the steel side plate. The steel side plate must be designed following the provisions for steel design.

Now the design process is to check the capacity of the rivets for failure and then to check the capacity of the connection from a wood failure standpoint. The lower value of either of these checks is the design value for the connection. There are two equations for checking the rivet strength. They are presented as Eqs. (6.24) and (6.25):

$$P_r = 280 \; p0.32 \; n_R n_c \tag{6.25}$$

where

P_r = reference design value for rivet failure in the parallel-to-grain direction

p = penetration of the rivet into the wood member (= rivet length − side-plate thickness [⅛ in.])

n_r = number of rows of rivets parallel to the direction of loading

n_c = number of rivets per row

$$Q_r = 160p^{0.32}n_R n_c \tag{6.26}$$

where Q_r is the reference design value for rivet failure in the perpendicular-to-grain direction.

The checks on the wood failure mechanism use tabular values because the equations are more complicated than most design equations. The reference design value for wood failure parallel to grain P_w is given in Tables 13.2.1A–F of ANSI/AF&PA NDS-2005, and the reference design value for wood failure perpendicular to grain Q_w is determined from:

$$Q_w = q_w p^{0.8} C_\Delta \tag{6.27}$$

where

q_w = value determined from Table 13.2.2A of ANSI/AF&PA NDS-2005

C_Δ = geometry factor that is given in Table 13.2.2B of ANSI/AF&PA NDS-2005

The reference values for P_r, P_w, Q_r, and Q_w must be multiplied by the appropriate end-use adjustment factors to obtain the allowable design values. If the load is at an angle to the grain other than 0 or 90 degrees, then the adjusted design value N' can be determined by Eq. (6.27):

$$N' = \frac{P'Q'}{P'\sin^2\theta + Q'\cos^2\theta} \tag{6.28}$$

Two other adjustments are possible for timber rivets. They include the *metal side-plate factor* for when the steel side-plate thickness is less than ¼ in., and the *end-grain factor* that has a value of 0.5 when the cut is perpendicular to the grain and can increases linearly as the cut approaches parallel-to-the-grain direction.

The spacing configurations for minimum end and edge distances are given in Table 13.3.2 of ANSI/AF&PA NDS-2005. The minimum spacing of the rivets in the connection is ½ in. perpendicular to the grain (s_q) and 1.0 in. parallel to the grain (s_p).

6.7 Shear Walls

Shear walls are the most common vertical element used for light-frame lateral force–resisting systems. In most discussions of timber design, diaphragms are discussed before the shear walls. However, the order is being reversed in this document because while the load is transferred to the shear wall by the diaphragm, the discussion of reentrant corners in the diaphragm section needs to be able to include the stiffness calculation for shear walls. Therefore, shear-wall design and determination of stiffness will be covered first to provide the basis for discussing irregular diaphragms in the following section.

The principal difference between the various types of shear walls is the amount of active overturning restraint provided by the designer. Determination of the forces required to be resisted is a controversial topic with design assumptions that either allow large portions of the dead load to be activated in resisting the uplift forces to those which do not allow any dead load to be used in resisting the uplift forces. This issue is important to consider because it can result in a large difference in the size of connectors and the expected stiffness of the wall. This is one of the areas where the designer needs to determine how well the design analog used represents reality in the final building. The designer should keep in mind, however, that while the real structural system is simplified to allow the design process to be completed, physics still works in the real world.

The use of adhesives to attach the sheathing to the framing in shear walls was banned in the mid-1990s due to the design rules not addressing the effects of added stiffness, strength, and brittleness when adhesives were used. Designers were following all the rules, but they were unknowingly designing buildings to fail in the anchorage, shear through the thickness, or tension failures in the studs. In the recent editions of the building code and ANSI/AF&PA SDPWS-08, the advantage of using adhesives has been acknowledged for high-wind events. Therefore, adhesives now can be used in regions where the seismic hazard is low to moderate (i.e., seismic design categories A, B, and C) when the seismic design parameters are changed to $R = 1.5$ and $\Omega_0 = 2.5$.[19] The reader should refer to Chap. 2 for determination of load effects in low-rise buildings.

6.7.1 Overturning Forces

There are two components to overturning forces that need to be considered when designing light-frame construction, global overturning (GO) and local overturning (LO). In this discussion, GO is the action of the entire building above the level being considered, trying to roll over as a rigid body like a cantilever beam on end. This action is illustrated in Fig. 6.9 and is determined using simple statics on the overall building above the level of interest. The induced overturning forces are located in the outer envelope of the building and sometimes are described as *envelop forces* or *shell effects*. The overall dead load of the building above the level of interest is usually used to counter the overturning effect.

For the building shown in Fig. 6.9, the GO forces at the ground floor would be computed as shown below. Subscript numbers in the following calculations indicate the respective floor level:

$$T = C = \frac{\sum F_i h_i - \sum W_i d_i}{\sum L_i}$$

$$= \frac{(F_2 h_2 + F_3 h_3 + F_4 h_4 + F_5 h_5 + F_R h_R) - (W_{w1} d_1 + W_2 d_2 + W_3 d_3 + W_4 d_4 + W_5 d_5 + W_R d_R)}{L_1 + L_2 + L_3}$$

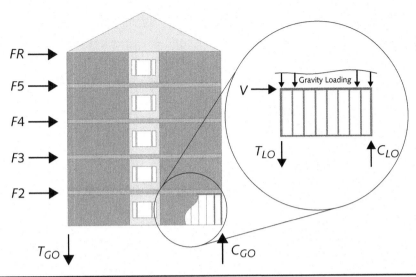

FIGURE 6.9 Forces applied to a multistory building: wind or seismic design loads applied to the individual levels of a building and the racking shear force applied to a given wall segment in the ground-floor wall line.

where

T = induced tension force at the end of the wall line

C = induced compression force at the end of the wall line

F_i = lateral design force imposed on the building at a given level due to either wind or seismic loading

h_i = height of the story level above the base of the structure (usually the top of the foundation)

W_i = weight associated with a given story level

d_i = horizontal distance from the compression end of the wall line to the center of action of gravity on the floor load

L_i = length of a given segment of the wall line

These forces are induced at the ends of the wall line and are due to the flexure or GO action of the building not the racking action of the wall line. They are superimposed on the LO forces that are induced into ends of each wall segment of a wall line due to the racking action of the wall segment, which are shown in the balloon in Fig. 6.9.

The LO forces are induced into the end studs of each wall segment as the wall is racked. These forces are illustrated in Figs. 6.9 and 6.10. The use of dead load to counter these tension forces is the area of controversy, and designers need to use judgment when deciding on how to determine the net uplift or compression force in the wall element.

Ignoring the dead load effects, when a shear wall or braced wall is subjected to a racking load as shown in Fig. 6.10, overturning forces are induced into the chords of the

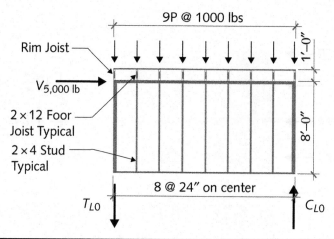

9P @ 1000 lbs

Rim Joist

$V_{5,000\,lb}$

2 × 12 Foor
Joist Typical

2 × 4 Stud
Typical

1'-0"

8'-0"

8 @ 24" on center

T_{LO} C_{LO}

FIGURE 6.10   Illustration of a typical light-frame wall with loads acting in both lateral and vertical directions.

wall (end studs). The induced force can be calculated using simple statics and summing the moments about the toe of the wall. If this is done, the tension force is calculated as follows:

$$\sum M_{@B} = 0 = Vh - TL$$

$$T = \frac{Vh}{L} = vh = C$$

where

$C =$ compression force that must be transferred to the platform or story below

$V =$ total applied racking load at the top of the wall coming from the diaphragm

$h =$ wall height

$T =$ uplift force that must be restrained for the wall to rack instead of roll over as a rigid body

$L =$ length of the wall segment

The simplified equation of the unit shear times the wall height usually is used by designers because the unit shear is the basis for all the tables used as design aides to determine the wall sheathing thickness and associated nail schedule. If the forces in the vertical direction are summed, then $T = C$.

This analysis ignores the effects of dead load on the performance of the wall because this is the controversial part of the analysis, and the options will be discussed now. The designer now must use engineering judgment to decide how much of the dead load can be activated to resist the uplift force in the end stud. The overturning tension and compression forces induced into the wall are located in the end stud due to the mechanism that transmits them from the shear load that is in the sheathing through the sheathing nails around the perimeter of the sheathing panels and into the end studs, as shown in the Fig. 6.11. Studs that have two sheathing panels attached have a net uplift force of zero due to the fact that since both panels rotate in the same direction, the forces on the common stud counter each other.

Now the dead load of the building above a given shear wall is being transmitted to the wall by the floors and walls above, along with the dead load of the wall being designed itself, and may be used to counter the racking-induced uplift force in the end stud if it can be physically activated. Regardless of the design analog used to simplify the design process, physics still controls reality. Therefore, the following three scenarios are given to allow the designer to compare the main methods used to determine the net uplift and compression forces that need to be accounted for in the design so that a rational decision can be reached as to which represents reality the best.

The first option is to ignore all the dead load when determining the tension force. This assumption results in the highest tension forces and therefore the most conservative result as far as sizing anchorage for the wall. Those who use this method assume that the dead load will not be effectively transferred horizontally to the end stud but rather will follow the nearest stud to travel to the next level down. Therefore, at best, the dead load associated with one stud spacing is the maximum that could be considered to counter the uplift, and unless load from a header is present, the magnitude of the dead load available is so small that it is not worth the effort to account for it in the design. This method usually also assumes that the magnitude of additional compression force for low-rise buildings is small with respect to the LO compression force and that the compression stud can be designed safely with the additional dead load being ignored.

The second most common method of accounting for the effects of dead load assumes that the wall is a rigid body. This method provides the most effect of dead load countering

FIGURE 6.11 Illustration of how forces are induced in studs from rotation of sheathing.

the uplift forces, but it also results in the maximum compression forces being determined. The free-body diagram shown in Fig. 6.10 is used in this analysis, and the summation of forces includes the dead load. Assume that the wall segment shown in Fig. 6.10 is the wall segment from the ground floor wall line that was shown in Fig. 6.9. One should assume that the loads P represent the joist reaction for the floor above and the cumulative loads from the walls above associated with one stud spacing. This also assumes that the framing is in-line framing (i.e., the studs and the joists are lined up). Consequently, this results in Eq. (6.28):

$$\sum M_{@B} = 0 = Vh - TL - \left(\frac{\sum P}{L}\right)\frac{L^2}{2} - (w_{DL})_w \frac{hL^2}{2}$$ (6.29)

The third term in this equation changes the point loads associated with the studs and floor joists to a distributed load and then sums the moments, and the last term uses the assumed dead load of the wall in force per square foot rather than linear load. Solving for T, Eqs. (6.29) and (6.30) are derived:

$$T = \frac{Vh - \left(\frac{\sum P}{L}\right)\frac{L^2}{2} - \left(\frac{(w_{DL})_w hL^2}{2}\right)}{L}$$ (6.30)

or

$$T = vh = \left(\frac{\sum P}{2}\right) - \left(\frac{(w_{DL})_w hL}{2}\right)$$ (6.31)

Next, if equilibrium is enforced in the vertical direction,

$$\sum F_y = 0 = -T - \left(\sum P\right) - \left[(w_{DL})_w hL\right] + C$$ (6.32)

Solving for C,

$$C = T + \left(\sum P\right) - \left[(w_{DL})_w hL\right] + C$$ (6.33)

or

$$C = +vh - \left(\frac{\sum P}{2}\right) - \left[\frac{(w_{DL})_w hL}{2}\right] + \sum P + (w_{DL})_w hL$$ (6.34)

or

$$C = vh + \left(\frac{\sum P}{2}\right) + \left[\frac{(w_{DL})_w hL}{2}\right]$$ (6.35)

If the lateral load associated with the wall segment is 12,000 lb, the loads due to the dead load above are each 3,500 lb, and the dead load of the wall is assumed to be 10 psf. The story height is 9 ft. If these assumed values are processed through the preceding equations, the LO tension force is –9,720 lb, which indicates that there is no tension uplift force, and the dead load will be sufficient to counter the uplift effects. The compression force will be 21,720 lb, and the compression stud will have to be designed for buckling and compression perpendicular to the grain on the platform and bottom plate of the wall.

If this same example were to be done with the tension and compression forces being determined without considering the dead load, they would both have a value of 6,000 lb. However, the dead load for the individual studs would be superimposed because the assumption is that the dead load follows the nearest stud to the next level down. This would result in a net tension force of 3,000 lb and a net compression force of 9,000 lb.

The third option for consideration is that the wall-floor system acts like a beam on an elastic foundation. The rim joist and wall plates along a wall segment act like a beam that is supported by the studs of the wall, similar to the way the rail on a railroad track is supported by the sleepers or ties. Now, to try to allow as much dead load to counter the uplift force induced in the end stud as possible, the beam should be as stiff as possible relative to the supporting foundation. Therefore, assume that the rim joist is a 2×12-in. nominal (1.5×11.25-in. actual dimensions), the bottom plate of a wall above and the double top plate of the wall below consist of 2×4-in. nominal (1.5×3.5-in. dressed dimensions), and the studs are 2×4-in. nominal lumber. This configuration is shown in Fig. 6.10. If the lumber is Douglas Fir–Larch that is graded as No. 2 and better, then the mechanical and physical properties would be computed as follows:

$$I_{beam} = \Sigma I_{members} = I_{2x12} + 3(I_{4x2}) = 177.98 + 3(0.98) = 180 \text{ in.}^4$$

$$E_{beam} = 1.6 \times 10^6 \text{ psi}$$

$$EI_{beam} = 288{,}000{,}000 \text{ lb-in.}^2$$

$K_{stud} = (E_{perp}A_{stud})/3.0 = (0.89 \times 10^6)(5.25)/(3.0) = 1.558 \times 10^6$ lb/in. The above assumes a 3.0-in. spring length, representing the thickness of the double top plate.

$$k_{foundation} = k_{stud}/s_{studs} = (1.558 \times 10^6)/24 = 64{,}896 \text{ psi}$$

$$s_{studs} = \text{stud spacing}$$

The preceding calculations consider the rim joist single bottom plate of the wall above the floor, and the double top plate of the wall below the floor to act as the beam. It should be noted that I_{beam} can alternatively be established as modulus of inertia of the composite section consisting of rim joist and the track plates, provided sufficient shear connection is provided at the interfaces of rim joist with the top and bottom track plates and at the interface of the two track plates below the rim joist. Following the beam-on-elastic-foundation analysis as outlined by Héteni[10] or Ugural and Fenster,[28] the decay length of the beam (i.e., the maximum distance any of the point loads associated with the dead load from above can have an effect on adjacent studs supporting the beam) is computed as follows:

$$L_{decay} = \frac{\lambda}{4} = \frac{2\pi}{4\left[\dfrac{K_{foundation}}{4(EI)_{beam}}\right]^{1/4}} = \frac{2\pi}{4\left[\dfrac{64{,}896}{4(288{,}000{,}000)}\right]^{1/4}}$$

$$= 18.1 \text{ in.}$$

This distance is less than the stud spacing, so the assumption that dead load can be transferred easily to the end post to be effective in resisting the induced uplift is false, and the designer should consider that only a single stud spacing of dead load is effective unless the post is also supporting a header beam above an opening, where the reaction of the header would counter the uplift forces. This also indicates that attributing more than one stud spacing of the dead load toward increasing the compression forces would not be justifiable. This concept is illustrated in Fig. 6.12, where the applied loads and the induced tension and compression loads in the chords are shown in the top illustration. The effective widths to which the gravity loads can be transferred are shown in the bottom segment of the figure. Notice that the gravity loads in the central section of the wall would not be able to counter the tension load induced on the left-hand chord in the top figure, and in fact, only the left two gravity loads would be effective in reducing the magnitude of the tension load.

6.7.2 Segmented Shear Walls

Segmented shear walls are the traditional "engineered" shear walls, where the end studs are restricted from uplift, and the design philosophy is to ensure that the yielding mechanism is in the sheathing nails and not the framing, hold-down connection, or the shear transfer connections. There is some discussion as to whether or not the tension and compression chords of the wall, as well as the hold-down connection, should be designed for the design seismic forces amplified by the overstrength factor Ω_o to ensure that the energy-dissipating mechanism occurs in the sheathing fastener connections.

While this design method provides the highest design racking capacity for the full-height segments, it ignores any segment of the wall line that contains an opening of any type. Essentially, the free-body diagram of the segmented shear wall uses the diagram of Fig. 6.11 for how the forces are transferred between the framing sheathing, and the tension and compression forces shown in the top figure of Fig. 6.12 are designed for as overturning forces.

The design process consists of the following steps:

1. Analyze the load path to determine the shear load applied to the wall line being considered.

2. Determine the length of full-height wall that has acceptable aspect ratios [sum of the lengths of the segments with aspect ratios less than 3.5:1 (height:length)].

3. Divide the shear load applied to the wall line by the length of the acceptable wall segments to determine the unit shear.

4. Determine the sheathing and fastening combination that provides an equal or greater unit shear value from ANSI/AF&PA SDPWS-08, Tables 4.3A–C.

5. Determine the required overturning connection resistance required using either simple statics for a given wall segment or the equation of $T = vh$, where T is the required hold-down strength, v is the unit shear, and h is the height of the wall segment.

6. Determine the deflection of the wall and compare it against the drift allowed by the building code.

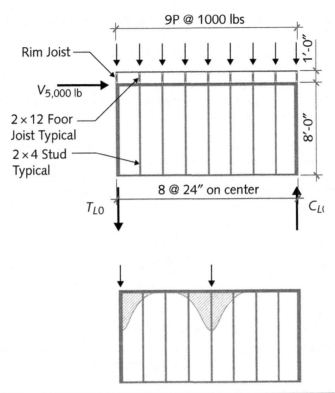

Figure 6.12 Illustration of how effective gravity loads may reduce the magnitude of induced tension overturning forces in a segmented shear wall.

The deflection of the shear wall is determined using one of two equations. The three-term equation that is presented in the ANSI/AF&PA SDPWS-08 is a modification of the four-term equation used historically that is presented as Eq. (C4.3.2-1) in the commentary of ANSI/AF&PA SDPWS-08. In essence, the three-term equation combines the panel shear deformation and the nail-slip term into a single term. The three-term equation for shear-wall deflections is shown as Eq. (6.35):

$$\delta_{SW} = \frac{8vh^3}{EAb} + \frac{vh}{1000G_a} + \frac{h}{b}\Delta_a \tag{6.36}$$

where

v = unit shear caused by the shear load on the wall line (lb/ft)

h = wall segment height (in.)

E = modulus of elasticity of the wall chord members (psi)

A = cross-sectional area of the wall chords (in.²)

b = wall segment length (in.)

G_a = effective or apparent shear stiffness of the wall segment (kip/in.) (Tables A4.3A and B)

Δ_a = total vertical elongation of the wall anchorage system (including the faster slip, device elongation, and so on and the compression deformation under the compression chord, which often includes the effect of compression perpendicular to the grain of the top and bottom plates of the wall and floor sheathing)

The first term of the three-term equation represents the effect of bending of the framing members. The second term represents the effective shear deformation of the wall segment. The third term represents the rigid-body deformations allowed by the anchoring of the wall segment.

The four-term equation for deflection was developed originally by the American Plywood Association (APA)—the Engineered Wood Association. This is shown as Eq. (6.36):

$$\delta_{SW} = \frac{8vh^3}{EAb} + \frac{vh}{G_v t_v} + 0.75 he_n + \frac{h}{b}\Delta_a \tag{6.38}$$

where

$G_v t_v$ = shear stiffness of the sheathing panel (lb/in.) and is tabulated in Tables C4.2.2A and B

e_n = nail slip at the load per nail applied to the wall

The load per nail is determined by dividing the unit shear by the nail spacing. The equation for calculating this term can be found in ANSI/AF&PA SDPWS-08, Table C4.2.2D. The nail slip term used in the four-term equation is provided by equation for only a few fastener types and sheathing classes rather than thicknesses. Many proprietary fastener manufacturers are now providing this information in their evaluation reports.

6.7.3   Force Transfer Around Openings

Walls designed for force transfer around openings, also known as perforated shear walls, typically are designed using rational analysis, which often is a coupled-wall analysis. All the aspect-ratio restrictions placed on segmented shear walls are imposed on the walls designed for force transfer around openings, except that the wall segment considered is the pier next to the opening rather than the entire height of the wall, and the length of the individual pier shall not be less than 2 ft.

The forces at the corners of the opening in the wall can become quite high due to the drag strut action required to provide equilibrium. These forces typically are transferred to the wall piers using blocking and straps, and often the full length of the wall pier needs to have both blocking and straps because of the reversing action of lateral loads.

The difference between the segmented (Fig. 6.13) and the force-transfer (Fig. 6.14) walls is that while both methods assume that the unit shear is distributed over the piers the same, the force-transfer allows the unit shear to be distributed over the entire wall length above and below the opening. Therefore, the unit shear in the top and bottom of the force-transfer method is lower than for the segmented method, but the unit shear in the pier regions are equal for the two methods. The connections required to transfer the loads into and out of the wall for the force-transfer method can be smaller and distributed over a longer length, and the number of overturning anchorage locations is reduced, but the added strapping around the window opening is the cost.

6.7.4 Perforated Shear Walls

Perforated shear walls are designed to take advantage of the added restraint provided to the full-height segments of the shear wall by the sheathing above and below the window and door openings in the wall. This wall option is based on an empirical design method that was proposed by Sugiyama and Matsumoto[25] and validated with full-scale tests by Johnson and Dolan[17] and Heine and Dolan.[13]

This type of construction is similar to the force transfer around openings, but without the additional strapping required. In essence, this type of wall functions as a segmented wall for the first element in the wall, and the rest of the elements function as prescriptive or "conventional construction" wall segments. Therefore, the shear force is not distributed uniformly as unit shear in reality. To simplify the design process, the concept of unit shear is maintained, but in reality, most of the force is transferred at the end of the wall line that produces tension in the end chord or stud.

The perforated-shear-wall method has the following requirements imposed on it:

1. Openings past the end stud can be included in the force analysis, but collectors need to be included to pull the forces into the wall line being considered. The wall length that can be considered as part of the perforated wall is the length from outside to outside of the end-most full-height wall segments that meet the aspect-ratio requirements.

2. The maximum unit shear allowed is 980 plf for seismic and 1,370 plf for wind for single-sided sheathing and no more than 2,000 plf for double-sided sheathing when wind is being considered.

3. Collectors for the shear transfer above openings shall be continuous for the entire wall length. The double top plate typically is used to make this transfer, but other methods to achieve this could be designed.

4. The top and bottom of the wall line must be at one level (i.e., no stepped walls or offset walls.), and the maximum height is 20 ft.

5. The shear anchorage provided to transfer the loads into and out of the wall line shall resist the maximum unit shear $v_{max} = V/(C_o L_i)$.

6. All full-height wall segments shall have anchorage provided to resist a uniform uplift force equal to v_{max}.

7. The end studs of the wall line must be held down with a hold-down device capable of resisting the entire uplift force induced when sufficient dead load to resist the induced uplift is not present.

The design process for designing a wall line using the perforated-shear-wall method follows these steps:

1. Determine the length of the wall line (from end stud to end stud).

2. Determine the unit shear for the wall line as if there were no openings in the length being considered.

3. Determine the height of the maximum-height opening in the wall.

4. Determine the percentage of the wall length that is full-height sheathing.

5. Determine the effective shear capacity ratio C_o from Table 4.3.3.5 in the ANSI/AF&PA SDPWS-08.

6. Divide the unit shear by C_o and determine the required sheathing and nail schedule from Table 4.3A of ANSI/AF&PA SDPWS-08.

7. Determine the end-chord forces $[T = C = Vh/(C_o \Sigma L_i)]$.

6.7.5 Example Wall Design

An example of these three options for design is illustrative. Therefore, consider the wall shown in Fig. 6.13. There is a 4,000-lb force applied at the top from a diaphragm, and assume that the force is due to wind load. The wall can be designed in three different ways: (1) as a segmented wall, (2) as a wall with detailing for force transfer around openings, and (3) as a perforated shear wall.

The difference among the three is the connection hardware used to detail them. The segmented wall will use overturning anchors at the ends of each full-height wall segment or a total of four hold-down devices. The force-transfer option will require only two hold-down anchors (at the ends of the wall line), but it will have to have strapping to transfer the collector forces generated around the opening. The perforated shear wall will require two hold-down anchors (at the ends of the wall line) and additional uplift anchorage along the length of each full-height segment equal to the adjusted unit shear. All will require shear-transfer anchorage along the top and bottom of the wall to transfer the forces into and out of the wall.

Segmented Option

For the segmented option, the design proceeds as follows:

Check aspect-ratio limits (for the narrowest full-height segment):

$$10 \text{ ft}/6.67 \text{ ft} \approx 1.5 < 3.5 \qquad \textit{Okay!}$$

Determine unit shear: V = 4,000 lb:

$$v = 4,000 \text{ lb}/(8 + 6.67 \text{ ft}) \approx 273 \text{ plf}$$

Determine the tension/compression forces in the chord:

$$T = C = vh = (273 \text{ plf})(10 \text{ ft}) = 2,730 \text{ lb}$$

Considering the gravity loads could reduce this force depending on the analysis assumptions that are made concerning the load path. The reduction will be ignored for this example. The resulting tension force would have to be resisted by using hold-down anchors at the end of each full-height wall segment.

Determine the sheathing and nail schedule to use: (See Table 4.3A and Table 4.3B of ANSI/AF&PA SDPWS-08.) Three options are

- Use 5/16-in. plywood siding with 6d galvanized casing nails spaced at 6 in. on center on the perimeter and 12 in. on center on all intermediate supports, or

- Use 5/16-in. wood structural panel–rated sheathing (plywood or OSB) with 6d common nails at 6 in. on center around the perimeter and 12 in. on center along all intermediate supports, or
- Use ½-in. gypsum wall board with No. 6 Type S or W drywall screws 1¼-in. long and spaced at 4 in. on center around the perimeter and 16 in. on center along all interior supports. The gypsum joints must be blocked.

Option for Detailing for Force Transfer Around Openings

The force-transfer option proceeds as follows:

Check aspect-ratio limits (for the narrowest wall pier next to the window opening):

$$5 \text{ ft}/6.67 \text{ ft} \approx 0.75 < 3.5 \qquad Okay!$$

Determine the tension/compression forces in the chord:

$$\Sigma M = 0 = (4{,}000 \text{ lb})(10 \text{ ft}) - T(20 \text{ ft})$$

$$T = C = 2{,}000 \text{ lb}$$

Considering the gravity loads could reduce this force depending on the analysis assumptions that are made concerning the load path. The reduction will be ignored for this example. The resulting tension force would have to be resisted by using hold-down anchors at the end of wall line (out to out of the two full-height segments).

Using the free-body diagram of the top half of the wall, the shear forces in the two piers can be determined using statics:

$$\Sigma F_x = 4{,}000 \text{ lb} - v(8 + 6.67 \text{ ft}) = 0$$

$$v = 273 \text{ plf} \qquad \text{(The same as found in the segmented method.)}$$

If the left-hand half of the wall is used as the free-body diagram, and the forces in the vertical direction are summed, the shear force for the two coupling-beam segments of the wall is found to be 1,000 lb for each beam (above and below the opening).

If the upper left-hand corner is used as the free-body diagram, and the forces are summed in the vertical direction, the vertical force in the pier is 1,000 lb. Now the wall can be divided into rectangles around the opening, as shown in Fig. 6.14b, and simple statics can be used to determine the forces on all edges of the rectangles. The unit shears can be determined simply by dividing the force on the edge of the rectangle by the appropriate dimension, and collector forces are determined as part of the statics analysis.

This option requires two hold-down anchors (one at each end of the wall line) and straps at the corners of the opening to transfer the collector forces. The highest shear is found to be 400 lb/ft. The options for sheathing and associated nailing schedule again are found in Table 4.3A of ANSI/AF&PA SDPWS-08. The options happen to be the same in this case as found in the segmented method if the wood structural panel–rated sheathing is the option chosen. If plywood siding is chosen, ⅜-in. siding with 8d galvanized casing nails spaced at 6 in. on center around the perimeter and 12 in. on center along interior supports would have to be chosen. There is no option available for using a single layer of gypsum wallboard sheathing. The load path characteristic for the Force Transfer Around Openings wall method of analysis is illustrated by Figure 6.15.

Perforated-Shear-Wall Option

For the perforated-shear-wall option, the design process proceeds as follows:

Check aspect-ratio limits (for the narrowest full-height segment the same as for the segmented option):

$$10 \text{ ft}/6.67 \text{ ft} \approx 1.5 < 3.5 \qquad Okay!$$

Determine the percentage of full-height sheathing:

$$\text{Total length } L = 20 \text{ ft}$$

$$\text{Length of full-height sheathing } \Sigma L_i = 8 + 6.67 = 14.67 \text{ ft}$$

$$\text{Percentage} = (14.67/20)100 = 73.3 \text{ percent}$$

Determine the shear capacity adjustment factor (C_o): Interpolating Table 4.3.3.5 of ANSI/AF&PA SDPWS-08 as shown below provides the value of C_o.

Percentage	Opening Height		
	2h/3 = 6 ft-8 in.	7 ft-6 in.	5h/6 = 8 ft-4 in.
70	0.77	0.73	0.69
73.3		0.75	
80	0.83	0.80	0.77

Therefore, $C_o = 0.75$.
Determine the overturning anchorage requirements:

$$T = C = \frac{Vh}{C_o \sum L_i} = \frac{(4,000 \text{ lb})(10 \text{ ft})}{0.75(14.67)} = 3,636 \text{ lb}$$

Determine the unit shear:

$$v_{max} = \frac{V}{C_o \sum L_i} = \frac{4,000 \text{ lb}}{0.75(14.67 \text{ ft})} = 364 \text{ plf}$$

Table 4.3A of ANSI/AF&PA SDPWS-08 shows that the same solution as used for the segmented option will provide sufficient resistance. Only wood structural panels can be used for this solution because plywood siding is not included in this option.

Additional distributed uplift anchorage capable of resisting 364 plf needs to be included in the detailing specification for all full-height segments for this wall, but only two overturning anchors are required instead of four for the segmented option.

6.7.6 Prescriptive Wall Construction

Prescriptive construction should be included in the discussion of timber design because while many residential buildings are designed and built following the rules of the 2009 *International Residential Code* (2009 IRC)[15], there are cases where either a building is being

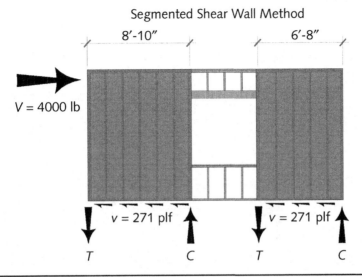

Figure 6.13   Anchorage forces using segmented wall approach.

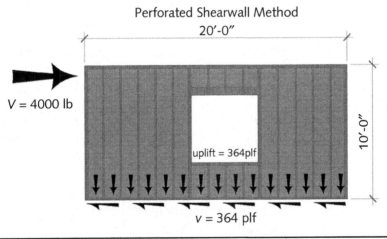

Figure 6.14   Anchorage forces using perforated wall approach.

modified or a component of the building is determined to be outside the scope of the 2009 IRC, and sufficient design needs to be completed to ensure that the component will perform safely and not adversely affect the building. In order to develop a design for this situation, a strength and stiffness assessment of the existing building needs to be completed. However, since usually no overturning anchorage or strong shear and uplift connections are included, the strength and stiffness cannot be calculated from information in the ANSI/AF&PA SDPWS-08.

(a)

Force Transfer Method
Coupling Beam Forces

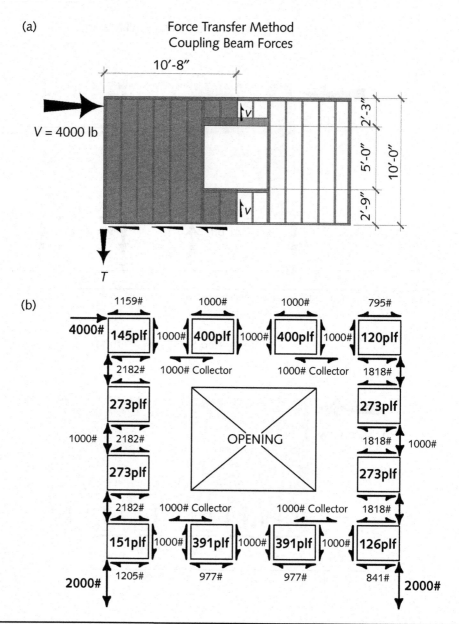

(b)

Figure 6.15 Internal force transfer around wall openings: (a) coupling-beam forces; (b) overall force distribution.

The reader is referred to other sources[8,9] for information that may be used to determine the design values to be considered. Historically, the lateral strength of conventional construction has been considered to be very low. Some committee decisions concerning the contribution of the surrounding structure on overturning resistance have made their way into the building codes and the ANSI/AF&PA SDPWS-08. However, the designer should seriously consider two items: (1) the analysis conducted earlier concerning how gravity loads are transmitted through a light-frame structure (beam on elastic foundation) and (2) walls oriented perpendicular to the wall of interest that are in fact resisting lateral loads as well. Just because the design is conducted using decoupled orthogonal analysis does not mean that the building is not loaded in both directions simultaneously. In fact, the peak load occurs in both directions within a few hundredths of a second of each other, and the uplift is induced on the same corner of the building at the same time by both walls. Therefore, if one is to assume that a return corner wall can provide uplift restraint, one should be sure that there is reserve capacity in the return wall after deducting the forces induced in the wall by the loads in that direction.

6.8 Diaphragms

Diaphragms are the horizontal elements that distribute the lateral loads applied in the interior of the building to the vertical components of the lateral force–resisting system. The system in wood diaphragms is assumed to act like a deep beam, as shown in Fig. 6.16. The difference between a deep beam and a flexural beam is that the shear is assumed to be distributed uniformly across the cross section in deep beams ($v = V/b$), whereas the flexural beams have a shear distribution that is parabolic ($v = VQ/Ib$). The shear is considered to be direct shear and therefore distributed uniformly across the cross section, similar to the way shear load is assumed to be uniformly distributed along a shear wall line. The unit shear is considered to be resisted by the web of the deep beam, which consists physically of the sheathing panels. The bending forces are assumed to be resisted by the flanges of the deep beam, which are physically the chords or edge framing members of the diaphragm (i.e., typically the band joist or end joist, but the bottom plate or top plates of the walls above or below can be included as well). The struts are the reactions of the diaphragms, which are again physically the edge framing. The same members serve as both struts and chords—the loading direction is the only difference. In actuality, they perform both actions simultaneously because the building is loaded in both directions simultaneously.

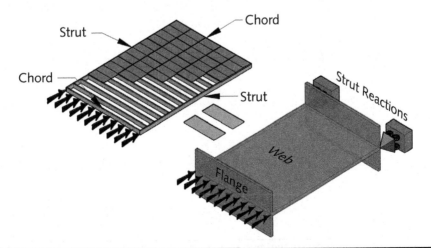

Figure 6.16 Illustration of deep-beam analog used for analyzing timber diaphragms.

Since the sheathing is required to resist the shear forces and transfer them from the middle of the beam to the struts (reactions), there needs to be a mechanism to transfer the shear forces from one sheathing panel to another. This is accomplished by the framing members in that the force in one sheathing panel is transferred to the sheathing nail, into the framing, into the sheathing nail in the next sheathing panel, and finally, into the adjacent sheathing panel. The concept is illustrated in Fig. 6.17. The framing member can be either the main floor or roof joists or blocking. The primary function of blocking is to transfer this force, and all blocking does is increase the number of nails available to transfer this force from one panel to another. Therefore, on larger, higher-aspect-ratio diaphragms, blocking is often used only in the high-unit-shear zones near the struts, and the unblocked condition is used in the central region of the diaphragm.

6.8.1 Typical Rectangular Diaphragms

Most timber diaphragms are rectangular and act as simple deep beams, as described earlier. The design process is relatively simple.

1. Analyze the load path to determine the distributed load on the diaphragm (beam), and determine the reactions.

2. Divide the reaction force by the depth of the diaphragm to determine the unit shear.

3. Determine the load case for each direction of loading. The cases are illustrated with any timber diaphragm design table. There are six different cases. The case depends on the orientation of the load with respect to the continuous joints in the sheathing and the direction of the framing or blocking. The cases are illustrated in Tables 4.2A and 4.2B of ANSI/AF&PA SDPWS-08. The most common load cases are Cases 1 and 3.

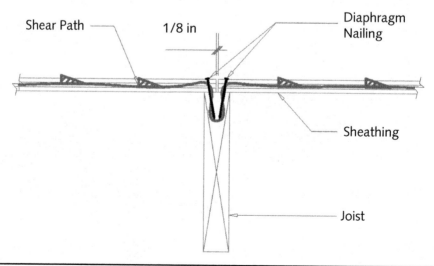

Figure 6.17 Illustration of the concept of transferring shear forces from one sheathing panel to an adjacent sheathing panel.

4. Determine the sheathing, nail schedule, and blocking requirements to resist the load by finding a configuration with equal or higher design values from Tables 4.2A, 4.2B, or 4.2C in the ANSI/AF&PA SDPWS-08, depending on whether blocking is required or not.

5. Determine the tension and compression forces in the chords, and determine the configuration of the framing members required to resist the force. (The compression check does not usually control due to the fact that the sheathing provides continuous bracing against buckling.)

6. Determine the connections required to transfer the forces (a) from the walls loading the diaphragm (perpendicular to the walls themselves) and (b) from the diaphragm strut to the vertical members of the lateral force–resisting system for the story below (usually shear walls).

7. Design collector members required to transfer forces around openings in the diaphragm and/or around openings in the vertical members of the lateral force–resisting system below.

8. Determine the diaphragm deformation.

Each segment of a diaphragm that spans between two lines of resistance for the vertical elements of the lateral force–resisting system on the level below is usually considered to be a simple diaphragm. Thus, for the building plan shown in Fig. 6.18, there is one diaphragm in the E-W direction and one or two diaphragms in the N-S

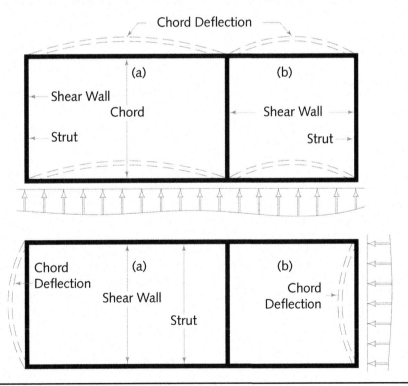

Figure 6.18   Determining the number and size of diaphragms in a building.

direction depending on whether the interior wall is detailed to take load from the diaphragm or represents only a partition wall. The figure shows the deflections of the diaphragm assuming that the interior wall is a shear wall, and the diaphragm therefore would be broken into two diaphragms for analysis.

The restrictions on aspect ratio for diaphragms are provided in Table 4.2.4 of ANSI/AF&PA SDPWS-08. However, for wood structural panel–sheathed diaphragms, the maximum aspect ratios are 3:1 and 4:1 for unblocked and blocked diaphragms, respectively. The aspect ratio is determined by dividing the length (distance between the struts) by the depth (distance between the chords).

The deflection calculation for a diaphragm is almost the same as that used to estimate the deflection of a shear wall. The three-term equation for diaphragm deflection is given by Eq. (6.37):

$$\delta_{diaphragm} = \frac{5vL^3}{8EAW} + \frac{0.25vL}{1,000G_a} + \frac{\sum(x\Delta_c)}{2W} \tag{6.38}$$

where

v = unit shear caused at the strut location (maximum value) (plf)

L = diaphragm length (ft)

E = modulus of elasticity of the wall chord members (psi)

A = cross-sectional area of the diaphragm chords (i.e., Σ of areas of all members associated with chord, possibly the rim joist and double top plate of the wall below) (in.²)

W = diaphragm width (ft)

G_a = effective or apparent shear stiffness of the sheathing and nail slip (kip/in.) (Table A4.2A and A4.2B of ANSI/AF&PA SDPWS-08)

x = distance from the chord splice connection to the strut or reaction for the diaphragm (ft).

Δ_c = chord slip connection slip at the design unit shear imposed on the diaphragm (in.)

The deflection also can be determined using the four-term equation that was developed originally by American Plywood Association (APA). This equation essentially breaks the term with G_a in it into two terms, one for shear deformation of the panel sheathing and one for nail slip. The four-term equation is presented in the commentary of ANSI/AF&PA SDPWS-08 as Eq. (C4.2.2-1).

An example design for the diaphragm shown in Fig. 6.18 in the N-S direction may be as follows: Consider the dimensions of the diaphragm to be 50 ft × 150 ft, with the interior shear wall being located 50 ft from the right end of the building. Assuming that the factored applied load due to wind is 600 plf, the reactions for the two diaphragms would be

$$R_R = WL/2 = (600 \text{ plf})(50 \text{ ft}/2) = 15,000 \text{ lb}$$

$$R_L = WL/2 = (600 \text{ plf})(100 \text{ ft}/2) = 30,000 \text{ lb}$$

The unit shears v then would be calculated as follows:

$$v_R = R_R/W = 15,000 \text{ lb}/50 \text{ ft} = 300 \text{ plf}$$

$$v_L = R_L/W = 30,000 \text{ lb}/50 \text{ ft} = 600 \text{ plf}$$

If the sheathing and framing are oriented so that the load case is Case 1, the options for sheathing, framing, and nail schedule for an unblocked diaphragm on the right portion of the building would be any listed in Table 4.2C of ANSI/AF&PA SDPWS-08. The left diaphragm could be sheathed with any combination of thicker sheathing and larger nails than ⅜-in. rated sheathing with 8d common nails spaced at 6 in. on center around the perimeter of the sheathing panels and 12 in. on center along all intermediate supports.

The struts would have to be connected to the tops of the walls to transfer the unit shear of 300 and 600 lb/ft on the right and left ends, respectively, but the strut above the interior wall would have to be connected sufficiently to transfer 900 lb/ft because both diaphragms contribute to the forces transferred to the interior wall. The chords and any splice connections would have to be designed to resist the induced tension and compression forces calculated by Eq. (6.38):

$$T = C = WL^2/8d \qquad\qquad (6.39)$$

For the right diaphragm,

$$T_R = C_R = (600 \text{ plf})(50 \text{ ft})^2/(8)(50 \text{ ft}) = 3,750 \text{ lb}$$

and for the left diaphragm,

$$T_L = C_L = (600 \text{ plf})(100 \text{ ft})^2/(8)(50 \text{ ft}) = 15,000 \text{ lb}$$

If the chords were made of a single 2×10 and the wall double top plate consisted of two 2×4s that were nailed together sufficiently, the cross sectional area of the chord would be computed as

$$A = (13.88 \text{ in.}^2) + (2)(5.25 \text{ in.}^2) = 24.38 \text{ in.}^2$$

If the lumber is Southern Pine graded as No. 2 and Better, $E = 1.6 \times 10^6$ psi (see Table 4B in NDS Supplement). From Table 4.2C of ANSI/AF&PA SDPWS-08, selecting diagonal lumber sheathing, $G_a = 6$ kips/in. Therefore, neglecting the slip of the chord connections, the deflection of the 100-ft diaphragm would be estimated as

$$\delta_{\text{diaphragm}} = \frac{5vL^3}{8EAW} + \frac{0.25vL}{1,000G_a} + \frac{\sum(x\Delta_c)}{2W}$$

$$= \frac{5(600)(100)^3}{8(1.6\times10^6)(24.38)(50)} + \frac{0.25(600)(100)}{1,000(6)}$$

$$\approx 0.19 + 2.50 = 2.69 \text{ in.}$$

If there were 9 connections in the chords (spaced at 10 ft on center) and each splice slipped ⅛ in. (bolted connections), the contribution of the connections to the diaphragm deflection would be

$$\delta_{\text{diaphragm connections}} = \frac{\sum (x\Delta_c)}{2W} = \frac{[(10+20+30+40)(2)+50](0.125)}{2(50)}$$

$$\approx 0.31 \text{ in.}$$

and the total deflection for the 100-ft diaphragm would be 2.69 + 0.31 = 3.00 in.

Flexible Diaphragms

Flexible diaphragms are diaphragms where the deflection of the diaphragm under the considered loading is equal to or greater than twice the deflection of the vertical lateral force–resisting system elements (for light-frame timber structures, this is usually the shear walls).[22] While the building code has a clause that states that wood-sheathed light-frame diaphragms may be assumed automatically to be flexible, it is rarely true. With the use of adhesives to reduce or eliminate floor squeaks, and with the typical aspect ratio of a wood-sheathed diaphragm, it is difficult to see how the deflections could be more than two times the deflection of the walls below when the walls typically are narrow and have many opening for windows and doors. While the flexible-diaphragm assumption does make the design process easier in apportioning the loads to the elements following tributary area concepts, it may not be the conservative solution, and some shear walls may be underdesigned because of it.

The issue is that if there is any torsional response, flexible diaphragms are not allowed to be used to redistribute the loads to other walls. Therefore, technically, if the floor plan of the building floor below the diaphragm is not symmetric in stiffness, the diaphragm should be rigid and should be analyzed and detailed as such.

Rigid Diaphragms

Rigid diaphragms are more typical in real buildings than flexible ones, whether the designer intentionally designed and detailed the diaphragm to be rigid or not. Rigid diaphragms are required if the building is configured such that the torsional response will occur, such as an open front or a large room in one corner of the building. In this case, the analysis must include consideration of the moment caused by the center of the loading having an eccentricity when compared with the center of stiffness (center of rigidity) of the supporting structure of the diaphragm (typically the position and relative stiffness of the wall lines on the story below the diaphragm).

The effect of torsion is that the walls oriented perpendicular to the loading direction being considered must resist significantly higher loads than calculated for the loading direction in their plane. The effect of the torsion moment must be superimposed on the ordinary lateral loads. The analysis uses standard statics to determine the reactions for the diaphragm, and the diaphragm is designed accordingly, but the deflection of the diaphragm must be kept lower than the associated deflections of the shear walls (or other lateral force–resisting system) below.

The designer often will find it easier to configure the lateral force–resisting system for the lower story to provide a more symmetric configuration with respect to stiffness. This can be accomplished with double-sheathed walls or other configurations. It should be noted that the ANSI/AF&PA SDPWS-08 does not allow timber diaphragms to be used to resist torsional response in masonry-wall buildings.

6.8.2   Irregular Diaphragms

Irregular diaphragms are quite common, but there is little guidance for the designer to follow. Shapes such as L or doughnuts to accommodate courtyards are not uncommon, and the best way to address these is to visualize the deformation patterns to start with. Consider the L-shaped diaphragm shown in Fig. 6.19. The deformation of the diaphragm will be (1) torsional due to the eccentricity between the center of stiffness and the center of load, (2) the difference in stiffness between the long depth of the E-W wing and the depth of the N-S wing will cause the interface to have significantly different lateral deformations, and (3) each wing of the diaphragm will deform similar to a shear wall in the plane of the diaphragm.

The torsional response requires that the diaphragm be designed and detailed as a rigid diaphragm, which most likely will result in the diaphragm being blocked and possibly glued to increase the stiffness. (It should be noted that while adhesives are banned from use in shear walls for buildings located in a high seismic region, there is no ban on the use of adhesives in floors or roofs.) The most effective method of stiffening a diaphragm is to add blocking and increase the nailing of the sheathing.

The effective shear-wall response of the wings of the L causes more serious problems. The effect is that the chords of the diaphragm not only have to resist the bending action associated with the moment imposed on the diaphragm, but they also must resist the overturning forces imposed by the shear-wall action. One could essentially visualize the diaphragm as a cantilevered wall that has elastic supports along the length, where any part of the lateral force–resisting system is located in the story below. Thus, if there were three walls in the building below the N-S wing, then the analog to be analyzed would be similar to that shown in Fig. 6.19. The stiffness of each spring support would be the stiffness of the shear wall located at that location on the floor below. This stiffness could be determined using the deflection equation for shear walls and would assume that the wall would be loaded at the allowable design unit shear.

The designer should be aware that this analysis will result in very large tension and compression forces that need to be resisted using collectors or other framing details. The issue is that one of the wings will be oriented such that the tension and compression forces will be oriented perpendicular to the framing in the common rectangular region of the building. This means that either special framing detailing is required so that there are continuous framing members to transfer these loads, or significant blocking and strapping would be required. This makes attaching the sheathing more difficult because the steel associated with the strapping will make driving nails difficult or impossible.

6.9 Serviceability

There are several issues concerning serviceability with which the designer of light-frame construction should be concerned. These include shrinkage, creep, and vibration.

Shrinkage is reasonable to accommodate by providing gaps between sheathing panels, calculating the shrinkage associated with perpendicular-to-grain orientations of the members, and detailing the interface with other components accordingly. If construction is going to use green lumber, these considerations should be considered mandatory because of the magnitude of the shrinkage that will occur.

However, one also should consider that engineered wood products typically are delivered at moisture contents that are considerably below the equilibrium moisture content expected in a completed structure. These products will swell over time as they equilibrate with the surrounding environment, and gaps should be left for them to swell into. One good example occurs when using tongue-and-groove floor sheathing. The tongue and groove is manufactured to provide out-of-plane continuity to the floor and prevent the differential movement of adjacent panels loaded by gravity loading.

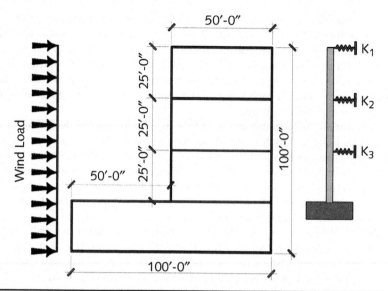

Figure 6.19　An irregular building diaphragm configuration and possible analogy for analysis of forces associated with shear wall deformation of the diaphragm wing.

However, the groove is cut at a different angle than the tongue so that the two merge to provide inherent gapping to allow swelling. If the builder lays the sheathing down and fits the tongue into the groove and then beats one sheet into the other with a sledge hammer and a 2×4 the inherent gapping is removed, and the floor may buckle. If the gaps between sheets of sheathing are not provided, the panels can buckle out of plane as much as 2 in. or more when they swell.

Creep is caused by the structure sustaining relatively high loads for long periods. Most people have seen the ridge of an old barn that is curved and sagging in the middle or a long bookshelf that sags over time. These are both illustrations of creep. There are a couple of options on how a designer might compensate for the fact that timber creeps. The first is to design for full dead load in all load combinations. This effectively means that the structure in most cases will resist a lower load than designed for, and therefore, the creep will be reduced because the magnitude of creep is proportional to the stress level.

Another option is to divide the loads causing deflection into long- and short-term loads and then amplify the long-term deflections and add them to the short-term deflections before comparing the total with the allowable deflection for the application. For green or wet lumber, glued-laminated beams, and wood structural panel construction, the recommendation is to double the long-term deflection and increase it by 50 percent for seasoned conditions. In equation form, this option has the form of Eq. (6.39):

$$Y_{total} = Y_{short\ term} + CY_{long\ term} \qquad (6.40)$$

where

Y = deflection being considered

C = 1.5 or 2.0 for seasoned or wet conditions, respectively

Annoying floor vibration is a common problem, and the United States is one of the few developed countries that does not have mandatory design requirements to prevent it. Humans are susceptible to annoyance from vibration in the range of "4 to 12 Hz". This is due to the fact that various organs in the human body resonate at frequencies in this range. The ANSI/AISC 360-05 Chapter L provides a specific requirement for considering the vibration effects in structural steel buildings. The methods for considering the vibration effects in structural steel floors are presented in AISC *Design Guide 11—Floor Vibrations Due to Human Activity*.[20] Structural steel and concrete floors typically vibrate in the range of 1 to 3 Hz, whereas light-frame floors typically resonates in the range of 10 to 15 Hz. Research by Dolan et al. (1999)[11] tested more than 80 floors with multiple people in occupied and unoccupied conditions to develop a simple method to check a design for annoying vibration. The issue is that the $L/360$ deflection criterion is not sufficient to provide more than protection from plaster cracking on the ceiling that might be attached to the joists. The longer the span of the floor, the more likely it is that the problem of annoying floor vibration will occur.

Therefore, a simple check is made to determine the natural frequency of the floor system and ensure that the floor is stiff enough to keep the floor fundamental frequency high enough to not be perceived as annoying by the occupants. The equation for checking the vibration of a simple beam is presented as Eq. (6.40):

$$f = (\pi/2)\sqrt{\frac{gEI}{WL^3}} \tag{6.41}$$

where

f = fundamental frequency of the floor system (Hz)

g = acceleration due to gravity (386.4 in./s²)

E = modulus of elasticity of the framing (psi)

I = moment of inertia of the framing member (in.⁴)

W = total dead load or dead plus realistic live loads associated with one joist (lb.)

L = span (in.)

For the floor supported directly on the foundation, the preceding equation is sufficient, and if the stiffness is sufficient to provide a fundamental frequency greater or equal to 14 or 15 Hz when dead load or dead plus live loads combined are considered, respectively, most people will consider the floor acceptable in performance. The target frequencies were chosen to eliminate all unacceptable and moderately acceptable rated floors so that only the floors rated acceptable by all the evaluators would be allowed.

If the floor system includes a girder, the combined frequency of the floor joists and girder must be determined. This involves calculating the fundamental frequency of the joist and girder independently and then combining the two frequencies using Eq. (6.41):

$$f = \sqrt{\frac{f_{joist}^2 \times f_{girder}^2}{f_{joist}^2 + f_{girder}^2}} \tag{6.42}$$

While this equation seems simple enough, one finds that the girder needs to be exceptionally stiff before the system will be rated as acceptable. The effect is that the girder must have either significant depth to provide the stiffness or a relatively short span.

There are other design criteria available for review. The British Standard Institution (BSI), *Eurocode 5: Design of Timber Structures, Part 1-1: General—Common Rules and Rules for Buildings* (BS EN 1955-1-1),[5] uses one developed in Sweden, and the Canadian practice employs another. Readers are referred to those standards for guidance if they wish to pursue this serviceability criterion further.

6.10   Seismic Design and Detailing Requirements

Seismic design and detailing are broken out from other types of loads because this is one type of design where the philosophy of design is different. A successful design is one that may be damaged, but the building did not collapse.

The idea is to have the framing members remain elastic and the connections yield. Therefore, the designer should design the anchorage to remain elastic and count on the other framing connections and the sheathing connections providing all the yielding and associated damping. This can be accomplished by designing all the framing connections to yield in Modes III and IV. These are the modes that cause the dowel to bend and yield, whereas Modes I and II result in the bolt remaining straight and elastic and the wood crushing and splitting.

While nailed connections almost always yield in Modes III and IV, split rings, shear plates, and large-diameter bolts yield in Modes I and II and result in brittle failure. Therefore, bolted connections should use slender bolts, even though this causes a larger shear-lag effect (group-action reduction). The improved ductility of the more slender bolt connections will provide a tougher building.

There has been some discussion as to whether wood light-frame design should be required to design collectors, chords, and anchorage while including the overstrength factor Ω_o. Cold-formed-steel light-frame design requires these types of elements be designed for the load combinations that include Ω_o as a way to ensure that the yield mechanism is the one that will provide a ductile or tough response. Designers who wish to ensure that the yield mechanism originates in the sheathing nails and other connections should consider designing the anchorage connections, chords, and collectors for the higher forces.

Finally, designers might consider using timber rivets with steel side plates that are cut to provide a necked-down region where yielding of the plate will be the failure mechanism. This will provide a moment frame with predictable response that could be repaired after the fact. If the framing is heavy timber or glued-laminated beams and columns, the yielded portion of the connection could be cut out, and a new piece of steel could be welded in its place after the frame is straightened.

6.11   International Building Code

Chapter 23 of the 2009 *International Building Code* (2009 IBC),[14] published by International Code Council (ICC), provides the model building code framework for the design of wood structures. Chapter 23 accommodates three design methodologies—allowable-stress design (ASD), load and resistance factor design (LRFD), and prescriptive conventional light-frame construction design. The ANSI/AF&PA NDS-2005 and ANSI/AF&PA SDPWS-08[2] are the primary design standards referenced by 2009 IBC Chap 23. This 2009 IBC chapter also references a number of other design standards, such as those of American Institute of Timber Construction (AITC) and Truss Plate Institute (TPI) for the design of various glulam elements and the design of preengineered metal plate–connected trusses, respectively. The 2009 IBC Chap. 23 also contains a complete set of design requirements that can under certain conditions be used for the design of conventionally light-framed buildings. Significance of the two NDS standards references earlier is that they represent bedrock design references capable of accommodating nearly all forms and methods of wood construction. However, readers are encouraged to familiarize themselves with 2009 IBC Chap. 23 to gain the full insight into the code-permissible design methods and the context within which the ANSI/AF&PA NDS-2005 design standards are used. Further discussion on relationships among building codes, design standards, and the load standard is provided in Chap. 1 of this book. Chap. 1 also provides a review of typical forms of construction associated with structural timber.

References

1. American Forest and Paper Association (AF&PA). 2005. *National Design Specification for Engineered Wood Construction—Commentary.* AF&PA, Washington, DC.
2. American Forest and Paper Association (AF&PA). 2005. *National Design Specification Supplement – Design Values for Wood Construction* (NDS Supplement), AF&PA, Washington, DC.
3. American Forest and Paper Association (AF&PA). 2008. *Special Design Provisions for Wind and Seismic* (ANSI/AF&PA SDPWS-08). AF&PA, Washington, DC.
4. American Institute of Steel Construction (AISC). 2005. *Specification for Structural Steel Buildings* (ANSI/AISC 360-05). AISC, Chicago, IL.
5. American Institute for Timber Construction (AITC). 2004. *Timber Construction Manual.* AITC, Englewood, CO.
6. Anderson, G. T. 2001. "Experimental Investigation of Group Action Factor for Bolted Wood Connections." Thesis submitted in partial fulfillment of the requirements of MS degree, Virginia Polytechnic Institute and State University, Blacksburg, VA.
7. Barrett, J. D., and Foschi, R. 1978. "Duration of Load and Probability of Failure in Wood," Parts I and II, *Canadian Journal of Civil Engineering* 5(4):505–532.
8. Breyer, D., Fridley, K., Cobeen, K., and Pollock, D. 2005. *Design of Wood Structures—ASD/LRFD.* McGraw-Hill, New York, NY.
9. British Standard Institution (BSI). 2004. *Eurocode 5: Design of Timber Structures,* Part 1-1: *General—Common Rules and Rules for Buildings* (BS EN 1955-1-1:2004). BSI, London, UK.
10. Crandell, J. H. 2007. "The Story Behind IRC Wall Bracing Provisions," *Wood Design Focus,* Summer, pp. 3–14.
11. Crandell, J. H., and Martin, Z. 2009. "The Story Behind the 2009 IRC Wall Bracing Provisions," Part 2: "New Wind Bracing Requirements," *Wood Design Focus,* Spring, pp. 3–10.
12. Hetényi, M. 1974. *Beams on Elastic Foundation.* University of Michigan Press, Ann Arbor, MI.
13. Dolan, J. D., Murray, T. M., Johnson, J. R., Runte, D., and Shue, B. C. 1999. "Preventing Annoying Wood Floor Vibrations," *American Society of Civil Engineers (ASCE) Structural Engineering Journal* 125(1):19–24.
14. Heine, C. P. 2001. "Simulated Response of Degrading Hysteretic Joints with Slack Behavior." Dissertation submitted in partial fulfillment of the requirements of PhD degree, Virginia Polytechnic Institute and State University, Blacksburg, VA.

15. Heine, C. P., and Dolan, J. D. 1998. "Effect of Tie-Down Restraint on the Performance of Perforated Wood Shear Walls," *Recent Advances in Understanding Full-Scale Behavior of Wood Buildings*. Forest Products Society, Madison, WI, pp. 23–32.

16. International Code Council (ICC). 2009. *International Building Code* (2009 IBC). ICC, Washington, DC.

17. International Code Council (ICC). 2009. *International Residential Code* (2009 IRC). ICC, Washington, DC.

18. Johansen, K. W., 1949: "Theory of Timber Connections". International association of Bridge and Structural Engineering, Publication No. 9, Bern, Switzerland, pp. 249 – 262

19. Johnson, A. C., and Dolan, J. D. 1997. "Performance of Long Shear Walls with Openings," in *Proceedings of the International Wood Engineering Conference*, Vol. 2, pp. 329–336.

20. Li, X. 1993. "Effect of Stiffness and Mass on the Dynamic Response of Wood Floors." Thesis submitted as partial fulfillment of the requirement for an MS degree at Virginia Polytechnic Institute and State University, Blacksburg, VA.

21. Line, P., and Russell, J. E. 2006. "2005 Special Design Provisions for Wind and Seismic (SDPWS)," *Wood Design Focus*, Spring, pp. 6–8.

22. Murray, T. M., Allen, D. E., and Ungar, E. E. 1997. *Design Guide 11—Floor Vibrations Due to Human Activity.* American Institute of Steel Instruction (AISC), Chicago, IL.

23. Structural Engineering Institute (SEI) of the American Society of Civil Engineers (ASCE). 2006. *Minimum Design Loads for Buildings and Other Structures, Including Supplements Nos. 1 and 2* (ASCE/SEI 7-05). ASCE, Reston, VA.

24. Structural Engineering Institute (SEI) of the American Society of Civil Engineers (ASCE). 2006. *Minimum Design Loads for Buildings and Other Structures* (ASCE/SEI 7-10). ASCE, Reston, VA.

25. Stark, J. W. 1993. "The Effect of Lateral Bracing on the Dynamic Response of Wood Floors Systems." Thesis submitted as partial fulfillment of the requirement for an MS degree at Virginia Polytechnic Institute and State University, Blacksburg, VA.

26. Stiess, T. M. 1994. "The Effect of Cutouts in Joists on the Vibrational Response of Wood Floors." Thesis submitted as partial fulfillment of the requirement for an MS degree at Virginia Polytechnic Institute and State University, Blacksburg, VA.

27. Sugiyama, H., and Matsumoto, T. 1994. "Empirical Equations for the Estimation of Racking Strength of a Plywood-Sheathed Shear Wall with Openings," *Mokuzai Gakkaishi*Vol. 39, No. 8:, pp. 924–929.

28. US Department of Agriculture (USDA). 2010. *Wood Handbook: Wood as an Engineering Material.* General Technical Reports FPL-GTR-190. USDA, Forest Service, Washington, DC.

29. American Concrete Institute (ACI). 2008. *Building Code Requirements for Structural Concrete* (ACI 318-08), ACI, Farmington Hills, MI.
American Forest and Paper Association (AF&PA). 2005. *National Design Specification for Engineered Wood Construction with 2005 Supplement* (ANSI/AF&PA NDS-2005). AF&PA, Washington, DC.

30. Wood, L. 1951. *Relation of Strength of Wood to Duration of Load*, Report R1916. Forest Products Laboratory, Madison, WI.

31. Ugural, A. C., and Fenster, S. K. 2003. *Advanced Strength and Applied Elasticity*, 4th ed. Prentice-Hall, Upper Saddle River, NJ.

CHAPTER 7

Open-Web Steel Joist
Systems

7.1 Introduction

Open-web joists are trusslike steel flexural members used in floor and roof structural framing. In terms of their application to low-rise buildings constructed of cold-formed steel, reinforced masonry, and structural timber, they are often found in bearing and building frame reinforced masonry applications, as illustrated in Chap. 1. Otherwise, open-web joists are used frequently in conventional structural steel framing applications both as roof and floor structural members. Otherwise, they are also used in tilt-up wall construction and occasionally in metal building systems with large bays. Aside from their primary usage as pin-connected floor and roof flexural members, open-web joists also can be employed in moment frames capable of resisting lateral forces and as drag and collector elements in roof and floor diaphragms. Primary benefits of open-web joist systems include weight economy compared with the competing conventional structural steel framing system, ease of construction, and the ability to absorb the plumbing, electrical, and mechanical conduits within their depths, as illustrated in Fig. 7.1. The ability of joist web openings to accommodate such conduits within the depth of the structural framing often can lead to decreased floor heights, resulting in additional economy. Use of open-web steel joists typically is very helpful in satisfying U.S. Green Building Council (USGBC) *Leadership in Energy and Environmental Design* (LEED) requirements because up to about 80 percent of the material currently used in fabrication of open-web joists comes from recycled steel.[33]

There are two primary types of open-web joist system—non-composite and composite open-web joists. The former typically are used in roof framing applications, although they also can be used for the support of floor decks, including those featuring concrete slabs. The latter typically are used in floor applications but also can be used in roof applications, where concrete topping is applied. These two systems are elaborated on in Secs. 7.2 and 7.3, respectively.

The first documented use of open-web joists dates back to 1923.[22] Significant variability in fabrication standards, design procedures, and types of manufactured open-web joists led to establishment of the Steel Joist Institute (SJI), which published its first specification in 1928.[22]

It is important to view open-web joist members as systems rather than as generic truss members with respect to both their analysis and their design. Specifically, the

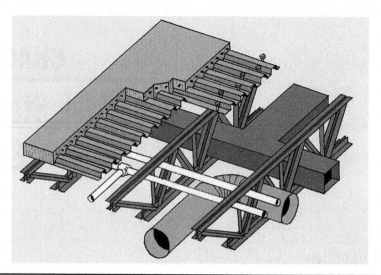

FIGURE 7.1   Floor framing with open-web joists. (*David Samuelson, P.E., Vulcraft.*)

behavioral and design considerations applicable to open-web joist systems, as presented herein, may not always be applicable to generic true trusses, designed and built with various structural shapes for custom applications. This is particularly true with consideration of truss node and element eccentricities. While specific analytical and experimental evidence embodied in the SJI specifications may permit that some eccentricities be neglected in design of an open-web joist element or a connection, this may not be a valid assumption in design of a generic composite truss, where all analytical elements require a strict element-level consideration. An SJI open-web joist is a preengineered structural system. Specifically, the structural engineer of record, responsible for the structural design of the overall building and its foundations, is not directly responsible for the design of the SJI open-web joist but rather for its specification on the basis of required load capacity and deflection performance. An SJI-certified open-web joist manufacturer performs the actual structural design.

With respect to SJI-certified manufacturers, the methodology employed in computing strength of composite joists is based on criteria established by the SJI specification. Any unique aspects pertaining to section and member analysis and design not specifically addressed by the specification may differ somewhat from manufacturer to manufacturer. However, as a condition of SJI membership and certification, the SJI technical staff (i.e., an SJI consulting engineer) checks each particular design methodology to ensure that a particular approach meets the intent of SJI specifications. It should be noted, however, that the SJI design review is not project specific, but rather it establishes that the design methodology of a particular manufacturer employed on a repetitive basis is consistent with the intent of the SJI specifications.

The 2009 *International Building Code* (2009 IBC)[12] references the following SJI specifications for the design of open-web joists:

- 2005 SJI *Standard Specification for Open Web Steel Joists, K-Series* (SJI-K-1.1-05),[26] for the design of short-span non-composite open-web joists

- 2005 SJI *Standard Specification for Longspan Steel Joists, LH-Series, and Deep Longspan Steel Joists, DLH-Series* (SJI-LH/DLH-1.1-05),[27] for the design of long-span and deep long-span non-composite open-web joists

- 2005 SJI *Standard Specification for Joist Girders* (SJI-JG-1.1-05)[25] for the design of non-composite open-web joist girders

- 2006 SJI *Standard Specification for Composite Steel Joists, CJ-Series* (SJI-CJ-1.0-06),[24] for the design of composite open-web joists

Capacities and deflection performance of open-web joist member design by the SJI specifications just listed are communicated to the specifier, most often the structural engineering of record on the overall project, in the form of load tables. The SJI published the first edition of its load tables for non-composite joists in 1929.[22] The 2005 SJI *42nd Edition Standard Specifications—Load Tables and Weight Tables for Steel Joists and Joist Girders, K-Series, LH-Series, DLH-Series, and Joist Girders* (2005 SJI Load Tables),[22] are based on SJI-K-1.1-05, SJI-LH/DLH-1.1-05, and SJI-JG-1.1-05. For a specific member span, the 2005 SJI Load Tables relate a particular SJI open-web joist designation to the corresponding member depth, unit weight, service and factored load capacities per unit length, and ratio of span-to-live-load deflection. Similarly, for the design of composite joists, the SJI *First Edition Standard Specifications for Composite Steel Joists—Weight Tables and Bridging Tables, Code of Standard Practice* (2005 SJI Composite Joist Load Tables),[23] based on SJI-CJ-1.0-06, relates specific joist spans, concrete slab compressive strength, number of headed stud shear connectors, diameter of headed stud shear connectors, and stability bridging pattern to the corresponding member depth, factored load capacity per unit length of the member, and ratio of span-to-live-load deflection. The structural engineer responsible for the overall structural design of the project will select a catalogued joist designation from the 2005 SJI Load Tables or 2005 SJI Composite Joist Load Tables as appropriate and indicate this designation in the construction documents. The joist manufacturer subsequently will perform an actual structural design of the open-web joist. The resulting design may differ from manufacturer to manufacturer. However, in all cases, a design from an SJI-certified manufacturer will result in the strength and deflection performance that will satisfy the intent of the applicable SJI specification and will support the strength and deflection criteria stipulated by the 2005 SJI Load Tables or the 2005 SJI Composite Joist Load Tables, as applicable, for the given joist designation. In short, structural drawings depicting joist layout, designations, and details can be viewed as performance specifications used to communicate the strength and deflection requirements to the joist manufacturer. The reader should note that the open-web joist selections from the 2005 SJI Load Tables and 2005 SJI Composite Joist Load Tables are based on uniform linear gravity load, simply supported condition, and uniform member depth. Any other load and member geometry conditions must be identified explicitly in the construction documents. Examples of non-uniform load requirements include the following:

- Concentrated and patch gravity forces and moments due to rooftop units, cladding elements and framing, support for other structural members, transitions in applied superimposed dead load, and live load in roofs and floors (i.e., open-web joist spanning through both a corridor and an office space)

- Sloping distributed loads due to snow drifts

- Wind and seismic component uplift forces

- Joist axial forces due to chord or collector action when joist chords are used as diaphragm elements in a wind or earthquake load path

- End-member moments, axial forces, and shear forces resulting from end-fixity when the joist is used as a part of a lateral force resisting system

Examples of non-uniform geometry typically are found in specific roof geometries. Examples of custom member depth configurations available from Vulcraft[33] are shown in Fig. 7.2.

Floor and roof framing systems featuring open-web joists contain, in addition to joist members themselves, additional elements. Although they can be used to support wood decking, open-web joist are used most commonly for support of cold-formed steel decking. Depending on whether a particular application is for roof or floor framing and whether a concrete slab tops the steel deck, non-composite, composite, and roof decks can be used. Deck attachments typically are performed with welds, screws, or power-actuated fasteners. Non-composite and composite floor joists typically feature a concrete slab, although some applications contain wood decking, either fastened to non-composite steel deck or fastened directly to the open-web joist. Concrete slabs typically are reinforced with mild wire mesh and deformed rebar reinforcement, although fiber reinforcement has seen increased use as a supplement to or replacement for the conventional mild steel reinforcement. Finally, composite joist systems contain headed studs used to resist shear forces developed at the interface of a steel deck and an open-web joist steel member. Subsequent sections provide further insight in the design of each of the constituent elements of open-web joist systems.

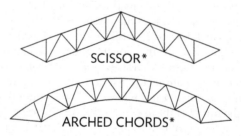

*Horizontal forces due to deflections of these types need to be considered by the design professional.

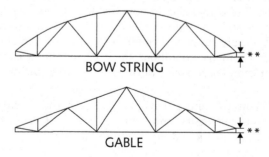

**Contact Vulcraft for minimum depth at ends.

Figure 7.2 Custom joist geometries. (*Vulcraft.[33]*)

7.2 Non-composite Open-Web Steel Joists

Non-composite steel joists can be used in either roof or floor framing, and they can support either a concrete slab on formed steel deck or only the deck itself. Typically, the applications not involving concrete slabs are seen in roof framing, whereas floor applications typically are associated with concrete slabs poured over steel deck. As noted in the preceding section, plywood deck also can be supported by non-composite open-web joists either by attachment to the non-composite deck or by direct attachment to the joist itself, although such applications are not nearly as common as those involving steel deck with or without a concrete slab.

The main characteristic of non-composite joists is that they do not rely on increased section properties and strength due to coupling of the concrete slab and the steel joist into a cohesive composite member, nor do they posses a mechanism at the interface of the concrete slab and the joist for such coupling to occur.

Typical non-composite open-web joist components and the associated terminology are illustrated in Fig. 7.3. As can be seen, horizontal top and bottom joist elements are referred to as *chords*. Their primary role is to resist axial forces derived from decoupling the flexural moment in the joist member. Additionally, since open-web joists are not true trusses, they support gravity and uplift loads applied along the entire top chord length, not just at the truss nodes. Consequently, flexural and shear resistance of the top chord is used in resisting gravity and uplift forces applied between truss nodes, known as *panel points*. A single open-web joist panel is the distance between two successive panel points. The primary role of web members is to provide resistance against shear forces. Bearing seats serve as pin-connected end-member anchorages to the supporting members. Joist depth is defined as the out-to-out depth of the entire member. The overall length of an open-web joist is defined as the end-to-end distance of the entire member, inclusive of bearing seats. The reader should not confuse this length with the joist span and design length. *Span*, or *base length*, is the term used to designate the *bay length*, or the length between support points of the two open-web joist ends. For a steel supporting member, the support point is defined as the center of the member. For a supporting wall, the support point is defined as the wall mid-depth. Depending on the specific conditions, span can be shorter or longer than the overall joist length. Finally, the *design length* is defined as the span used in calculations of joist moment and shear demands. The 2005 SJI Load Tables define this length as span minus 4 in. The design

JOIST DEFINITIONS

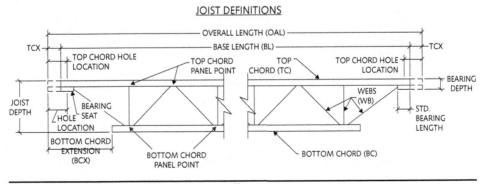

FIGURE 7.3 Custom joist geometries. (*Vulcraft*.[33])

length represents the actual flexural length of the member. The 4-in. reduction in length reflects the design assumption that the centroids of the web and the chord members coincide at 2 in. from the end-span points. The joist design length and span definitions are illustrated in Fig. 7.4. Finally, it is often necessary to extend the joist members past the end of the span to accommodate cladding details, short cantilevers, or floor segments extending past the span points. This is accomplished by providing top chord extension, as designated by the dimension *TCX* in Fig. 7.3.

The types of steel shapes used in fabrication of top chord, bottom chord, web members, and bearing supports depend on the type of joist, the load and span demands, and the particular manufacturer involved in the design and fabrication. The most typical joist element types are illustrated in Fig. 7.5. Top and bottom chords typically consist of hot-rolled double-angle sections. Short-span joists are in some cases also fabricated with cold-formed double-angle sections. Hot-rolled sections sizes are not

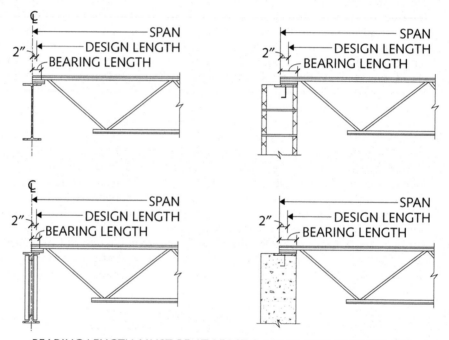

BEARING LENGTH MUST BE AT LEAST 2 1/2 IN. FOR BEARING ON STEEL MEMBERS.

BEARING LENGTH MUST BE AT LEAST 4 IN. FOR BEARING ON CONCRETE AND MASONRY MEMBERS.

FOR JOISTS BEARING ON SUPPORTS WITH DIFFERENT ELEVATIONS RESULTING IN A SLOPE OF AT LEAST 1/2:12, THE SPAN AND DESIGN LENGTHS ARE DEFINED AS LENGTHS ALONG THE SPAN.

FIGURE 7.4 Definitions of joist spans.

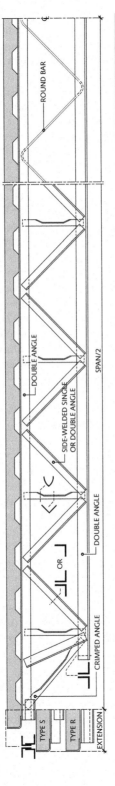

FIGURE 7.5 Typical joist elements.

necessarily the standard American Society for Testing and Materials (ASTM) shapes found in the American Institute of Steel Construction (AISC) *Manual of Steel Construction* (AISC 325-05)[3] and typically are treated as proprietary data of the manufacturer. Other members, such as hot-rolled T-sections, also are used for chord members, although they are not common in the United States. Web members typically consist of side-welded single or double angles or single angles concentrically crimped in between chord angles. For applications involving short-span joists, web members can be fabricated using a continuously bent round bar. End bearing seats in most cases are configured by side welding an inverted double-angle section to the top chord. Open-web joist elements are joined together entirely via shop welding. Open-web joists typically are bolted to column supports and welded to beam and joist girder supports. Although not guaranteed in each case, non-composite open-web joists in most situations feature a single size of the top chord throughout the joist length. The same is true with the bottom chord. Multiple types of web members can occur within a single open-web joist, maximizing the overall economy.

The first number in a joist designation indicates the joist depth, whereas the letter or letters following this number indicate the joist series. Finally, the last number in a joist designation indicates a chord size within a particular series. It is a relative measure of the weight of the chords within a particular series—the higher the number, the heavier is the chord size. For example, an 18K3 joist is an 18-in. deep K-Series joist with a #3 chord size. Also, the number relates the chord sizes across different depth series. For example, a 16K4 joist and an 18K4 joist will have a similarly sized chord. Non-composite SJI open-web joists come in several types, each targeting a specific application. These types can be summarized as follows:

- *K-Series* joists range from 10 to 30 in. deep, with intermediate depths available in 2-in. increments, and available size designations from 10K1 to 30K12. The K-Series joists apply to spans ranging from 8 to 60 ft in length, with the required ultimate load capacity of 190 to 825 lb/ft and service live load causing deflection of 1/360 of the span ranging from 63 to 550 lb/ft. The K-Series joists have a maximum span-to-depth-ratio of 24, as stipulated in Sec. 5.2 of SJI-K-1.1-05. The standard depth of K-Series joists bearing a seat is 2.5 in. They are the most frequently employed joist type in both floor and roof applications. By definition, K-Series joists are available only in parallel-chord configurations. Therefore, when a sloped-roof configuration is required, it can be achieved by an adjustment in support member elevations. The K-Series joists can be used in moment-resisting frames. When this is done, the bottom chords axially restrain only after all dead loads have been applied, eliminating moments due to dead load at end restraints. The K-Series joists are designed using SJI-K-1.1-05. The 2005 SJI Load Tables contain the allowable stress design (ASD) and load and resistance factor design (LRFD) load capacities for K-Series joists designed using SJI-K-1.1-05.

- *KCS-Series* joists are similar to K-Series joists. The distinction between the two types is that KCS-Series joists are designed for constant shear capacity across the entire span. The main applications of KCS-Series joists are in situations involving non-uniform forces and highly varying load diagrams among members, whereby the benefits of fabrication repetition outweigh the benefits of the design for shear based on highly variable shear diagrams changing on a given project on a member-to-member basis. The KCS-Series joists are designed according to SJI-K-1.1-05. The 2005 SJI Load Tables for KCS joists are expressed in terms of moment and shear capacities. The tables provide joist moments of inertia for a direct consideration of deflection by the structural engineer. Although a KCS-Series joist is designed for uniform shear strength, the designer still must indicate all non-uniform load conditions in the construction documents. The reason for this is the fact that the joist manufacturer must have the accurate definition of loading for flexural design of the joist top chord. The KCS-Series joists are designed for a full stress reversal. Specifically, all web members in tension are also designed for compression of the same magnitude, and vice versa. The only exception to this are the end web members, which are always in tension under gravity loads. The available factored moment capacities range from 258 to 2,749 kip-in., and KCS-Series joists are designed for a factored uniform load of 825 lb/ft and a concentrated factored force of up to 13.8 kips. As noted previously, as a matter of fabrication convenience, top and bottom chords typically do not vary in size along the joist span, resulting, as a consequence, in a uniform moment capacity along the span. In KCS-series joists, the design for constant moment along the span is deliberate. The moment capacity is constant within all interior panels, and it tapers to zero from the first interior panel point to support.

- *Joist substitutes* are auxiliary joist members conforming to the SJI-K-1.1-05. As depicted in Fig. 7.6, they are used in short-span and short-cantilever applications where application of the smallest available K-Series joist (10K1) would provide capacity far exceeding the required strength and therefore would be impractical and uneconomical. The joist substitutes also can be used as lateral braces to steel columns and masonry walls. When used as flexural members, joist substitutes are used in spans not exceeding 10 ft, with ultimate load capacities ranging from 267 to 825 lb/ft. Joist substitutes can be used in cantilevers not exceeding 6.5 ft in length, with ultimate load capacities ranging from 215 to 825 lb/ft. The joist substitutes are 2.5 in. deep, making them convenient with use in combination with K-Series joists. The joist substitutes bear the designation 2.5K1, 2.5K2, and 2.5K3. The section moment of inertia is provided for each of the available joist substitutes, allowing the structural engineer to evaluate the deflection directly. Joist substitutes can come in the form of hot-rolled C-sections, hot-rolled angles welded into a C-shape, cold-formed tubes, cold-formed C-sections, and double shapes combining the aforementioned shapes in a back-to-back fashion.[11] The 2005 SJI Load Tables provide the load capacities and section properties for joist substitute sections.

- *Joist extensions*, depicted in Fig. 7.5 and discussed earlier in this text, also conform to SJI-K-1.1-05. Two types of joist extensions, Type S and Type R, exist. The former is formed by extending the top chord angle over the bearing support for up to 4.5 ft in length. The latter consists of extending the continuously connected top chord and bearing angle support for lengths of up to 6 ft. Both the extension types are illustrated in Fig. 7.5. The available Type S extension designations are tabulated in the 2005 SJI Load Tables, range from S1 to S12, and provide uniform factored load capacities of between 150 and 825 lb/ft, with permissible service live load causing the deflection of 1/360 of the span ranging from 41 to 550 lb/ft. Similarly, available Type R extension designations range from R1 to R12, provide uniform factored load capacities of between 175 and 825 lb/ft, with permissible service live load causing the deflection of 1/360 of the span of 32 to 550 lb/ft. Joist extensions and the loads supported thereby must be communicated to the joist manufacturer by noting both in the construction documents.

- *LH-Series* and *DLH-Series* non-composite open-web joists are used in long-span applications exceeding the capacity of *K-Series* and *KCS-Series* joists. Unlike K-Series and KCS-Series joists, LH-Series and DLH-Series joists are available in custom non-parallel and non-horizontal chord configurations. Several custom configurations are depicted in Fig. 7.2. Ranging from 18 to 48 in. in depth, LH-Series joists can span up to 96 ft and yield factored load capacities from 261 to 1,602 lb/ft. Service live loads causing deflection of 1/360 of the span range from 83 to 569 lb/ft. Deep long-span joists under the DLH-Series feature member depths ranging from 52 and 72 in. and an ultimate load capacity range of 312 to 1,059 lb/ft. DLH-Series joists can support spans of up to 144 ft. The bearing seat height for the #18 and #19 LH-Series and DLH-Series joists is 7.5 in. For other joists within these type series, bearing seats are 5 in. high. The long-span joists comprised by the LH-Series and DLH-Series conform to SJI-LH/DLH-1.1-05. Other derivatives of long-span joists, such as the Vulcraft SLH-Series joist capable of spanning up to 240 ft, are also found in use. These joists conform to specifications developed by individual manufacturers.[33] Although not specifically acknowledged by the building code, the use of such products and corresponding specifications can be justified under the guidance of 2009 IBC, Sec. 104.11, which permits alternative materials, design and methods of construction. Implications of this 2009 IBC provision are explored further in Chap. 1 of this book.

- *JG-Series* joists cover joist girders. These members are employed as primary framing members charged with supporting other open-web joists conforming to one of the categories just listed. Unlike preceding joist types, joist girders are not in direct contact with decking. Instead, they provide support for point reactions transferred from bearing seats of other joists. These supports are located at discrete joist girder panel points. Figure 7.7 shows typical details involving a system of open-web joists supported by joist girders. As can be seen, the supported open-web secondary joists can serve as brace support points for both girder chords, eliminating the need for supplemental bridging bracing. Since joist girders are located primarily along the main framing column lines, they can be integrated into the lateral force resisting system of the building by forming a moment connection at the column. Typical moment connection details are shown in Fig. 7.8. As can be seen, moment connections involving joist girders are formed by axially restraining both chords at the support. Under applied gravity load, the bottom chord of an end-fixed joist girder is in compression at the chord ends. Consequently, to facilitate a more efficient design, the bottom chord in end-fixed joist girders typically is restrained at the column only after the dead loads have been applied, eliminating moment due to dead load at member ends.[25] Joist girders are analyzed and designed as true trusses, with all loads applied at panel points and with no joist elements resisting flexural effects. As a result, the weight of the joist girder itself is lumped into point loads at individual panel points at which supported joist reactions are applied. The 2005 SJI Load Tables provide weights of joist girders per linear foot for a given joist girder depth, span, magnitude of a typical supported point

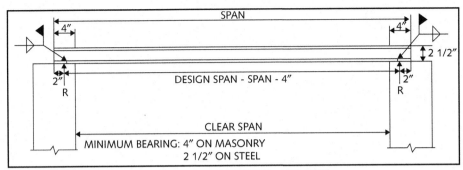

SPAN

4"

4"

2 1/2"

DESIGN SPAN - SPAN - 4"

2"
R

2"
R

CLEAR SPAN

MINIMUM BEARING: 4" ON MASONRY
2 1/2" ON STEEL

NOTE: 2.5K SERIES NOT U.L. APPROVED.

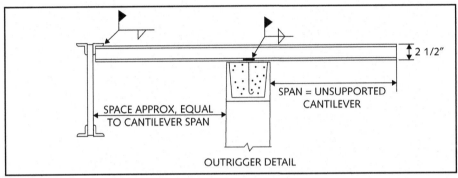

2 1/2"

SPACE APPROX, EQUAL
TO CANTILEVER SPAN

SPAN = UNSUPPORTED
CANTILEVER

OUTRIGGER DETAIL

NOTE: 2.5K SERIES NOT U.L. APPROVED.

FIGURE 7.6 Joist substitute applications. (*Vulcraft.*[33])

load, and spacing of the supported load. These weights must be incorporated in the supported point load reactions communicated to the joist manufacturer. The required depth, the number of the supported loads, and their magnitude and spacing are communicated to the joist manufacturer via joist girder designation in the form of $xGyNzF$ or $xGyNzK$, where the value x indicates the joist girder depth in inches, the value y indicates the number of the spacing of the supported joist members, and z indicates the magnitude of the supported joist reaction. The value z is followed by the letter F if the indicated reaction is a factored load and the letter K if the indicated reaction is a service load. For example, 16G4N12F indicates a 16-in. deep joist girder supporting three joist reactions located at four equal spaces between joist girder ends. The supported joist reactions are factored forces of 12 kips. When unequal joist reactions or multiple spacing is present, a diagram indicating joist reactions and their locations along the span must be shown in structural drawings. Joist girders are available in a number of non-parallel-chord geometries, facilitating many roof slope geometries without the need of sloping the roof framing. The primary standard for the design of JG-Series joists is SJI-JG-1.1-05. Unlike other joist series, the 2005 SJI Load Tables do not provide service live loads causing deflection of 1/360 of the span for serviceability considerations. However, deflection of a joist girder can be computed directly using an approximate moment of inertia, as illustrated later in this chapter.

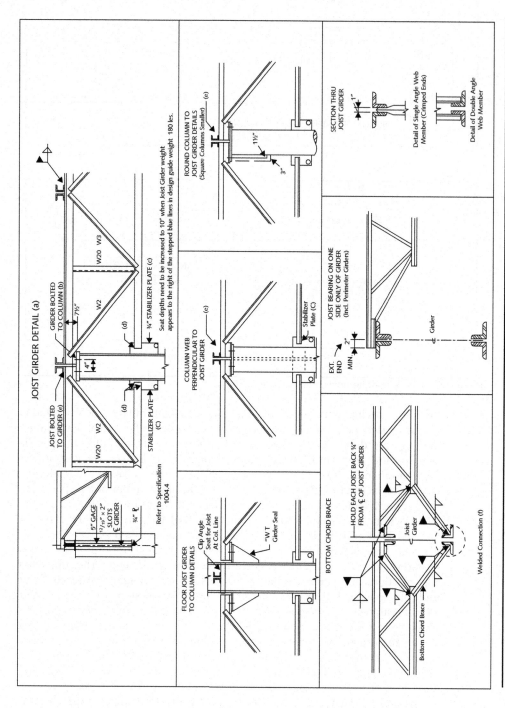

FIGURE 7.7 Joist girder details. (*Vulcraft.*[33])

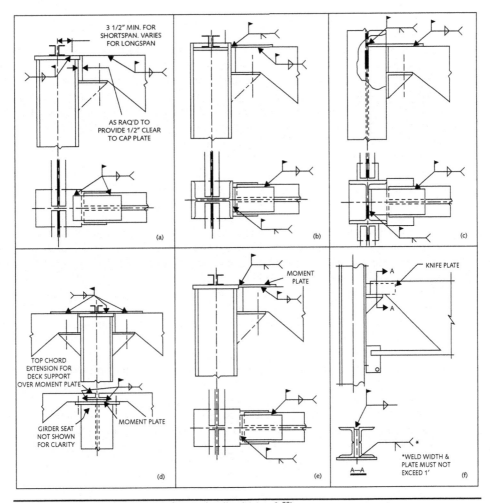

FIGURE 7.8   Joist girder moment frame details. (*Vulcraft.*[33])

The SJI specifications used in the design of the types of joists just described are similar in structure. Each respective specification defines the scope of the joist series it applies to and stipulates permissible materials for joist design and fabrication, explicit strength calculations and criteria for individual joist elements such as chords, web members, and connections, and stability bracing requirements for the overall member. Furthermore, each specification lays out serviceability criteria, manufacturing requirements, applicability guidelines, and any specific considerations required for the joist series covered by a particular specification. The latest editions of the IBC-referenced open-web steel joist design specifications and accompanying design load tables that can be obtained from the SJI at www.steeljoist.org. The SJI specifications and accompanying load tables typically are also incorporated in the manufacturers' catalogues. Manufacturers' joist product catalogues contain additional helpful design information, such as construction details, manufacturer-specific design information, and other auxiliary design aids.[33] The following is an overview of the design requirements for the most common non-composite open-web joist—K-Series joist—embodied in SJI-K-1.1-05.

Section 4.1 of SJI-K-1.1-05 permits both ASD and LRFD as the design basis of K-Series joists. This review focuses on the latter. The K-Series joists can be designed using either cold-formed or hot-rolled elements. Per Secs. 2 and 3.2 of SJI-K-1.1-05, the joist chords must be designed with a material with the nominal yield strength of 50 ksi, whereas web elements can be of either 36 or 50 ksi yield strength. Chapter 3 of this book provides a review of material requirements and a list of the materials for joist design and fabrication permitted by Sec. 3.1 of SJI-K-1.1-05.

Applicable load combinations are those prescribed by the applicable building code. The design professional with overall responsibility for the project must communicate the applicable load combinations to the joist manufacturer either by direct notation of each load combination to be considered or by reference to the applicable building code, such as 2009 IBC. When loading combinations are not stipulated in the construction documents, Sec. 4.1 of SJI-K-1.1-05 stipulates that the load combinations of the American Society of Civil Engineers (ASCE) *Minimum Design Loads for Buildings and Other Structures, Including Supplements Nos. 1 and 2* (ASCE/SEI 7-05)[29] be used by the structural engineer responsible for the joist design.

As noted in Sec. 4.1 of SJI-K-1.1-05, for any design provisions not explicitly provided in SJI-K-1.1-05, the user must refer to the AISC *Specification for Structural Steel Buildings* (ANSI/AISC 360-05)[2] for hot-rolled steel elements and to American Iron and Steel Institute (AISI), *North American Specification for the Design of Cold-Formed Steel Members* (ANSI/AISI S100-07)[4] for cold-formed steel elements.

Flexural strength of a non-composite open-web joist is determined from Eq. (7.1). It is determined simply by coupling the axial strengths of the top and bottom chords at the joist mid-span. This flexural model is shown in Fig. 7.9. For gravity loading, the bottom chord is in tension, and the top chord in compression. Obviously, for a loading case with net uplift, the top chord is in tension, and the bottom chord is in compression.

$$\phi M_n = \phi F(d_j - y_{tc} - y_{bc}) \tag{7.1}$$

where

ϕF = lesser of the top and bottom chord axial strength

d_j = overall joist depth, as illustrated in Fig. 7.9

y_{tc} = top chord geometric centroid measured from the outer surface of horizontal chord legs, as illustrated in Fig. 7.9

y_{bc} = bottom chord geometric centroid measured from the outer surface of horizontal chord legs, as illustrated in Fig. 7.9

For chords in tension, the axial strength is determined using SJI-K-1.1-05, Sec. 4.2(a), as follows, provided by Eq (7.2):

$$\phi F = \phi_y F_y A_g \tag{7.2}$$

where

ϕ_y = strength-reduction factor for tension yielding

= 0.90

F_y = tension chord yield stress

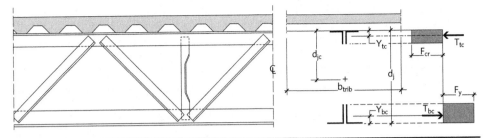

FIGURE 7.9 Non-composite open-web joist flexural model.

$= 50$ ksi

A_g = tension chord gross cross-sectional area

No criteria for net fracture limit state are explicitly provided in SJI-K-1.1-05. However, when a tension chord possesses holes, the effect of net section fracture should be considered. Holes can occur due to attachment of architectural ceiling and mechanical, plumbing, and electrical components. When the top chord is in tension due to net uplift, net fracture should be considered if the deck is attached using mechanical fasteners. Reduction in cross-sectional area typically is significant only in very light joists with small top chord cross-sectional areas. Also, since chords generally are fabricated with a single size of chord over its entire length, the farther the tension-chord holes are from the joist mid-span, the less critical they are in the consideration of flexural strength. Tension chord net fracture strength can be considered using ANSI/AISC 360-05 Sec. D3 and computed using Eq. (7.3). The same equation can be used when considering the net fracture strength of a cold-formed tension chord with $\frac{3}{16}$ in. or less in thickness per ANSI/AISI S100-07 Chap. E and App. A.

$$\phi F = \phi_u F_u A_e \tag{7.3}$$

where

ϕ_u = strength-reduction factor for tension fracture

$= 0.75$ for net fracture strength calculated per ANSI/AISC A360-05

$= 0.65$ for net fracture strength calculated per ANSI/AISI S100-07

F_u = tension-chord yield stress

$= 50$ ksi

A_e = tension-chord effective net cross-sectional area

$= A_n$ for chords in tension

A_n = tension-chord net effective area

$= A_g - A_{hole}$

A_{hole} = area of the chord material displaced by the hole

The determination of hole diameter in net fracture checks depends on the type of connector for which the hole is intended. As required by ANSI/AISC 360-05 Chap. B, the hole diameter in net fracture check equals the nominal diameter of the hole, given in ANSI/AISC 360-05 Chap. J for various bolt diameters plus $\frac{1}{16}$ in. When net fracture is considered using ANSI/AISI S100-07, bolt-hole sizes are given in Table E3a of ANSI/AISI S100-07. For screws, the nominal diameter of the screw can be used as the hole diameter. Research has shown that power-actuated fasteners with nominal diameters of up to 0.146 in. do not have an adverse effect on the load-carrying ability of open-web joists with chord sizes as small as No. 2, such as 12K2, 14K2, or 16K2.[9] For bigger fastener diameters or for No. 1 chord sizes, such as 10K1, 12K1, or 14K1, the user can perform the net fracture check using hole diameter equal to 1.1 times the nominal diameter of the power actuated fastener employed.[7] Other than for deck attachments via mechanical fasteners to hot-rolled chord angles, holes should be avoided whenever possible in tension chords. To this end, the structural engineer with overall responsibility for the project may choose to restrict the use of screw- and bolt-based attachments in the bottom chord in favor of welds, clamps, and power actuated fasteners. Deck attachments are performed using small diameter fasteners that typically do not reduce a hot rolled tension chord cross section to the point where net fracture strength would govern over gross yielding, and as discussed earlier, power actuated fasteners have been shown not have an adverse effect on joist flexural strength even in least favorable configurations of chord size and fastener diameter. Net fracture, however, may govern in cold-formed tension chords where the net fracture strength of the chord is calculated per ANSI/AISI S100-07. Consequently, the construction documents should indicate the type and location of any penetrations in joist tension chords, allowing the joist design engineer to consider them in design.

The SJI-K-1.1-05 Sec. 4.3 mandates that the maximum slenderness ratio l/r of the tension chord not exceed 240. In considering this limitation, the length l is the distance between successive panel points, and r is the radius gyration corresponding to the horizontal chord axis.

Axial strength of hot-rolled chords in compression is determined per SJI-K-1.1-05, Sec. 4.2(b) using Eq. (7.4). The compressive strength of cold-formed steel chords is determined in accordance with ANSI/AISI S100-07 Chap. C. Chapter 4 of this book provides additional background on the design of cold-formed steel members, including those in axial compression. As noted previously, either the top or the bottom chord can be in axial compression depending on the magnitude of loads and load combination under consideration.

$$\phi F = \phi_c F_{cr} A_g \tag{7.4}$$

where

ϕ_c = strength-reduction factor for compression buckling
    = 0.90

F_{cr} = compression buckling stress

    = $Q\left(0.658^{QF_y/F_e}\right)F_y$ when $l/r \le 4.71[E/(QF_y)]^{1/2}$

    = $0.877F_e$ when $l/r > 4.71[E/(QF_y)]^{1/2}$

Q = slenderness-reduction factor prescribed by ANSI/AISC 360-05, Sec. E7

F_y = compression chord yield strength
    = 50 ksi

F_e = elastic buckling stress

    $= \dfrac{\pi^2 E}{(l/r)^2}$

l = length between joist panel points

r = radius of gyration corresponding to the chord horizontal axis

A_g = compression chord gross cross-sectional area

E = steel modulus of elasticity

= 29,000 ksi

Although a double-angle chord possesses a single axis of symmetry, research shows that its compressive strength can be calculated considering only flexural buckling, as provided by Eq. (7.4), and that flexural-torsional buckling, ordinarily applicable to singly symmetric cross sections, need not be considered.[17] As noted earlier, open-web joist members should be viewed as systems for which various design aspects such as this are established through extensive research based on consideration of specific framing methods employed therein. Consequently, they may not necessarily apply to other forms of truss assemblies.

The SJI-K-1.1-05 Sec. 4.3 stipulates that the l/r ratio for the compression chord, as defined in Eq. (7.4), not exceed 90 in. interior panels and 120 in. end panels. Furthermore, SJI-K-1.1-05, Sec. 4.4(a) requires that the slenderness ratio l_v/r_v of the compression chord does not exceed 145. The variable r_v indicates radius of gyration about the vertical chord axis, whereas l_v indicates the unbraced length in the direction perpendicular to the joist span. Discrete bridging lines, discussed later in this chapter, form the bracing points. In the case of a top chord, bracing also can be accomplished by the deck attachments to joist via puddle welds, screws, or power-actuated fasteners. The SJI-K-1.1-05, Secs. 4.4(a) and 5.8(e) state that the top chord can be considered fully laterally braced when the deck attachments are not spaced further than 36 in. This requirement is satisfied routinely because deck attachments required for diaphragm resistance typically are spaced at increments of less than 18 in. for commercially available formed steel decks.

As noted previously, open-web steel joists are not true trusses. This is particularly true for the top chord, whose flexural resistance is relied on for resisting the loads imposed by the steel deck between the joist panel points. As provided by SJI-K-1.1-05 Sec. 4.4(a), the bending due to uniformly distributed load between panel points can be ignored when the panel points are spaced at no more than 24 in. When the top chord panel-point spacing exceeds 24 in., or when concentrated forces occur between panel points, top chords must be checked for bending. Consequently, they also must be checked for combined axial force due to joist flexure and bending moment due to the loads supported between chord panel points. Finally, the portion of the top chord located between the support and the panel end must be designed for bending and shear resulting from the eccentricity of the support reaction with respect to the first panel point. Consideration of element eccentricities in the joist design is discussed later.

Bottom chords also can experience combined axial and bending effects. While the bending effects due to uniformly distributed loads applied between panel points by the steel deck are anticipated in the joist design, any concentrated forces or patch loading in excess of uniformly distributed live loads must be communicated to the joist manufacturer in the construction documents. Such concentrated forces can occur due to the support of mechanical units or miscellaneous architectural features such as parapet braces. As is the case with the top chord, the bottom chord also can experience combined bending and axial force, where bending typically is caused by the attachments of architectural features or mechanical, plumbing, and electrical elements between the panel points. The practical aspects of considering concentrated forces are discussed later.

Regardless of whether ASD or LRFD is used, SJI-K-1.1-05 Sec. 4.2(c) stipulates that the moment strength of a joist element be determined using elastic section modulus. The moment strengths of both the top and bottom chords are determined using Eq. (7.5):

$$\phi M = \phi_b F_y S_x \qquad (7.5)$$

where

ϕ_b = strength-reduction factor for bending

= 0.90

F_y = chord yield stress

= 50 ksi

S_x = chord section modulus with respect to its horizontal axis

At panel points, combined bending and axial force in the chord must satisfy the interaction provided by Eq. (7.6). Half-way between panel points, combined bending and axial force must satisfy the interaction provided by Eq. (7.7).

$$(P_u/A_g) + (M_u/S_x) \leq \phi F_y \qquad (7.6)$$

where

P_u = factored axial force in the chord

A_g = chord gross cross-sectional area

M_u = chord factored moment at the panel point

S_x = chord section modulus with respect to its horizontal axis

ϕ = strength-reduction factor

= 0.90

F_y = chord yield strength

= 50 ksi

When $P_u/(\phi_c F_{cr} A_g) \geq 0.2$, Eq. (7.7a) applies. Otherwise, the interaction of moment and axial force should be considered using Eq. (7.7b).

$$\frac{P_u}{\Phi_c A_g F_{cr}} + \frac{8}{9}\left[\frac{C_m M_u}{\left(1 - \dfrac{P_u}{\Phi_c A_g F_e}\right) Q\Phi_b S_x F_y}\right] \leq 1.0 \qquad (7.7a)$$

$$\frac{P_u}{2\Phi_c A_g F_{cr}} + \left[\frac{C_m M_u}{\left(1 - \dfrac{P_u}{\Phi_c A_g F_e}\right) Q\Phi_b S_x F_y}\right] \leq 1.0 \qquad (7.7b)$$

where

P_u = factored axial force in the chord

F_{cr} = buckling stress, as defined by Eq. (7.4)

ϕ_c = strength-reduction factor for compression

  = 0.90

A_g = chord gross cross-sectional area

C_m = moment diagram shape factor

  = $1 - 0.3P_u/(\phi_c A_g F_e)$ for end panels

  = $1 - 0.4P_u/(\phi_c A_g F_e)$ for interior panels

Q = slenderness-reduction factor prescribed by ANSI/AISC 360-05, Sec. E7

S_x = chord section modulus with respect to its horizontal axis

ϕ_b = strength-reduction factor for bending

  = 0.90

F_e = elastic buckling stress, as defined by Eq. (7.4)

M_u = chord factored moment at the panel point

F_y = chord yield strength

  = 50 ksi

The web elements are charged with resisting flexural shear. Based on their position along the span and actual loading condition, web members can be in compression and tension. The most typical condition pertaining to the consideration of shear in web elements is illustrated in Fig. 7.10. As can be seen, the role of the design is to ensure that the shear resistance envelope provided by the web members does not exceed the shear

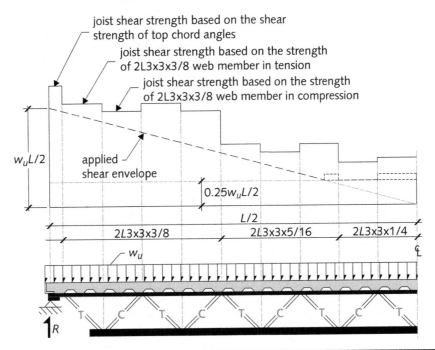

FIGURE 7.10 Joist web forces.

diagram of the member. For a distributed load, the maximum shear at the support V_u equals $w_u L/2$, where w_u is the uniform load per unit length, and L is the design length. For the short top chord segment located between the reaction support and the first panel point of the span, vertical legs of the top chord angles provide shear resistance. For the remainder of the half-span, shear strength is provided using axial strength of alternating diagonal web members in tension and compression. As shown in Fig. 7.10, web member sizes can be decreased gradually away from the support. SJI-K-1.1-05 Sec. 4.4(b) requires that the minimum provided shear strength along the span be at least $0.25V_u$. Although idealized as uniform linear loads, live loads, and especially superimposed dead loads, may occur as a series of small point loads applied as several locations along the joist span. The requirement for the minimum shear strength of the joist equal to $0.25V_u$ takes into account the fact that the actual applied shear envelope may not be linear, as achieved in the analysis based on the assumption of perfectly uniform applied load. In other words, for a concentrated force occurring near the joist mid-span, the resulting joist shear would be higher than calculated assuming uniform loads.

The tension and compression strength of a web member is calculated as provided by Eqs. (7.2) and (7.4), respectively, except that F_y can be either 36 or 50 ksi depending on the material used. SJI-K-1.1-05 Sec. 4.3 requires that the slenderness ratio l/r of a web member in compression be limited to 200. The previously given slenderness limit of 240 applies to web members in tension. When the first web member is a crimped single angle whose end-welded connection is not designed as fixed, SJI-K-1.1-05 Sec. 4.2(b) mandates that the slenderness ratio l/r used in calculations of compressive strength must be taken as $1.2l/r_x$, where r_x is the angle radius of gyration about the axis perpendicular to the joist span. In calculating buckling strength, joist web members typically will be considered as end pinned with effective length equal to the clear distance between the joist chords.

With respect to local flexural effects in top and bottom chords (i.e., bending between panel points), concentrated forces of up to 200 lb typically can be accommodated without a significant impact on chord size. To this end, structural engineers with overall responsibility for the project commonly specify the requirement that the chords be designed to resist a minimum concentrated force of 200 lb occurring anywhere along the chord. This force typically is treated as a superimposed dead load along the bottom

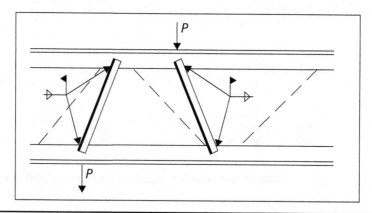

FIGURE 7.11   Joist reinforcement for concentrated forces. (*Vulcraft.*[33])

chord and either wind, live, or superimposed dead load, whichever is most critical, along top chord. For forces over 200 lb, construction documents typically mandate the chord reinforcement. As shown in Fig. 7.11, such reinforcement typically consists of field welding an additional web member in the form of a single angle that translates an intermediate concentrated force directly into a panel point. The 200-lb force is not a load in addition to what the open-web joist actually was designed for. Instead, this consideration is simply made to facilitate field attachment of superimposed dead loads because superimposed dead loads, while idealized as uniform loads along the joist span, actually occur as a series of small, concentrated forces occurring along the joist span.

When considering the impact of concentrated and other non-uniform forces on the overall shear and flexural capacity of the joist member, all concentrated forces and their known locations must be reported to the joist manufacturer by indicating such loads in the construction documents. When locations of concentrated forces greatly vary from member to member, the design may consider specifying a KCS-Series joist, ensuring constant shear and constant moment capacity along the joist span and eliminating the need to consider discontinuities in the joist shear diagram on a joist-by-joist basis while also removing the burden of providing specific load diagrams for each unique joist in the construction drawings. When concentrated forces exist and the decision is made not to specify a KCS-Series joist, the user may elect to call out a K-Series joist on structural plans on the basis of an equivalent distributed load, with the goal of properly communicating the required member flexural strength to the joist manufacturer. A load diagram must in such cases accompany the K-Series joist selection. For example, for a joist with the uniform loading w_u and a concentrated force P_u applied at mid-span, the equivalent uniform $w_{u,eq}$ load used for the selection of a joist member from the 2005 SJI Load Tables can be computed using Eq. (7.8):

$$w_{u,eq} = w_u + 2P_u/L \tag{7.8}$$

where L is the joist design length.

The final as-installed configuration of joist loading incorporating all concentrated forces must be verified by the structural engineer with overall responsibility for the project against the joist shear and moment capacity envelope of the joist.

Interconnectivity of joist elements is achieved through shop welds designed using Chap. E of ANSI/AISI S100-07 when attaching cold-formed steel members with up to $\frac{3}{16}$ in. thickness and Chap. J of ANSI/AISC A360-05 for all other cold-formed steel and all hot-rolled members. SJI-K-1.1-05 Sec. 4.5(b) stipulates that all element-to-element welded joints must be designed for a force equal to 1.35 times the factored force calculated in the analysis. For splicing two chord section segments, the same SJI–K-1.1-05 section stipulates that the butt weld splicing the two chord segments must be capable of resisting the axial chord force corresponding to $F_y = 57$ ksi.

Although centroidal axes of the chord and web members often do not coincide, as illustrated in Fig. 7.12, SJI-K-1.1-05 Sec. 4.5(c) permits that the eccentricity e_w be neglected in the design of web members as long as it does not exceed 0.75 times the width of the web angle leg parallel to the joist span. The eccentricity e illustrated in Fig. 7.12 represents the top-chord-to-web eccentricity at interior panel points. It is typically neglected in the design of top chord, which is thereby treated as a continuous beam on knife supports. The eccentricity of the top-chord-to-web connection with repect to the support reaction must be considered in all cases. This condition is shown in Fig. 7.13. Therefore, the joist top chord must be designed for the moment $M = Re$ and the shear R applied immediately to the left of the web attachment.

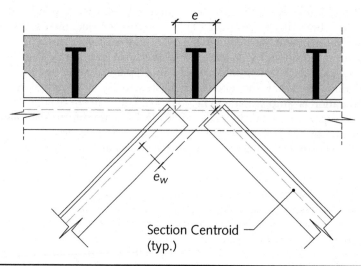

FIGURE 7.12 Top-chord-to-web attachment eccentricity.

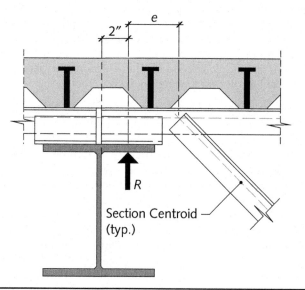

FIGURE 7.13 Top-chord eccentricity at support.

Open-web joists are laterally and torsionally stabilized using top chord deck attachments and discrete continuous bridging, typically in the form of small angles and steel rods. Typical joist bridging details are shown in Fig. 7.14. Bridging need not be provided where the top chord is continuously braced by the deck attachment. Bracing must be positively anchored to either a wall, as shown in Fig. 7.14, a diaphragm, or another structural element capable of resisting cumulative bracing forces. As stipulated by SJI-K-1.1-05 Sec. 5.4, bridging connections to the joist chords must be designed for the minimum service force in the direction of the bridging of 700 lb. The maximum slenderness ratio of horizontal bridging lines may not exceed 300. The maximum slenderness ratio of diagonal bridging lines may not exceed 200. The number of required bracing lines perpendicular to joist span depends on the joist span, size of the chord, and joist depth. The number of required rows of bridging, given by SJI-K-1.1-05, Table 5.4-1, ranges from one to five rows for K-Series joist spans of up to 60 ft. The ultimate responsibility for determining the required bridging on a specific project rests with the joist manufacturer. However, the structural engineer with overall responsibility for the project should review and approve the proposed bridging arrangement to ensure compatibility with the supporting structure, especially as it relates to anchoring the bracing forces into a supporting member. As shown in Fig. 7.14, in order to prevent accumulation of gravity or uplift forces in the bridging system due to deflection of the nearest joint, diagonal bridging forming a truss should not be located between the first joist and a wall or another more rigid component to which the bridging is anchored.

Section 1604.3.3 of 2009 IBC requires that the open-web joist deflections for K-Series joists be evaluated in accordance with SJI-K-1.1-05. SJI-K-1.1-05 Sec. 5.9 requires that the service live-load deflection in floors not exceed $L/360$, where L is the joist span. Similarly,

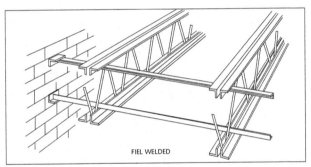

HORIZONTAL BRIDGING
SEE SJI SPECIFICATION 5.5 AND 6.

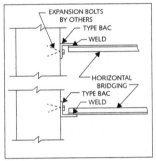

BRIDGING ANCHORS
SEE SJI SPECIFICATION 5.5 AND 6.

NOTE: DO NOT WELD BRIDGING TO JOIST WEB MEMBERS.
DO NOT HANG ANY MECHANICAL, ELECTRICAL, ETC. FROM BRIDING.

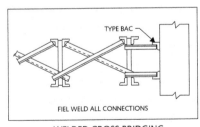

WELDED CROSS BRIDGING
SEE SJI SPECIFICATION 5.5 AND 6.

HORIZONTAL BRIDGING SHALL BE USED IN
SPACE ADJACENT TO THE WALL TO ALLOW FOR
PROPER DEFLECTION OF THE JOIST NEAREST
THE WALL

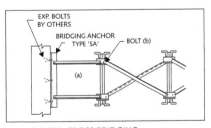

BOLTED CROSS BRIDGING
SEE SJI SPECIFICATION 5.5 AND 6.

(a) HORIZONTAL Bridging units shall be used in the space adjacent to the wall to allow for proper deflection of the joist nearest the wall.

(b) For required bolt size refer to bridging table on page 136.
NOTE: Clip configuration may vary from that shown.

FIGURE 7.14   Joist stability bridging. (*Vulcraft.*[33])

the same deflection limit applies to roof joists supporting plaster ceilings. For other roof joists, deflection under the service load may not exceed $L/240$.

The SJI-K-1.1-05, Sec. 4.6 states that the approximate camber of K-Series joists ranges from 0.25 in. for 20-ft top chord length to 1.5 in. for 60-ft top chord length. When the structural engineer with overall responsibility for the project wishes to impose total load deflection criteria, it is necessary to estimate a modulus of inertia of the joist section. A typical situation where a deflection limit under total load is imposed involves long open-web non-composite joists supporting concrete slabs, where the approximate cambers prescribed by SJI-K-1.1-05, Sec. 4.6 are considered insufficient to prevent significant concrete ponding effects. The moment of inertia of K-Series joist sections can be estimated using Eq. (7.9)[22] or Eq. (7.11).[16]

$$I_j = 80 w_{LL} L^3 / [3.45(10^6)] \tag{7.9}$$

where

I_j = joist section moment of inertia (in⁴)

w_{LL} = service live load causing deflection of $L/360$ from 2005 SJI Load Tables (lb/ft)

L = design span corresponding to w_{LL} (ft)

Consideration of vibration control typically is required on non-composite open-web joists supporting concrete slab floors. This can be accomplished using SJI *Technical Digest No. 5—Vibration of Steel Joist-Concrete Slab Floors*[28] as a guide. Alternatively, the user may choose to use the method presented by the AISC *Design Guide 11—Floor Vibrations Due to Human Activity*.[14] Whether a floor consists of composite or non-composite open-web joists, it is evaluated in the same manner with respect to the vibration performance using either of the preceding referenced methods. For the method presented in SJI *Technical Digest No. 5*, the moment of inertia used in computing the floor frequency is calculated assuming a fully composite section and transformed slab area, defined as I_{comp} in Eq. (7.10). For the method presented in the AISC *Design Guide 11*, the composite effective moment of inertia is used in frequency computations. The effective moment of inertia, which takes into account the distortion of the joist cross section due to web openings, is calculated using Eq. (7.10):

$$I_{eff} = [(\gamma/I_{chords}) + (1/I_{comp})]^{-1} \tag{7.10}$$

where

$\gamma = (1/C_r) - 1$

$C_r = 0.90(1 - e^{-0.28L/d})^{2.8}$ for joists with $6 \leq L/d \leq 24$ and single- or double-angle web members

$\quad = 0.721 + 0.00725(L/d)$ for joists with $10 \leq L/d \leq 24$ and continuous-rod web members

d = overall joist depth

L = joist span

I_{chords} = moment of inertia of bare joist

$\quad = I_{tc} + I_{bc} + d_e^2(A_{tc} A_{bc})/(A_{tc} + A_{bc})$

d_e = effective joist depth

$\quad = d - y_{tc} - y_{bc}$, where y_{tc} and y_{bc} are as defined in Eq. (7.1)

I_{tc} = top chord moment of inertia about its centroidal axis

I_{bc} = bottom chord moment of inertia about its centroidal axis

A_{tc} = top chord cross-sectional area

A_{bc} = bottom chord cross-sectional area

I_{comp} = moment of inertia assuming a fully composite section using transformed slab section

The main challenge of calculating I_{chords} and I_{comp} in applications involving non-composite joists is the inability to positively verify the chord section and sizes and effective joist depth d_e at the design stage. For this reason, d_e can be taken as approximately $0.95d$.[16] The I_{chords} also can be taken as I_j computed by Eq. (7.9) or estimated as given by Eq. (7.11).[16] When I_{chords} is computed as shown in Eq. (7.10), A_{tc} and A_{bc} can be estimated using Eqs. (7.12)[16] and (7.13).[16] In open-web non-composite SJI K-Series joists subjected to gravity loads, top chords typically are fabricated with a slightly higher size than bottom chords. This is done in an attempt to maximize the utilization of the bottom tension chords because a compression chord will provide a lower capacity per unit area than a tension chord. Conversely, when the joist design is dominated by uplift, the bottom chord generally will be fabricated larger than the top chord. When A_{tc} and A_{bc} are known and I_{chords} is computed as shown in Eq. (7.10) and used in the evaluation of bare joist properties, the computed value of I_{chords} should be reduced by a factor of 1.15 to take into account shear deformations of the joist. This reduction is not necessary if I_j computed by Eqs. (7.9) and (7.11) is used because the effects of shear deformations already were incorporated therein. Similarly, when Eqs. (7.9) and (7.11) are used to estimate I_{chords} for the purpose of computing I_{eff} with Eq. (7.10), the values of I_{chords} so computed should be increased by a factor of 1.15. When indicating a required moment of inertia of a bare joist as a requirement to the joist manufacturer, the design professional with responsibility for the overall project must indicate whether the shear deformations were incorporated into the indicated required value of I_j or whether shear deformations are to be taken into account separately by the joist designer.

$$I_j = 0.0247 w_{ASD} L^2 d_e / 1,000 \tag{7.11}$$

$$A_{tc} = \frac{1.5 w_{ASD} L^2}{23.79 d_e} \tag{7.12}$$

$$A_{bc} = \frac{1.5 w_{ASD} L^2}{29.15 d_e} \tag{7.13}$$

where

$\quad I_j$ = joist section moment of inertia (in.[4])

$\quad w_{ASD}$ = uniform service load capacity from the 2005 SJI Load Tables (lb/ft)

$\quad L$ = joist design length (ft)

$\quad d_e$ = effective joist depth (in.)

$\quad\quad = 0.95d$

A_{tc} = top chord cross-sectional area (in.^{2})

A_{bc} = bottom chord cross-sectional area (in.^{2})

Although SJI-K-1.1-05 does not preclude the design of end-fixed web members, in a typical K-Series joist configuration, web members will be analytically considered as end-pined.

In determining the joist loads for comparison with the capacities from the 2005 SJI Load Tables, the user should pay particular attention to the direction and distribution of the applied loads in sloped roofs. As illustrated in Fig. 7.15, the roof live and snow loads are specified on the basis of a projected flat surface, whereas dead loads are specified on the basis of distribution along the surface of the roof sloped at an angle Ψ. Component wind pressures and vertical components of the main lateral force resisting system wind pressures are applied as forces perpendicular to the roof surface. The vertical and horizontal seismic component loads F_{pv} and F_{ph}, respectively, are applied along the absolute vertical and horizontal directions, respectively. Only the loads or components of the loads perpendicular to the roof surface are supported by the joist member. The components of the loads occurring in the plane of the roof surface are resisted by the diaphragm. As can be seen, only wind loads occur perpendicular to the surface. It is therefore necessary to isolate the load components perpendicular to the roof surface before performing the comparison of the unit applied to loads to the unit load capacities from the 2005 SJI Load Tables. Similarly, when it is necessary to report loads to the joist manufacturer, as is the case with non-uniform loads, wind uplift, and joist girders with unequal joist reaction spacings, it is necessary to isolate and report only the force components resisted by the joist itself.

For example, for the strength load combination given by 2009 IBC Eq. (16-3), $1.2D + 1.6L_r + 0.8W$, the actual load for comparison with the unit load capacities from the 2005 SJI Load Tables is shown in Eq. (7.14). The load resulting from this load combination to be resisted by the diaphragm is shown in Eq. (7.15). Similarly, considering the load

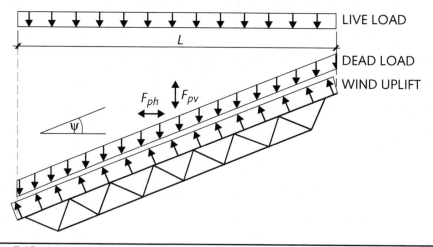

FIGURE 7.15 Joist loads in sloping roofs.

combination given by 2009 IBC Eq. (16-7), $0.9D + 1.0E$, and seismic component loads F_{pv} and F_{ph}, the uniform loads on the joist are given by the Eq. (7.16), whereas the loads resisted by the diaphragm are as provided by Eq. (7.17).

$$w_u = 1.6w_{LL} \cos^2(\Psi) + 1.2w_{DL} \cos(\Psi) + 0.8w_W \tag{7.14}$$

$$w_u = 1.6w_{LL} \cos(\Psi) \sin(\Psi) + 1.2w_{DL} \sin(\Psi) \tag{7.15}$$

$$w_u = w_{Fph} \sin(\Psi) + w_{Fpv} \cos(\Psi) + 0.9w_{DL} \cos(\Psi) \tag{7.16}$$

$$w_u = w_{Fph} \cos(\Psi) + w_{Fpv} \sin(\Psi) + 0.9w_{DL} \sin(\Psi) \tag{7.17}$$

Example 7.1 illustrates application of SJI-K-1.1-05 by the joist manufacturer. Example 7.2 illustrates application of SJI-K-1.1-05 and the 2005 SJI Load Tables by the structural engineer with overall responsibility for the project.

Example 7.1

For the given properties and load conditions, determine whether a typical non-composite K-Series joist spaced at 36 in. on center possesses adequate flexural strength if it supports an office floor with a uniform superimposed dead load of 20 psf. Assume that the slab supports the entire superimposed dead load. Also, check the adequacy of the top chord for support of the gravity loads between panel points, and check chord bending due to eccentricity at the support assuming eccentricity of 7 in., as illustrated in Fig. 7.13. Find the required arrangement of bridging. Assume that web-to-chord eccentricity, e_w, illustrated by Fig. 7.12, is 0.75 in. Check the adequacy of the web member closest to the joist mid-span using web member sizes given in Fig. 7.10. Web members conform to ASTM A36. Check whether the live load deflection limit of $L/360$ is satisfied. The project is governed by the 2009 IBC.

Given

$d_j = 24$ in.; span = 37 ft, 6 in.

ASTM A572 Grade 50

Top chord:

2L 3 × 3 × ½; $A_{tc} = 5.50$ in.²; $y_{tc} = 0.929$ in.; $r_x = 0.895$ in.; $r_y = 1.58$ in.

Bottom chord:

2L 4 × 4 × ⅝; $A_{bc} = 9.21$ in.²; $y_{bc} = 1.22$ in.; $r_x = 1.20$ in.; $r_y = 2.00$ in.

Gap between top chord angles is ¾ in.

Diagonal web members are oriented at 45 degrees from the chords.

Weight of the slab and non-composite deck = 40 psf.

The attachment pattern for the specific non-composite deck used in this example requires attachments to support at 12 in. on center at both interior and exterior zones.

Sections 2 and 3.2 of SJI-K-1.1-05 stipulate that chords have nominal yield strength of 50 ksi, which is met, as shown earlier. Section 3.1 of SJI-K-1.1-05 permits ASTM A572 Grade 50 for chord fabrication.

$$\text{Design length, } L = 37.5 \text{ ft} - 0.33 \text{ ft} = 37.17 \text{ ft}$$

The first step is to tabulate all the applicable loads before determining whether the design flexural strength is adequate. Dead load due to the self-weight of deck and slab is computed as follows:

$$w_{slab} = 3(40) = 120 \text{ lb/ft}$$

Since web sizes vary along the span, the weight of the chord is calculated slightly conservatively by considering the heaviest of the three angle sizes, using 2L3×3×⅜. Unit weight of 2L3×3×⅜ is obtained from AISC 325-05 as 7.2 lb/ft. Using joist depth of 24 in. and 45-degree member orientation, the weight of the web members per unit length of joist is $7.2(2^{1/2}) = 10.12$ lb/ft. The total weight of the joist is

$$w_{joist} = w_{tc} + w_{bc} + w_{web} = 490(5.50 + 9.21)/144 + 10.12 = 60.17 \text{ lb/ft}$$

With the weight of the deck and superimposed dead load included, total dead load is calculated:

$$w_{dead} = 120 + 60.17 + 3(20) = 240.17 \text{ lb/ft}$$

The 2009 IBC, Sec. 1607.3, Item 25 stipulates a uniform live load of 50 psf for office occupancy. The live load may be reduced per 2009 IBC, Sec. 1607.9.2 if the minimum area supported by the member is at least 150 ft².

$$A = 3(37.17) = 111.51 \text{ ft}^2 \le 150 \text{ ft}^2$$

Therefore, live-load reduction cannot be applied. Calculated live load per unit length is then

$$w_{live} = 3(50) = 150 \text{ lb/ft}$$

It should be noted that for office buildings, the 2009 IBC Sec. 1607.4 requires consideration of a 2000-lb concentrated forces distributed over an area confined by a 2 ½ by 2 ½-ft square, and positioned such as to cause the most critical effect for the member and the limit state being considered. This concentrated force is to be considered acting non-concurrently with the uniform live load specified by the 2009 IBC Sec. 1607.3. With respect to this particular example, it is determined by inspection that the consideration of the required concentrated live load will not govern the flexural strength of the joist or the size of the joist top chord section subjected to the combined effects of axial force due to joist flexure and top chord bending under the loads applied between joist panel points. Section 4.1 of SJI-K-1.1-05 stipulates load combinations for situations where loading combinations are not given. Since 2009 IBC applies to this problem, the load cases given by Sec. 1605.2.1 of 2009 IBC govern. It should be noted that the load cases of SJI-K-1.1-05 Sec. 4.1 are effectively identical to the 2009 IBC load cases. Factored line loads are computed as follows using the 2009 IBC Eqs. (16-1) and (16-2), respectively:

1.4D:

$$w_{u,1.4D} = 1.4(240.17) = 336.24 \text{ lb/ft} = 0.34 \text{ k/ft}$$

1.2D + 1.6L:

$$w_{u,1.2D+1.6L} = 1.2(240.17) + 1.6(150) = 528.20 \text{ lb/ft} = 0.53 \text{ k/ft}$$

Therefore, the load combination given by 2009 IBC Eq. (16-2) governs. The required moment strength finally is computed as

$$M_u = \frac{w_u L^2}{8} = \frac{(0.53)(37.17)^2}{8} = 91.53 \text{ kip-ft}$$

It is now necessary to determine the joist capacity. Factor yield strength of the bottom chord is

$$\phi T_{bc} = A_{bc} F_y = 0.90(9.21)(50) = 414.45 \text{ kips}$$

Next, it is necessary to determine the top chord strength. Since the distance between points exceeds 24 in., SJI-K-1.1-05, Sec. 4.4(a) mandates that the top chord be designed for combined compression and bending due to load between panel points.

$$F_e = \frac{\pi^2 E}{(l/r)^2} = \frac{\pi^2(29,000)}{(48/0.895)^2} = 99.51 \text{ ksi}$$

Referring to ANSI/AISC 360-05, Sec. E7.1, and considering the top chord angle geometry,

$$b/t = 3/0.50 = 6.0 \qquad \text{and} \qquad \left\{ 0.56 \sqrt{\frac{E}{F_y}} = \sqrt{\frac{29,000}{50}} = 13.49 \right\} => 6.0$$

Thus $Q = 1.0$.

$$\{l/r = 48/0.895 = 53.63\} \le \left\{ 4.71 \sqrt{\frac{E}{QF_y}} = 4.71 \sqrt{\frac{29,000}{1.0(50)}} = 113.43 \right\}$$

Therefore,

$$F_{cr} = Q(0.658^{QF_y/F_e})F_y = 1.0(0.658^{1.0(50)/99.51})50 = 40.52 \text{ ksi}$$

$$\phi T_{tc} = \phi_c F_{cr} A_{tc} = 0.90(40.52)(5.50) = 200.57 \text{ kips}$$

As expected, the strength of the top chord is smaller than the strength of the bottom chord, and it therefore governs the flexural strength of the joist. However, since it is necessary also to consider the local bending of the top chord due to the load between panel points, the full axial strength of the top chord cannot be exhausted for the joist flexural strength. The available reduced top chord strength is found from the interaction equation provided by SJI-K-1.1-05 Sec. 4.4(a). Using moment diagrams from AISC A325-05 and the joist panel length L_p of 48 in., the top chord moment at the mid-panel point coinciding with the joist mid-span is calculated as follows:

$$M_{u,tc} = 0.036 w_u L_p^2 = 0.036(0.53)(4)^2 = 0.31 \text{ kip-ft}$$

Assuming initially that $P_u/(\phi_c F_{cr} A_g) \ge 0.2$, Eq. (7.7a) applies. Also, the reduced magnitude of P_u from the Eq. (7.7) must be determined iteratively because the magnitude of parameter C_m and the P-δ amplifier in Eq. (7.7) depends on P_u. Initially, it is assumed that $P_u = 200$ kips.

$$C_m = 1 - 0.4 P_u/(\phi_c A_g F_e) = 1 - 0.4(200)/[0.90(5.50)(99.51)] = 0.84$$

Chord section modulus $S_{x,tc}$ can be backcalculated from the given values of r_x and A_{tc} as follows:

$$S_{x,tc} = \frac{A_{tc}r_x^2}{\text{leg width} - y_{tc}} = \frac{5.50(0.895)^2}{3 - 0.929} = 2.13 \text{ in.}^3$$

$$\frac{P_u}{\Phi_c A_g F_{cr}} + \frac{8}{9}\left[\frac{C_m M_u}{\left(1 - \dfrac{P_u}{\Phi_c A_g F_e}\right)Q\phi_b S_x F_y}\right]$$

$$= \frac{200}{0.90(5.50)(40.52)} + \frac{8}{9}\left\{\left[\frac{0.84(0.31)}{\left[1 - \dfrac{200}{0.90(5.50)(99.51)}\right]1.0(0.90)(2.13)50}\right]\right\} = 1.00$$

Therefore, the interaction is satisfactory. It is now necessary to check the interaction equation at the panel point. Since the panel point is 24 in. removed from the joist mid-span, the smaller axial force in the top chord applies as calculated below:

$$P_{u@24 \text{ in.}} = 200 - 0.53(2)^2/[2(2)] = 199.47 \text{ kips}$$

Applying Eq. (7.6),

$$(P_u/A_g) + (M_u/S_x) = (199.47/5.50) + (0.31/2.13) = 36.41 \text{ ksi} \le \phi F_y = 0.90(50) = 45 \text{ ksi}$$

Thus, accounting for the top chord bending due to the load between panel points, the resulting governing chord force is $\phi T_{tc} = 200$ kips. Finally, the factored moment strength can be computed:

$$\phi M_n = \phi T_{tc}(d_j - y_{tc} - y_{bc}) = 200(24 - 0.929 - 1.22)/12 = 364.18 \text{ kip-ft} > 91.53 \text{ kip-ft}$$

Therefore, the moment capacity is satisfactory. Alternatively, this can be presented in terms of factored load capacity per unit length:

$$\phi w_n = 8\phi M_n/L^2 = 8(0.90)(364.18)/37.17^2 = 1.90 \text{ kip/ft} = 1,900 \text{ lb/ft} > 528.20 \text{ lb/ft} \quad Okay!$$

Per the problem statement, it is now necessary to check the shear capacity of the first web member near the mid-span, 2L3×3×¼. The following properties are retrieved from AISC 325-05 for this member: $A_g = 2.87$ in.^{2} and $r_x = 0.926$ in.

The computed factored joist shear for this member is $V_u = 0.53(2) = 1.06$ kips. However, the minimum required shear resistance per SJI-K-1.1-05 Sec. 4.4(b) is $0.25V_{u,max}$, where $V_{u,max}$ is the shear at support.

$$V_{u,\text{min}} = 0.25(37.17/2)0.53 = 2.46 \text{ kips}$$

Therefore, the minimum shear governs. The force in the diagonal web member is calculated as follows:

$$V_{u,2L3\times3\times1/4} = 2.46(2)^{1/2} = 3.48 \text{ kips}$$

$$\phi F = \phi_y F_y A_g = 0.90(36)(2.87) = 92.99 \text{ kips} > 3.48 \text{ kips} \quad Okay!$$

Therefore, the shear strength at the first web member is satisfactory. The next step is to check the bending due to member eccentricity. At intermediate panels, the eccentricity of 0.75 in. may be neglected in the design of web members because $(\tfrac{3}{4})3 = 2.25$ in. > 0.75 in., as provided by SJI-K-1.1-05, Sec. 4.5(c). For the top chord design, the eccentricity at the support must be considered. The net factored moment in the top chord at the first panel point due to the 7-in. support reaction eccentricity is computed as follows:

$$M_{u,\text{ecc.}} = V_{u,\max}(e) - w_u(e)^2/2 = (0.53)(37.17/2)7 - 0.53(1/12)(7)^2/2 = 68.95 - 1.08$$
$$= 67.87 \text{ kip-in.}$$

$$\phi M_n = 0.90 S_{x,tc} F_y = 0.90(2.13)(50) = 95.85 \text{ kip-in.} > 67.87 \text{ kip-in.} \qquad \textit{Okay!}$$

Finally, live-load deflection is to be checked:

$$d_e = d - y_{tc} - y_{bc} = 24 - 0.929 - 1.22 = 21.85 \text{ in.}$$

$$I_{\text{chords}} = I_{tc} + I_{bc} + d_e^2(A_{tc} A_{bc})/(A_{tc} + A_{bc})$$
$$= 0.924 + 2.76 + (21.85)^2(5.50 \times 9.21)/(5.50 + 9.21) = 1647.72 \text{ in.}^4$$

$$\Delta_{LL} = \frac{5w_{LL}L^4}{384 E I_{\text{chords}}/1.15} = \frac{5(0.40/12)\left[12(37.17)\right]^4}{384(29{,}000)(1647.72)/1.15} = 0.41 \text{ in.}$$

The deflection ratio under service live load is $37.17(12)/0.41 = 1088 > 360$. Therefore, live-load deflection is satisfactory. An unsatisfactory check can be improved by using larger chords or by using a composite joist.

Example 7.2

For the non-composite open-web joist roof system shown in Fig. 7.7, the roof snow load p_f as defined in Chap. 2 of this book is 25 psf. The roof superimposed dead load is 10 psf, including the weight of the suspended ceiling, and the deck weight is 2 psf. The 2 ft, 10 in. by 2 ft, 10 in. by 6 ft, 0 in. mechanical unit (Unit A in Fig. 7.16) weighs 700 lb and is suspended from the joist bottom chords at its four corner points. For the joists, the applicable component wind suction is 20 psf, whereas the component wind positive pressure is 7 psf. The wind edge zone is 3.7 ft wide and corresponds to the wind uplift of 31 psf. Specify the proper SJI joists for the members labeled "JOIST X" and "JOIST Z" and the joist girder along gridline B. The roof structure is not occupiable, but it is subject to periodic maintenance.

The snow load exceeds the roof live load of 20 psf stipulated by 2009 IBC Sec. 1607.3 and should therefore be used in the design. It can also be shown by inspection that the load cases incorporating a 300-lb concentrated live load as stipulated by 2009 IBC Sec. 1607.4 will not govern the design. It is consequently determined that the two governing load combinations, one for gravity and another for uplift, respectively, are 2009 IBC Eqs. (16-3) and (16-6) as follows:

$$1.2D + 1.6S + 0.8W$$

$$0.9D + 1.6W$$

$$\text{Joist tributary width} = [26.42 - 2(0.33)]/12 = 2.15 \text{ ft}$$

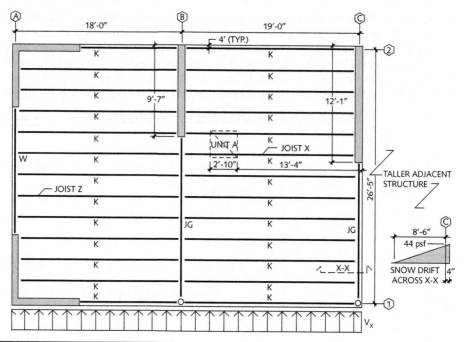

FIGURE 7.16 Roof framing plan.

For the gravity design, assuming initially 5 lb/ft for the joist weight, the distributed factored load is calculated considering the 2009 IBC Eq. (16-3) as follows:

$$w_u = 1.2(5) + 2.15[1.2(10 + 2) + 1.6(25)] = 122.96 \text{ lb/ft}$$

The 2009 IBC Sec. 1603.1.4 requires that component and cladding wind pressures be provided in the construction documents in pounds per square foot for the exterior building component and cladding elements not specifically designed by the design professional with overall responsibility for the project. Engineers typically provide the values of positive and negative wind pressure, incorporating internal pressure coefficients, for various wind tributary areas ranging from 10 to 500 ft² in either tabular or graphic form. Widths and locations of edge and corner zones also are indicated. When a joist is subjected to a net uplift load, such as the one calculated by 2009 IBC Eq. (16-3), and wind loads consist of only ordinary uniformly applied wind pressures, in the absence of point loads resulting from parapet braces, supported equipment, or supported miscellaneous framing, the non-composite joists typically are not specified by providing a joist load diagram or by requiring a KCS-Series joist. Instead, the joist design engineer is referred to the component and cladding wind tables provided in the construction documents. Furthermore, the values indicated in the 2005 SJI Load Tables only represent gravity loads and cannot be used in specifying joists for uplift resistance.

Therefore, for an 18-ft span, the 2005 SJI Load Tables indicate that a 10K1 joist has a factored load capacity of 369 lb/ft, which is comfortably more than the required capacity of 122.96 lb/ft; thus the capacity is satisfactory.

SJI-K-1.1-05 does not stipulate a deflection limit under snow load. However, using SJI-K-1.1-05, Sec. 5.9 live load deflection criteria and Table 1604.3 snow deflection criteria as a guide, $L/360$ can be applied as applicable criteria.

$$w_{snow} = 2.15(25) = 53.75 \text{ lb/ft}$$

As can be seen in the 2005 SJI Load Tables, for the 10K1 joist just selected and the span of 18 ft, the service load causing a deflection of span/360 is 134 lb/ft. Since 134 lb/ft > 53.75 lb/ft, a deflection check is satisfactory. Also, since the 10K1 joist weighs 5 lb/ft, the initial assumption of the joist self-weight is confirmed.

Therefore, the final design communicated through the construction documents will indicate 10K1 near the line designating member joist Z. Also, it will stipulate a requirement for the joist design engineer to proportion the joist for the component and cladding wind forces per the building code governing the project, such as 2009 IBC, which may result in net uplift. To accurately combine the loads, the joist manufacturer must have access to all other applicable loads, such as snow, dead, and live loads. These typically are reported on the general notes sheet of the structural drawings. The stated design requirement for the typical joist Z of this example could be in the form of the following note:

> Unless noted otherwise, all steel joists shall be designed for all applicable 2009 IBC load combinations incorporating component and cladding positive or negative pressures, as appropriate, indicated in the drawings. As required by a particular load combination, superimposed dead loads, roof live loads, and snow loads shall be as stipulated in the general notes. The manufacturer shall determine the appropriate bridging configuration to provide stability against effects resulting from all applicable load combinations.

For joist X, the structural engineer has two options. One is to specify a K-Series joist and augment the plan joist designation with a schedule of all applicable loads, as shown in Fig. 7.17, providing the joist manufacturer with all the necessary information to design the joist for all applicable loads and load combinations. The first step is to determine the equivalent uniformly distributed gravity load. By inspection, the governing load combination is given by 2009 IBC Eq. (16-3): $1.2D + 1.6S + 0.8W$.

$$\text{Design length} = 18.67 - 0.33 = 18.33 \text{ ft}$$

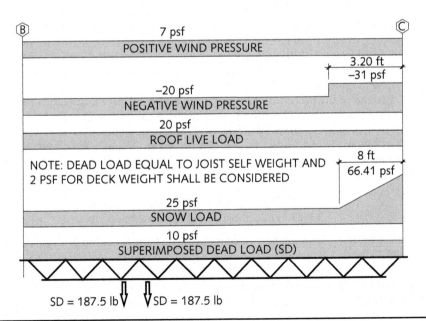

Figure 7.17 Joist X loading diagram.

Assuming initially a 5 lb/ft joist weight, reaction R_B at the support corresponding to gridline B is calculated as follows:

$$R_B = [0.8(7)(2.15)(18.33^2/2) + 1.6(25)(2.15)(18.33^2/2) + 1.6(8^2/6)(41.41) \\ + 1.2(5)(18.33^2/2) + 1.2(12)(18.33^2/2) + 187.5(13+15.83)]/18.33 = 1{,}418.96 \text{ lb}$$

The maximum moment and the corresponding equivalent uniform load are computed subsequently as follows:

$$w_u = 1.2[(2.15)(2 + 10) + 5] + 1.6(25) + 0.8(7) = 82.56 \text{ lb/ft}$$

$$M_u = 3.01(1{,}418.96 + 1{,}170.45)/2 + 2.83(945.45 + 711.81)/2 + 5.90(711.81/2) \\ = 8{,}341.92 \text{ lb-ft}$$

$$w_{u,eq} = 8M_u/L^2 = 8(8{,}341.92)/18.33^2 = 198.62 \text{ lb/ft}$$

As can be seen from the 2005 SJI Load Tables, for a 19-ft span, a 10K1 joist has a factored uniform load capacity of 331 lb/ft, exceeding the required factored distributed load of 198.62 lb/ft. Since a 10K1 joist weighs 5 lb/ft, the initial assumption of the self-weight therefore is adequate. The equivalent distributed service snow load is computed next.

$$R_{B,snow} = 2.15[25(18.33^2/2) + 8(41.41/2)(8/3)]/18.33 = 544.42 \text{ lb}$$

$$w_{snow} = 2.15(25) = 53.75 \text{ lb/ft}$$

$$M_{snow} = (544.42/53.75)544.42/2 = 2{,}757.15 \text{ lb-ft}$$

$$w_{snow,\,eq} = 8M_{snow}/L^2 = 8(2{,}757.15)/18.33^2 = 65.65 \text{ lb/ft}$$

For a 10K1 joist and a 19-ft span, the 2005 SJI Load Tables indicate that a deflection equal to $L/360$ is caused by a uniform service load of 113 lb/ft. Since 113 lb/ft exceeds 65.65 lb/ft, the deflection is satisfactory. In this scenario, the design consists of indicating a joist designation 10K1-SP on the framing plan for joist X. The designation "SP" typically is used to alert the joist design engineer that the K-Series joist is subjected to non-uniform loading condition and requires special design consideration. In such cases, a loading diagram similar to the one shown in Fig. 7.17 is provided, indicating actual load requirements. Although shown in Fig. 7.17, the wind component and cladding wind pressures and uniform loads typically are not provided in the individual joist loading diagrams.

Instead of the preceding approach, the user may elect to specify a KCS-Series joist. It is first necessary to identify the maximum shear and moment under the governing gravity load combination. From the preceding calculations, $M_u = 8{,}341.92$ lb-ft and $R_B = 1{,}418.96$ lb. To determine the maximum shear, it is also necessary to compute the reaction R_C corresponding to the gridline C. The joist self-weight of 5 lb/ft is assumed for the analysis.

$$R_C = 18.33[1.2(5) + 1.2(2.15)(12) + 1.6(25)(2.15) + 0.8(7)] + 2(187.5) \\ + 1.6(41.41)(2.15)(8)/2 - 1{,}418.96 = 1{,}882.35 \text{ lb}$$

Therefore, the governing shear occurs near the support at the gridline C, and $V_u = R_C = 1{,}882.35$ lb.

It can be seen from the 2005 SJI Load Tables that a 10KCS1 joist has a factored shear capacity of 3,000 lb, which significantly exceeds the required capacity of 1,882.35 lb. Also, the 10KCS1 joist has a factored moment capacity of 258 kip-in., or 21,500 lb-ft, which significantly exceeds the required moment strength of 8,341.92 lb-ft. Although the 10KCS1 joist weight is given as 6 lb/ft, the difference from the initially assumed weight is, by inspection, negligible. The 10KCS1 joist moment of inertia is 29 in.[4], and the resulting deflections under the snow load are computed as follows:

$$R_{B,snow} = 2.15[25(18.33^2/2) + 8(41.41/2)(8/3)]/18.33 = 544.42 \text{ lb}$$

$$R_{C,snow} = 2.15[25(18.33) + 8(41.41/2)] - 544.42 = 796.94$$

$$w_{snow} = 2.15(25) = 53.75 \text{ lb/ft}$$

$$M_{snow} = 796.94x - 53.75(x^2/2) - 2.15(41.41)(4)(x - 8/3)$$

$$EI\theta = 398.47x^2 - 8.96x^3 - 356.13(x^2/2 - 2.67x) + C_1$$

$$EI\Delta_{snow} = 73.47x^3 - 2.24x^4 + 475.43x^2 + C_1x + C_2$$

Since $\Delta_{snow} = 0$ at $x = 0$, $C_2 = 0$. Similarly, at $x = 18.33$ ft, $\Delta_{snow} = 0$, and $C_1 = -19,604.30$ lb-ft^2

The maximum deflection occurs when the slope of the deflected shape θ is 0. Therefore, $8.96x^3 - 220.41x^2 - 950.87x + 19,604.30 = 0$, with maximum deflection occurring at $x = 8.90$ ft from the support C. Therefore, increased 15 percent to account for the effect of shear deformations, the maximum deflection due to the service snow load is computed as follows:

$$\Delta_{snow} = 1.15[73.47(8.90)^3 - 2.24(8.90)^4 + 475.43(8.90)^2 - 19,604.30(8.90)]/$$
$$[29,000(29)1,000/144] = -113,941.60/5,840,277.78 = -0.02 \text{ ft} = 0.23 \text{ in.} \quad \downarrow$$

Since $18.33(12)/360 = 0.61$ in. > 0.23 in., the deflection check is satisfactory.

With the option involving a KCS-Series joist, the design for wind pressures is identical to the manner in which this is handled when a K-Series joist is specified. The joist manufacturer is instructed via a design note to use the component and cladding wind pressures from the construction documents to size the joist for load cases involving wind effects. However, a load diagram depicting non-uniform gravity loads is not necessary with the approach using KCS-Series joists because they are designed for a constant moment and shear capacity.

To specify a joist girder along gridline B, it is first necessary to find the joist reactions. Given the presence of the joist concentrated loads at joist X, it is clear that the joist reactions will be different for the typical joist not supporting a non-uniform load compared with joist X. To specify different loads, the user would need to provide a joist girder load diagram indicating the location and magnitude of each reaction. However, for the sake of simplicity, the structural engineer may elect simply to use the larger joist reaction at all joist locations and thus eliminate the need for a joist girder load diagram. Therefore, from the preceding calculations, the typical factored joist reaction is calculated as follows:

$$R_u = 122.96(18/2) + 1,418.96 = 2,525.60 \text{ lb} = 2.5 \text{ kips}$$

Since there are eight equal joist reaction spaces, the following joist designation can be indicated on the framing plan for the joist girder along the gridline B: 16G8N2.5F. The depth of 16 in. was selected arbitrarily. If the approximate joist girder weight is desired for the purposes of structural design of the supporting columns and walls, the joist girder weight can be found in the 2005 SJI Load Tables. Specifically, the joist girder span in this case is approximately 17.25 ft. As can be seen from the 2005 SJI Load Tables, for a 20-ft joist girder span, 16 in. of depth, and 10 uniformly spaced factored reactions of 6 kips, a joist girder weight of 28 lb/ft is reported. Although the precise configuration of loads as used in this example cannot be found in the tables, the closest, more conservative scenario can be used safely for the purpose of estimating the girder weight.

For the purpose of computing deflections, the joist girder moment of inertia can be estimated using the Eq. (7.18). The moment of inertia given by Eq. (7.18) incorporates a 15 percent reduction for the effect of shear deformations.

$$I_{jg} = 0.016NPLd \tag{7.18}$$

where

I_{jg} = joist girder effective moment of inertia (in.^{4})

N = number of spaces between supports and joist reactions

L = joist girder span (ft)

P = factored typical joist reaction (kips)

d = joist girder depth (in.)

7.3 Composite Open-Web Steel Joists

This section provides a review of common behavioral models and assumptions used in composite joist strength calculations. It focuses on flexural behavior because the aspects pertaining to shear design, consideration of eccentricities, and connection strength are similar to those in non-composite open-web joists presented in the preceding section. Composite joists derive their flexural strength through interface shear connection, whose strength couples the concrete slab and the steel joist member into a cohesive composite cross section with a continuous strain diagram. The standard method of achieving shear connection is through the use of headed shear studs, as illustrated in Fig. 7.18. The material properties pertaining to headed shear studs are reviewed in Chap. 3 of this book.

With respect to SJI-certified manufacturers, the methodology employed in computing the strength of composite joists is based on criteria established by SJI-CJ-1.0-06. Any unique aspects pertaining to section and member analysis and design not specifically addressed by the specification may differ somewhat from manufacturer to manufacturer. However, as a condition of SJI membership and certification, the SJI technical staff (i.e., an SJI consulting engineer) checks each particular design methodology to ensure that a particular approach meets the intent of SJI-CJ-1.0-06 and can justify the values given in the 2005 SJI Composite Joist Load Tables. As noted previously for the open-web joist systems in general, it is important to view a composite joist member as a system. Specifically, the reader is cautioned that the behavioral and design considerations applicable to composite joists systems, as presented herein, may not always be applicable to generic composite trusses designed, built, and connected to various structural shapes for custom applications. This is particularly true in consideration of truss node and element eccentricities. While specific analytical and experimental evidence (Fig. 7.19) embodied in SJI-CJ-1.0-06 may permit some eccentricities to be neglected in the design of a composite joist element or a connection, this may not be a valid assumption in the design of a generic composite truss, where all analytics require strict element-level consideration. For further reading on composite trusses, the reader is referred to Chien and Ritchie[8] and Viest et al.[32]

SJI-CJ-1.0-06, Sec. 100 stipulates the use of LRFD. The moment strength of a simply supported joist is calculated at its mid-span cross section, the location of governing

Figure 7.18 Shear connection in composite joists. *(Source: D. Samuelson, P.E., Vulcraft.[15])*

Figure 7.19 Testing of a composite joist. *(Source: W. Samuel Easterling, Ph.D., P.E., Virginia Tech.)*

ultimate moment. The approach for performing these calculations is specified by SJI-CJ-1.0-06, Sec. 103.5(a)(2) and consists of summing the moments due to plastic strengths of individual components within the cross section about an arbitrary point. With this model, the contribution of the top chord to the flexural strength of a fully plastic composite section is neglected. This simplification has little effect on the computed strength given the relative proximity of the top chord angles to the location of counteracting force in the concrete slab and typically smaller size of the top chord

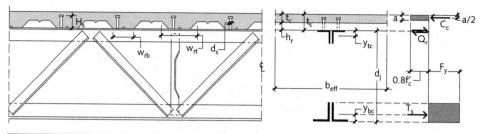

Figure 7.20 Flexural model for composite joists.

compared with the bottom chord. The strength model pertaining to this method is illustrated in Fig. 7.20. Composite joists typically are designed as fully composite, which in the context of the typical method means that sufficient shear connection is provided for the bottom chord to achieve full yielding.

Referring to the flexural model depicted in Fig. 7.20, the nominal moment strength of a composite joist then can be computed using Eq. (7.19):

$$M_n = T_s(d_j + t_s - a/2 - y_{bc}) = C_c(d_j + t_s - a/2 - y_{bc}) \tag{7.19}$$

where

T_s = bottom chord force corresponding to yield stress (kips)

$\quad = A_{bc}F_y$

A_{bc} = bottom chord cross-sectional area (in.²)

F_y = bottom chord yield stress (ksi)

d_j = depth of steel joist section (in.)

t_s = overall thickness of the concrete slab (in.)

a = effective depth of the concrete slab stress block (in.)

$\quad = T_s/(0.85f'_c b_{eff})$

y_{bc} = bottom chord cross-sectional geometric centroid (in.)

t_c = thickness of slab above metal deck (in.)

$\quad = t_s - h_r$

h_r = height of deck or height of concrete rib (in.)

f'_c = concrete compressive strength (ksi)

b_{eff} = concrete slab effective width (in.)

$\quad$ = the smaller of $L/4$ and $S/2$ on either side of member, but not more than the length of a cantilever

L = design length of the joist (in.)

S = joist spacing (in.)

C_c = concrete block force (kips)

$\quad = T_s$

The required shear connection strength Q_r is determined simply from the conditions of equilibrium, which dictate that $Q_r = T_s = C_c$. The factored design strength $\phi_t M_n$ then is computed by applying a strength-reduction factor based on bottom chord yielding $\phi_t = 0.90$ to M_n, as stipulated by SJI-CJ-1.0-06, Sec. 103.5(a)(2), and SJI Eq. (103.5-7).

Obviously, to achieve section equilibrium, and as can be seen from Eq. (7.19), T_s must equal C_c. This relationship determines the depth of the concrete stress block c and, consequently, its effective depth a. Where a very large bottom chord is used with a relatively thin slab, it is possible that the balancing force C_c would require a concrete stress block depth in excess of what is provided by the thickness of the slab above the deck t_c. For this reason, SJI-CJ-1.0-06, Eq. (103.5-4) limits the depth a to a maximum of t_c. Since the concrete effective depth a corresponds only to the rectangular geometric approximation of the actual concrete stress block depth c, the authors believe that it is prudent to further restrict a to no more than $\beta_1 t_c$, where β_1 is the compression block factor, as defined by American Concrete Institute (ACI), *Building Code Requirements for Reinforced Concrete* (ACI 318-08)[1] Sec. 10.2.7.3. It is equal to 0.85 for $f'_c = 4$ ksi and 0.80 for $f'_c = 5$ ksi. In short, Eq. (7.19) is invalid if the computed depth a exceeds $\beta_1 t_c$.

When a situation arises where the force C_c is limited by t_c (i.e., when the computed depth a exceeds $\beta_1 t_c$), a thicker slab or denser joist spacing with smaller bottom chords could be used. A second option is for the joist flexural strength to be based on a reduced value of T_s limited by C_c, as provided by SJI-CJ-1.0-06, Sec. 103.5(a)(2) and its Eq. (103.5-9), and shown in Eq. (7.20), effectively resulting in a partially composite joist. The boundary of these two cases is the balanced condition where $a = \beta_1 t_c$. SJI-CJ-1.0-06 treats this condition as fully composite and permits the design strength to be computed using $\phi_t = 0.90$. Since bottom chord yielding, concrete slab crushing, or simultaneous bottom chord yielding and concrete slab crushing are possible at $a = \beta_1 t_c$, we believe that $\phi_{cc} = 0.85$, as provided by Eq. (7.20), is more appropriate for this case because it takes into account increased statistical variability characteristic of concrete failures. It is recommended that the joist be configured such that bottom chord yielding governs whenever possible because this is important in ensuring member and shear connection ductility. Where composite joists are configured such that bottom chord yielding governs, it also becomes easier to perform retrofit redesign after a member is placed in service.

SJI-CJ-1.0-06, Sec. 103.5(a)(2) provides criteria for two additional partially composite scenarios. The first, embodied in Eq. (7.21) (SJI Eq. 103.5-8), concerns situations where the bottom chord force is limited by tension fracture strength. Given the fact that the ratio of tensile-to-yield strength of the materials used in fabrication of composite joists given by SJI-CJ-2010, Sec. 102.1(a) exceeds the ratio of resistance factors for gross yielding and net fracture, the net fracture will not govern the design unless the chord contains a hole of sufficient enough size to cause the net fracture strength to govern the design. This typically occurs due to the fastened attachments of non-structural components to the bottom chords. Wherever possible, flexural strength governed by chord fracture should be avoided. This can be achieved by specifying that no fastened attachments to the bottom chord occur within the two truss panels adjacent to the mid-span, where the composite flexural strength is calculated. Alternatively, welded, clamped, or wire attachments can be specified instead. Finally, Eq. (7.22) (SJI Eq. 103.5-10) indicates a flexural strength governed by the strength of shear connection. SJI-CJ-1.0-06, Sec. 103.5(a)(2) also requires that when the bottom chord force is limited by the strength of the shear connection, the bottom chord force reaches the minimum of 50 percent of its yield strength, which is aimed at maintaining minimum member and shear connection ductility. This requirement effectively sets the minimum degree of composite action in the member at 50 percent. As noted earlier, wherever possible, the composite member should be configured such that the member flexural strength is limited by bottom chord yielding.

$$\phi M_n = \phi_{cc} 0.85 \beta_1 f'_c b_{eff} t_c (d_j + t_s - a/2 - y_{bc})$$ (7.20)

$$\phi M_n = \phi_{tr} A_{n,bc} F_u (d_j + t_s - a/2 - y_{bc})$$ (7.21)

$$\phi M_n = \phi_{stud} N Q_n (d_j + t_s - a/2 - y_{bc})$$ (7.22)

where

ϕ_{cc} = resistance factor for concrete crushing

= 0.85

ϕ_{tr} = resistance factor for chord net fracture

= 0.75

ϕ_{stud} = resistance factor for shear stud fracture

= 0.90

It is important to recognize that composite joist floors are constructed without temporary shoring. The consequence of this is that from the time of concrete slab casting until the time the concrete hardens and the joist becomes composite (referred to as pre-composite stage), the weight of the steel deck and concrete, self-weight of the joist, as well as a construction live load, must be supported by the bare non-composite strength of the joist itself. Therefore, a joist member will require a pre-composite strength check. The SJI-CJ-1.0-06 Sec. 103.5(a)(1) stipulates the design requirements for non-composite (or in this case pre-composite) design. Pre-composite dead loads can be determined easily by tabulating individual weights of slab, deck, and joist of known geometries and properties. Pre-composite live load of 20 psf generally is used for ordinary building construction conditions, as stipulated by both SEI/ASCE 37-02[30] and Steel Deck Institute (SDI), *Standard for Composite Steel Floor Deck* (ANSI/SDI-C1.0-06).[19]

The general approach for considering pre-composite strength is to check the fully composite joist section under full factored load and then, under a separate check, establish the adequacy of the bare joist section under pre-composite loads. In so doing, the chord forces are simply calculated by decoupling the pre-composite ultimate moment into the chord forces and comparing those with the respective compressive and tensile axial capacities. Consequently, SJI-CJ-1.0-06 Sec. 103.5(a)(1) mandates that adequate pre-composite moment strength $\phi M_{n,pc}$ must be provided under factored pre-composite loads by coupling the compressive and tensile axial strength of the top and the bottom chords T_c and T_s, respectively, over the pre-composite effective depth distance. The pre-composite effective depth is defined as shown in Eq. (7.23):

$$d_{pc} = d_j - y_{bc} - y_{tc}$$ (7.23)

where y_{tc} is the bottom chord cross-sectional geometric centroid (in.).

General requirements of SJI-CJ-1.0-06 for compression member design are presented in Chap. 2. However, an aspect of top chord compression buckling due to pre-composite joist bending is further explored here. Namely, previous research shows that top chords in compression are subject only to flexural buckling.[17] Considering a double-angle top chord section, where yy is the axis of symmetry of the double-angle shape and xx is the horizontal axis through the centroid of the shape, as depicted in Fig. 7.21, the unbraced length for flexural buckling about the xx axis L_x is the distance between adjacent nodes at which web members attach to the top chord. Consideration of flexural buckling about the yy axis is somewhat more complex. For flexural buckling about the yy axis, SJI-CJ-1.0-06 Sec. 103.4 stipulates an unbraced length L_y of 36 in. This distance, based on the ANSI/AISC 360-05 Sec. I8.2d requirement for minimum shear stud spacing, adequately covers the condition of top chord bracing via initial layout of steel deck on supporting joist members prior to concrete placement. In addition to its weight, the bare joist is also supporting the weight of deck and construction live load at this stage. By the stage where the weight of wet concrete is added, all deck attachments by either studs or puddle welds will be in place. Currently commercially available composite deck profiles prescribe attachments of 6 to 14 in. depending on location along the deck surface, deck profile, or manufacturer. Furthermore, ANSI/AISC 360-05 Sec. I3.2c.(1)(4) limits the maximum distance between adjacent deck attachments either by puddle welds or by studs to 18 in. In view of these circumstances, the distance L_y can be limited to either 18 in. or the maximum permitted deck attachment distance.

The SJI-CJ-1.0-06 Sec. 103.4 requires that the largest section slenderness of an individual angle ($K_z L_z / r_z$) does not exceed the governing slenderness of the double-angle section (typically $K_x L_x / r_x$). Furthermore, SJI-CJ-1.0-06, Table 103.4-1 requires that the governing slenderness of the double-angle top chord, as well as the governing slenderness of each individual angle within the chord, does not exceed 120 in. the joist panels adjacent to supports and 90 elsewhere. As illustrated in Fig. 7.21, either fillers and ties or deck attachments can prevent buckling of individual angles about either yy and zz axes; thus $L_{yi} = L_{zi}$. The individual angle slenderness $K_{zi} L_{zi} / r_{zi}$ will govern in most cases. However, since $L_{xi} = L_x$, and since L_{xi} typically is larger than L_{zi}, there may be cases

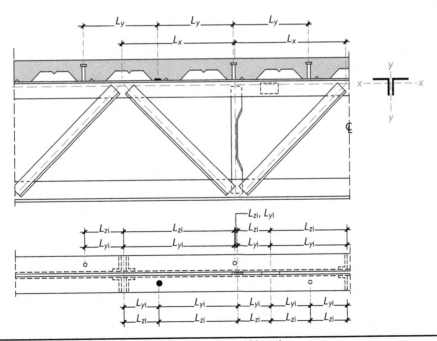

Figure 7.21 Top-chord compression buckling unbraced lengths.

where K_xL_x/r_x governs. It is therefore necessary to check the slenderness with respect to both the zz and yy axes while determining the governing slenderness value. It should be understood that where the nature of structural details is such that the location of an individual deck attachment relative to an adjacent filler or a web member attachment cannot be determined with certainty; the L_{yi} and L_{zi} should be determined without considering deck attachments. The unbraced length so determined corresponds to the term l_s in SJI-CJ-1.0-06 Sec. 103.4.

The provisions for computing strength of shear connection with headed shear studs are given by SJI-CJ-1.0-06 Sec. 103.6(d) and are virtually identical to those given by ANSI/AISC 360-05.

Flexural strength calculations per SJI-CJ-1.0-06 are illustrated in the following example.

Example 7.3

For the given properties of shear studs, slab, and joist, determine whether a typical composite joist spaced at 96 in. on center possesses adequate flexural strength if it supports an office floor with a uniform superimposed dead load of 20 psf. Assume that the slab supports the entire superimposed dead load. The project is governed by 2009 IBC. The given variables are as depicted in Fig. 7.20.

Shear Studs	Slab	Joist
H_s = 5 in.	t_s = 6¼ in.	d_j = 24 in., L = 360 in.
D = 0.75 in.	h_r = 3 in.	ASTM A572 Grade 50
$F_{ut,sc}$ = 65 ksi	w_{rb} = 5 in.	Top chord:
2 studs in each rib, located in strong position and welded at midwidths of top chord angle widths	w_{rt} = 7 in.	2L 3 × 3 × ½
	w_c = 150 lb/ft³	A_{tc} = 5.50 in.²
	f'_c = 4 ksi	y_{tc} = 0.929 in.
Steel Deck	w_{deck} = 2 psf	r_x = 0.895 in.; r_y = 1.58 in.
The attachment pattern for the specific composite deck used in this example requires attachments to support at 12 in. on center at both interior and exterior zones.		Bottom chord:
		2L 4 × 4 × ⅝
		A_{bc} = 9.21 in.²
		y_{bc} = 1.22 in.
		r_x = 1.20 in.; r_y = 2.00 in.
		Gap between top chord angles is ¾ in.
		Diagonal web members are oriented at 45 degrees from the chords.
		Fillers are provided at chord and web members at points halfway between joints.

Section 103.3 of SJI-CJ-1.0-06 stipulates that chords have a nominal yield strength of 50 ksi, which is met as shown above. Section 102.1(a) of SJI-CJ-1.0-06 permits ASTM A572 Grade 50 for chord fabrication.

The first step is to tabulate all the applicable loads before determining whether the design flexural strength is adequate. Dead load due to the self-weight of the slab is computed as follows:

$$w_{slab} = (S/2)(w_c)(t_s - h_r/2)/12 = (8)150(3.25 + 3/2)/12 = 475 \text{ lb/ft}$$

Since the web members are not known, it will be assumed that their weight is 15 lb/ft. This assumption can be verified and any correction made when the web members are designed. Thus total dead line load due to the weight of the steel joist itself is the sum of the weights of the top chord, bottom chord, and web members, that is,

$$w_{joist} = w_{tc} + w_{bc} + w_{web} = 490(5.50 + 9.21)/144 + 15 = 65.05 \text{ lb/ft}$$

With the weight of the deck and the superimposed dead load included, total dead load is calculated:

$$w_{dead} = 475 + 65.05 + 8(2) + 8(20) = 716.05 \text{ lb/ft}$$

The 2009 IBC Table 1607.1, Item 25 stipulates a uniform live load of 50 psf for office occupancy. The live load may be reduced per 2009 IBC Sec. 1607.9.2. The live load reduction percentage for the composite joist flexural strength is as follows:

$$A = 8(30) = 240 \text{ ft}^2$$

$$R = 0.08(A - 150) = 0.08(240 - 150) = 7.20\% \qquad \text{(2009 IBC, Eq. 16-23)}$$

$$W_{dead} = w_{dead}/S = 716.05/8 = 89.51 \text{ psf}$$

$$R_{max} = 23.1(1 + D/L_0) \qquad \text{(2009 IBC, Eq. 16-24)}$$

$$= 23.1(1 + 89.51/50) = 64.45\% \qquad 40\% > 7.20\% \qquad \leftarrow Okay!$$

Therefore, the reduced live load $w_{live} = 8(50)(1 - 0.072) = 371.2 \text{ lb/ft}$.

The 2009 IBC Sec. 1607.4 also requires consideration of a 2000-lb non-concurrent concentrated live load distributed over a 2 ½ by 2 ½ square and positioned as to cause the maximum effect for the member and limit state being considered. With respect to the flexural strength of composite joist considered in this example, it can be established by inspection that the consideration of the concentrated live load required by 2009 IBC Sec. 1607.4 will not govern the design. It should be noted, however, that the concentrated forces stipulated by 2009 IBC Sec. 1607.4 should be considered when proportioning the joist top chord for bending under loads applied between joist panel points. Section 103.2(b) of SJI-CJ-2010 stipulates load combinations for situations where loading combinations are not given. Since 2009 IBC applies to this problem, the load cases given by Sec. 1605.2.1 of 2009 IBC govern. It should be noted that the load cases of SJI-CJ-1.0-06, Sec. 103.2(b) are effectively identical to the 2009 IBC load cases. Factored line loads are computed as follows:

1.4*D*:

$$w_{u,1.4D} = 1.4(716.05) = 1,002.47 \text{ lb/ft} = 1.00 \text{ k/ft} \qquad \text{(2009 IBC, Eq. 16-1)}$$

1.2*D* + 1.6*L*:

$$w_{u,1.2D+1.6L} = 1.2(716.05) + 1.6(1 - 0.072)(50)(8)$$

$$= 1,453 \text{ lb/ft} = 1.45 \text{ k/ft} \qquad \text{(2009 IBC, Eq. 16-2)}$$

Therefore, the load combination given by 2009 IBC Eq. (16-2) governs. The required moment strength finally is computed as

$$M_u = \frac{w_u L^2}{8} = \frac{(1.45)(30)^2}{8} = 163.13 \text{ kip-ft}$$

It is now necessary to determine the joist capacity. First, the effective slab width is determined. It is the minimum of joist spacing and one-quarter of joist analytical span, that is,

$$b_{eff} = \min\{S = 96 \text{ in.}, L/4 = 360/4 = 90 \text{ in.}\} = 90 \text{ in.}$$

Next, the composite joist flexural strength and required shear connection are determined:

$$T_s = A_{bc} F_y = 9.21(50) = 460.50 \text{ kips}$$

Although the flexural capacity of the composite joist in the final analysis may be far greater than the required flexural strength, an attempt will be made first to design the joist as fully composite. Therefore, the required strength of the shear connection $Q_r = T_s = 414.45$ kips.

Referring to the Sec. 103.6(d) of SJI-CJ-1.0-06,

$$d_{stud}/t_{top\ chord} = 0.75/0.50 = 1.5 \leq 2.7$$

Thus SJI-CJ-1.0-06 Eq. (103.6-1) applies, and we get

$$Q_n = \min\left\{0.5 A_{stud} \sqrt{f_c' E_c},\ R_p R_g A_{stud} F_{u,stud}\right\}$$

Per SJI-CJ-1.0-06 Table 103.5-1, the minimum leg width to accommodate a ¾-in. stud is 2.50 in., and the minimum thickness is 0.250 in. For $L3 \times 3 \times \frac{1}{2}$, both stipulations are satisfied.

For a ¾-in. diameter headed stud, $A_{stud} = 0.44$ in.^{2}. For $f_c' = 4$ ksi and normal-weight concrete, $E_c = 3,834$ ksi.

Average rib width $w_r = (w_{rb} + w_{rt})/2 = (5 + 7)/2 = 6$ in. For $w_r = 6$ in., stud diameter of 0.75 in., and $h_r = 3$ in., SJI-CJ-1.0-06 Table 103.6-1 yields $R_p = 0.50$.

For two studs per rib in strong position, side by side, as illustrated in Fig. 3.1, SJI-CJ-1.0-06 Sec. 103.6(a)(d) yields $R_g = 0.85$. The shear strength per stud is then computed as follows:

$$Q_n = \min\left\{0.5(0.44)\sqrt{(4)(3834)},\ (0.50)(0.85)(0.44)(65)\right\}$$

$$Q_n = \min\{27.24 \text{ kips},\ 12.16 \text{ kips}\} = 12.16 \text{ kips}$$

The required number of studs per one-half joist span for a fully composite joist is then

$$T_s/Q_n = 460.50/12.16 = 38.16 \rightarrow 39 \text{ required studs per one-half joist span}$$

With $w_r = 6$ in. and a span of 30 ft, there are approximately 15 ribs in each half of the span. The 39 required studs for a fully composite joist would, with two studs positioned side by side in each rib, warrant 20 ribs in each half-span, which is unavailable. As discussed previously, all attempts should be made to configure the composite joist such that bottom chord yielding governs. In such a case, this could be accomplished by decreasing the size of the bottom chord. However, for illustrative purposes, an attempt will be made to consider the member as 68 percent partially composite. Thus,

$$0.68T_s/Q_n = (0.68)460.50/12.16 = 25.75$$

Therefore, 13 ribs are comfortably available for placement of 26 studs at two per rib, starting from each joist support. The governing horizontal component force then is $NQ_n = 26(12.16) = 316.16$ kips.

The 2005 SJI Composite Joist Load Tables stipulate a minimal transverse stud spacing of $4d_s$ when studs are spaced in a single rib, side by side. In this particular example, the stud center-to-center spacing is $1.5 + 0.75 + 1.5 = 3.75$ in.

$$4d_s = 4(0.75) = 3.00 \text{ in.} \leq 3.75 \text{ in.}$$

Thus the transverse stud spacing is satisfactory.

Next, focus is on computation of the composite flexural strength. Instead of the bottom chord yielding force, the strength of the shear connection governs the concrete effective stress block depth, that is,

$$a = \Sigma Q_n/(0.85f'_c b_{eff}) = 316.16/[(0.85)(4.0)(90)] = 1.03 \text{ in.}$$

Per ACI 318-08 Sec. 10.2.7.3, for $f'_c = 4$ ksi, $\beta_1 = 0.85$. Since $a/\beta_1 = 1.03/0.85 = 1.21$ in. and $a/\beta_1 \leq t_c$ (1.21 in. $\leq$ 3.25 in.), concrete crushing does not govern. Also, since $\Phi_{tr}F_u = 0.70(65) = 45.50$ ksi $\geq \Phi_t F_y = 0.90(50) = 45$ ksi, and since $A_n = A_{bc}$, net fracture will not govern.

$$M_n = \Sigma Q_n(d_j + t_s - a/2 - y_{bc})$$
$$= 316.16(24 + 6.25 - 1.03/2 - 1.22)$$
$$= 9{,}015.30 \text{ kip-in.} = 751.28 \text{ kip-ft}$$

Finally, the factored flexural strength is computed as

$$\phi_t T_s = 0.90(316.16) = 284.54 \text{ kips} \quad \text{and} \quad \phi_t M_n = 0.90(751.28) = 676.15 \text{ kip-ft}$$

Therefore, $\phi_t M_n \geq M_u$ (676.15 kip-ft $\geq$ 163.13 kip-ft); thus the composite flexural strength is adequate.

The final step with respect to evaluation of composite joist flexural strength is to check its adequacy at the pre-composite stage. This will be accomplished by performing a separate check involving the strength of only the bare joists necessary to establish the factored pre-composite load immediately after the concrete is cast. The dead load at this stage will be the same as at the composite stage, except for the absence of superimposed dead load. Therefore,

$$w_{dead, pc} = 716.05 \text{ lb/ft} - 8(20) = 556.05 \text{ lb/ft}$$

The pre-composite live load simply will consist of the construction live load in the magnitude of 20 psf, as discussed earlier; thus

$$w_{live, pc} = 8(20) = 160 \text{ lb/ft}$$

The governing factored load then can be established:

$1.4D_{pc}$:

$$w_{u,1.4Dpc} = 1.4(556.05) = 778.47 \text{ lb/ft} = 0.78 \text{ k/ft} \qquad \text{(2009 IBC, Eq. 16-1)}$$

$1.2D_{pc} + 1.6L_{pc}$:

$$w_{u,1.2Dpc+1.6Lpc} = 1.2(778.47) + 1.6(160) = 1,190.16 \text{ lb/ft} = 1.19 \text{ k/ft} \qquad \text{(2009 IBC, Eq. 16-2)}$$

Therefore, the load combination given by 2009 IBC, Eq. (16-2) governs. The required pre-composite moment strength finally is computed as

$$M_{u,pc} = \frac{w_u L^2}{8} = \frac{(1.19)(30)^2}{8} = 133.88 \text{ kip-ft}$$

Next, it is necessary to establish the pre-composite flexural depth d_{pc} and the resulting moment strength $\phi M_{n,pc}$.

$$d_{pc} = d_j - y_{bc} - y_{tc} = 24 - 1.22 - 0.929 = 22.85 \text{ in.}$$

$$\phi_t T_s = A_{bc} F_y = 0.9(9.21)(50) = 414.45 \text{ kips}$$

Since fillers are provided per SJI-CJ-1.0-06 Sec. 103.4 in. the top chord angles and located at middle panel points, SJI-CJ-1.0-06 Table 103.4-1 gives $K_x = 0.75$ and $K_y = 1.0$ for top chord interior panels. Referring to ANSI/AISC 360-05 Sec. E7.1 and considering the top chord angle geometry,

$$b/t = 3/0.50 = 6.0 \qquad \text{and} \qquad 0.56\sqrt{\frac{E}{F_y}} = \sqrt{\frac{29,000}{50}} = 13.49 \geq 6.0$$

Thus $Q = 1.0$ (ANSI/AISC 360-05 Eq. E7-4).

In the direction perpendicular to the yy axis, the deck is attached at 12 in. on center; thus $L_y = 12$ in. In the direction perpendicular to the xx axis, unbraced length L_x is the distance between adjacent web member attachments; thus, with a 45-degree diagonal member orientation, $L_x = 24$ in. On this basis, slenderness ratios are computed next.

$$\frac{K_x L_x}{r_x} = \frac{0.75(24)}{0.895} = 20.11 \qquad \text{and} \qquad \frac{K_y L_y}{r_y} = \frac{1.0(12)}{1.58} = 7.59$$

Therefore, the slenderness about the xx axis governs. SJI-CJ-1.0-06 Table 103.4-1 mandates that slenderness ratios of chord members not exceed 90 in the interior panel and 120 in. exterior panels. As can be seen above, and assuming that the same top chord size is used along the entire member, this requirement is met. Furthermore, SJI-CJ-1.0-06, Sec. 103.4 requires that the individual angle slenderness not exceed that of the overall section. To that end, the principal radius of gyration of an individual L3×3×½ angle section r_z is 0.580 in., as found in AISC 325-05. The unbraced length for the individual angle check is the distance between adjacent fillers, deck attachments, or web member attachments. Assuming that at the most conservative location the distance between deck attachment, filler, or center of a web member attachment is 10 in.,

$$\frac{K_z L_z}{r_z} = \frac{1.0(10)}{0.580} = 17.24 \leq 20.11 \qquad \leftarrow Okay!$$

Compressive strength is computed per SJI-CJ-2010 Sec. 103.3:

$$4.71\sqrt{\frac{E}{QF_y}} = 4.71\sqrt{\frac{29,000}{1.0(50)}} = 113.43 \geq 20.11$$

Therefore, SJI-CJ-1.0-06 Eq. (103.3-5) applies. On this basis, elastic buckling stress, inelastic buckling stress, and compressive strength are computed.

$$F_e = \frac{\pi^2 E}{(KL/r)^2} = \frac{\pi^2 (29,000)}{(20.11)^2} = 707.74 \text{ ksi}$$

$$F_{cr} = Q\left(0.658^{(QF_y/F_e)}\right)F_y = 1.0\left(0.658^{[1.0(50/707.74)]}\right)50 = 48.54 \text{ ksi} \qquad \text{(SJI, Eq. 103.3-3)}$$

$$\phi_c T_c = \phi_c A_{tc} F_{cr} = 0.90(5.50)(48.54) = 240.27 \text{ kips} \leq \phi_t T_s$$

Finally, the pre-composite design moment is computed based on the governing chord force, in this case the top force:

$$\phi M_{n,pc} = \phi_c T_c d_{pc} = 240.27(22.85) = 5,490.17 \text{ kip-in} = 457.51 \text{ kip-ft}$$

Since $\phi M_{n,pc} \geq M_{u,pc}$ (457.51 kip-ft $\geq$ 133.88 kip-ft), the pre-composite joist strength immediately following concrete casting is satisfactory.

For the pre-composite check prior to concrete placement, $L_y = 36$ in. Then,

$$\frac{K_y L_y}{r_y} = \frac{1.0(36)}{1.58} = 22.78$$

$$4.71\sqrt{\frac{E}{QF_y}} = 4.71\sqrt{\frac{29,000}{1.0(50)}} = 113.43 \geq 22.78$$

$$F_e = \frac{\pi^2 E}{(KL/r)^2} = \frac{\pi^2 (29,000)}{(22.78)^2} = 551.56 \text{ ksi}$$

$$F_{cr} = Q(0.658^{(QF_y/F_e)})F_y = 1.0(0.658^{[1.0(50/551.56)]})50 = 48.14 \text{ ksi} \qquad \text{(SJI, Eq. 103.3-3)}$$

$$\phi_c T_c = \phi_c A_{tc} F_{cr} = 0.90(5.50)(48.14) = 238.29 \text{ kips} \leq \phi_t T_s$$

$$\phi M_{n,pc} = \phi_c T_c d_{pc} = 238.29(22.85) = 5,444.93 \text{ kip-in.} = 453.74 \text{ kip-ft}$$

Since this moment capacity is significantly higher than the factored applied moment even with the weight of concrete included, there is no need to compute the factored applied moment not including the weight of concrete; by inspection, it won't govern.

It should be noted that the top chord strength in final analysis will be evaluated with due consideration of combined axial effects due to member flexure and local top chord flexural effects due to gravity load between panel points. These design considerations are identical to those presented in the preceding section dealing with non-composite open-web joists.

As is the case with non-composite open-web joists, the structural engineers with overall responsibility for a project do not engage in routine computations of the composite joist flexural strength or of shear strength or other limit states applicable to the joist individual elements. Instead, use is made of the 2005 SJI Composite Joist Load Tables. Using these tables, for a given span, geometry of the slab, compressive strength of the concrete, joist spacing, joist depth, number and size of shear-headed studs, the support bearing height, and the required uniform factored load capacity, the user can select an SJI CJ-Series composite joist designation of certain weight and effective moment of inertia. CJ-Series joists are specified in the form of the following designation: $(X)CJ\ (A)/(B)/(C)(N)–(d)$, where (X) indicates the joist depth, (A) indicates the total factored uniform composite design load, (B) designates the total factored uniform composite live load, and (C) shows the total uniform factored composite dead load. Obviously, in all cases, $(A) = (B) + (C) +$ non-composite dead load. The variable (N) indicates the number of shear studs per member, and (d) is the diameter of the stud. As an example, 28CJ 1000/640/360/76–0.75 is a 28-in. deep joist with a factored uniform load capacity of 1,000 lb/ft, 640 lb/ft of which is composite factored live load and 360 lb/ft of which is composite factored dead load. Furthermore, the joist requires a total of 76 (38 per each half-span) 0.75-in. diameter studs. The 2005 SJI Composite Joist Load Tables also provide the required size and spacing of stability bridging for a given composite joist configuration.

Example 7.4

For the loads, span, and member depth from Example 7.3, select the appropriate CJ-Series joist and preliminary bridging configuration.

From Example 7.3, $d = 24$ in., and $w_u = 1.45$ k/ft and $w_{LL} = 371.2$ lb/ft. The design length from the preceding example is 30 ft, meaning that the span is 30.33 ft. However, it is only negligibly unconservative to use the table for the 30-ft span. As can be seen from the 2005 SJI Composite Joist Load Tables, using a 1.5-in. deck and total slab thickness of 4 in., with $f'_c = 4$ ksi, 5-in. high joists bearing a seat, and a minimum joist spacing of 5.5 ft, a uniform factored load capacity of 1,600 lb/ft can be achieved with thirty ½-in. diameter studs, yielding an effective moment of inertia of 773 in.^{4} and a service live load causing a deflection of 1/360 times the span of 1,229 lb/ft. Since 1,229 lb/ft > 371.2 lb/ft, the live-load deflection is satisfactory. A single row of L1.25×1.25×0.109 bridging angles is required.

Consideration of vibration control typically is required on composite open-web joists supporting floors. This can be accomplished using the SJI *Technical Digest No. 5—Vibration of Steel Joist–Concrete Slab Floors* as a guide. Alternatively, the user may chose to use the method presented by the AISC *Design Guide 11—Floor Vibrations Due to Human Activity*. Whether a floor consists of composite or non-composite open-web joists, it is evaluated in the same manner with respect to the vibration performance using either of the preceding methods. For the method presented in SJI *Technical Digest No. 5*, the moment of inertia used in computing the floor frequency is calculated assuming a fully composite section and transformed slab area, defined as I_{comp} in Eq. (7.10). For the method presented in AISC *Design Guide 11*, the composite effective moment of inertia is used in frequency computations. The composite effective moment of inertia I_{eff}, which takes into account the distortion of the joist cross section due to web openings, is calculated using Eq. (7.10).

I_{eff} is provided in the 2005 SJI Composite Joist Load Tables for a variety of composite joist configurations. When it is necessary to compute I_{eff} for a composite joist configuration not provided by the 2005 SJI Composite Joist Load Tables, the main challenge of calculating I_{chords} and I_{comp} in applications involving non-composite joists is the inability to positively verify the chord section and sizes and effective joist depth d_e at the design stage. For this reason, d_e can be taken as approximately $0.91d$.[15] Similarly, A_{bc} can be taken as $1.67A_{tc}$.[15] The A_{tc} can be estimated using the joist capacities from the 2005 SJI Load Tables using

$$A_{tc} = w_{PC}L^2/(\phi 8F_y d_e) \tag{7.24}$$

where

W_{PC} = uniform factored load supported by the pre-composite joist section

L = joist design length

ϕ = strength-reduction factor

= 0.90

F_y = top chord yield stress

= 50 ksi

d_e = effective joist depth

= 0.91d

d = overall joist depth

7.4 Cold-Formed Steel Deck

The most typical applications of composite and non-composite open-web steel joists involve cold-formed steel decks. As noted earlier, steel decks can come in three different types—composite floor decks, non-composite floor decks, and roof decks. The former two are used when an open-web joist system is used to support a concrete slab. The main distinction between the non-composite and composite floor decks is that in former, the cold-formed steel decking is used simply as a flexural member for the support of wet concrete. Once the concrete hardens, the role of the deck diminishes because the supported slab is designed to act as a continuous reinforced-concrete member. In composite floor decks, on the other hand, the steel deck serves not only as a form supporting the weight of the wet concrete but also as a cohesive composite unit with the concrete slab, supplementing the slab mild reinforcement. The composite action between the steel deck and the concrete slab in composite deck systems is achieved through a combination of chemical adhesion of concrete to steel deck, steel deck embossments that create a mechanical bond, and mechanical attachments consisting of headed shear studs. It should be noted that although composite joists typically feature composite slabs, taking advantage of the slab-to-deck bond created by the headed shear studs welded through the deck, composite joist members could just as well be designed with non-composite floor decks. Similarly, non-composite open-web joists systems may be used in support of composite deck slabs. Finally, roof decks are used in open-web joist roof framing applications that do not involve concrete slab topping. Where concrete slab is required in roof framing applications, either for the purposes of better durability or for the support of heavier loads, such as those due to mechanical units, terraces, or green roofs, either composite or non-composite decks can be used. Chapter 3 provides a discussion on material properties and depicts profiles of typical commercially available composite floor, non-composite floor, and roof decks.

The 2009 IBC–referenced standard for the design of roof steel decks is the SDI *Standard for Steel Roof Deck* (ANSI/SDI-RD1.0-06).[21] Similarly, for the design of non-composite floor decks, 2009 IBC references the SDI *Standard for Non composite Steel Floor Deck* (ANSI/SDI-NC1.0-06).[20] For the design of composite slabs, 2009 IBC references the American Society of Civil Engineers (ASCE) *Standard for the Structural Design of Composite Slabs* (ANSI/ASCE 3-91).[6] This standard is widely perceived as outdated and not adequate in addressing the needs of the current design practice. As a result, ANSI/ASCE 3-91 was removed from Chap. 35 of the 2012 IBC.[13] Although not an IBC-referenced standard, the provisions of ANSI/SDI-C1.0-06 most typically are consulted for the design of composite slabs. It is expected that the next edition of the IBC will reference a consensus ASCE/SDI standard for the design of composite slabs currently in development. When a concrete slab is used, SJI-K-1.1-05 Sec. 5.8(b) requires that the minimum slab thickness be at least 2 in.

The flexural and shear strength of any of the three types of deck discussed earlier can be established either by test or by calculation, although the former is more prevalent for all three types of deck. The design of an untopped steel deck, both roof and non-composite floor, is by reference performed using the provisions of ANSI/AISI S100-07. Concrete slabs supported by a non-composite floor deck are by reference designed as one-way reinforced-concrete slabs using the provisions of ACI 318-08. The strength of composite decks is calculated using an analytic approach provided in ANSI/SDI-C1.0-06. In terms of practical day-to-day application, generic calculations using the provisions of the above-referenced standards typically are not performed. Instead, design aids in the form of section properties, strength properties, and span tables provided in manufacturers' catalogs are used commonly. Such values, in turn, are established based on the building code–referenced design standards. The most common process is to select a roof deck or a slab thickness and floor deck as appropriate based on fire protection, durability, vibration, economy, constructability, and other criteria and then to use the allowable span tables to limit the spacing of the framing joist members to achieve the required uniform load capacity. To aid the economy of design, a single type of deck typically is chosen for the entire roof surface or the entire floor framing as appropriate on a project. As an example, Figure 7.22 shows a common roof deck model from Vulcraft. If the designer were to verify whether this deck is sufficient for use for the roof framing analyzed in the Example 7.2, the first step would be to establish the service gravity load. The most critical case in this example results from 2009 IBC Eq. (16-10), providing the load case $D + S$ at the peak of the snow drift load. The governing service load at this location, therefore, equals $10 + 2 + 25 + 44 = 81$ psf. Of this, 69 psf is due to the snow load. By inspection, the load cases involving roof live load will not govern. As can be seen from Fig. 7.22, for the minimum continuity of three spans, as is the case with the deck from Example 7.2, one can safely use 1.5B Ga. 24 deck, which is capable of resisting a service load of 154 psf with a center-to-center of support span of 5 ft or less. Similarly, the maximum deflection under the snow load permitted by 2009 IBC Table 1604.3 for roof members not supporting a plaster ceiling is $L/240$, where L is the member span. As can be seen from Fig. 7.22, the service load of 120 psf will cause a deflection of $L/240$ for spans of 5 ft or less. Since the span in this example is only 2.15 ft and the snow load is 69 psf, the deflection check is satisfactory. The deck also must be checked for construction loads. As can be seen from Fig. 7.22, the maximum permissible construction span for this deck with a minimum three-span continuity is 5 ft-10 in., which by far exceeds the actual span of 2.15 ft. In an unlikely situation where the deck is shored during construction, a construction span check would not be necessary. The final consideration is to determine the width of bearing. Decks with insufficient support bearing width, that is, the support width, may fail by web crippling. The manufacturer's design aids, such as the one presented in Fig. 7.22, typically are generated assuming a certain bearing width. As can be seen, for the deck shown in Fig. 7.22, the minimum bearing width is 1.5 in. The smallest top chord angle sizes employed in the fabrication of very light joists typically do not feature a leg width of less than 1.25 in. Consequently, for even the smallest joist sections, the minimum bearing requirements are satisfied routinely. However, to avoid scenarios whereby the capacity of a larger deck profile is limited by the bearing on a smaller joist section, the designers should stipulate the

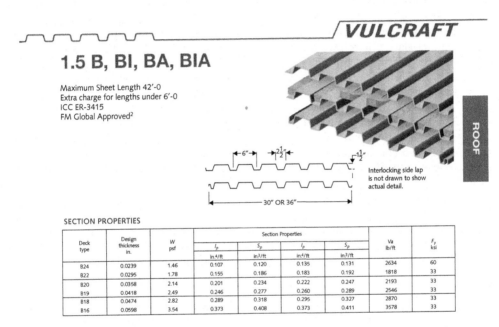

1.5 B, BI, BA, BIA

Maximum Sheet Length 42'-0
Extra charge for lengths under 6'-0
ICC ER-3415
FM Global Approved[2]

Interlocking side lap is not drawn to show actual detail.

30" OR 36"

ROOF

SECTION PROPERTIES

Deck type	Design thickness in.	W psf	Section Properties				V_a lb/ft	F_y ksi
			I_p in.⁴/ft	S_p in³/ft	I_p in⁴/ft	S_p in³/ft		
B24	0.0239	1.46	0.107	0.120	0.135	0.131	2634	60
B22	0.0295	1.78	0.155	0.186	0.183	0.192	1818	33
B20	0.0358	2.14	0.201	0.234	0.222	0.247	2193	33
B19	0.0418	2.49	0.246	0.277	0.260	0.289	2546	33
B18	0.0474	2.82	0.289	0.318	0.295	0.327	2870	33
B16	0.0598	3.54	0.373	0.408	0.373	0.411	3578	33

VERTICAL LOADS FOR TYPE 1.5B

No. of Spans	Deck Type	Max. SDI Const. Span	Allowable Total (PSF) / Load Causing Deflection of L/240 or 1 inch (PSF) Span (ft/in.) ctr to ctr of supports										
			5-0	5-6	6-0	6-6	7-0	7-6	8-0	8-6	9-0	9-6	10-0
1	B24	4'-8	115/56	95/42	80/32	68/26	59/20	51/17	45/14	40/11	35/10	32/6	29/7
	B22	5'-7	98/81	81/61	68/47	58/37	50/30	44/24	38/20	34/17	30/14	27/12	25/10
	B20	6'-5	123/105	102/79	86/61	73/48	63/38	55/31	48/26	43/21	38/18	34/15	31/13
	B19	7'-1	146/129	121/97	101/75	86/59	74/47	65/38	57/31	51/26	45/22	40/19	36/16
	B18	7'-8	168/152	138/114	116/88	99/69	85/55	74/45	65/37	58/31	52/26	46/22	42/19
	B16	8'-8	215/196	178/147	149/113	127/89	110/71	96/58	84/48	74/40	66/34	60/29	54/24
2	B24	5'-10	124/153	103/115	86/88	74/70	64/56	56/45	49/37	43/31	39/26	35/22	31/19
	B22	6'-11	100/213	83/160	70/124	59/97	51/78	45/63	39/52	35/43	31/37	28/31	24/27
	B20	7'-9	128/267	106/201	89/155	76/122	66/97	57/79	51/65	45/54	40/46	36/39	32/33
	B19	8'-5	150/320	124/240	104/185	89/145	77/116	67/95	59/78	52/65	47/55	42/47	38/40
	B18	9'-1	169/369	140/277	118/213	101/168	87/134	76/109	67/90	59/75	53/63	48/54	43/46
	B16	10'-3	213/471	176/354	149/273	127/214	110/172	95/140	84/115	74/96	66/81	60/69	54/59
3	B24	5'-10	154/120	128/90	108/69	92/55	79/44	69/35	61/29	54/24	48/21	43/17	39/15
	B22	6'-11	124/167	103/126	87/97	74/76	64/61	56/50	49/41	43/34	39/29	35/24	31/21
	B20	7'-9	159/209	132/157	111/121	95/95	82/76	72/662	63/51	56/43	50/36	45/31	40/26
	B19	8'-5	186/250	154/188	130/145	111/114	96/91	84/74	74/61	65/51	58/43	52/37	47/31
	B18	9'-1	210/289	174/217	147/167	126/132	108/105	95/86	83/71	74/59	66/50	59/42	54/36
	B16	10'-3	264/369	219/277	185/214	158/168	136/135	119/109	105/90	93/75	83/63	74/54	67/46

Notes: 1. Minimum exterir bearing length required is 1.50 inches. Minimum interior bearing length required is 3.00 inches. If these minimum lengths are not provided, web crippling must be checked.
2. FM Global approved numbers and spans available on page 21.

FIGURE 7.22 Design data for a typical Vulcraft roof deck profile. (*Vulcraft*.[34])

minimum required bearing width for the selected deck profile in the construction documents. When bearing width isn't sufficient, the designer has an option of checking the web crippling strength for the actual load using Chap. C of ANSI/AISI S100-07. Web crippling of cold-formed decks is discussed further in Chap. 5.

The design process just illustrated for a roof deck is similar to that used for composite and non-composite floor decks. The most notable exception is the need to select the appropriate mild reinforcement. Although fiber reinforcement has seen an increased use, welded wire fabric remains a primary means of composite and non-composite slab reinforcement. In the case of non-composite slabs, the user must verify the adequacy of the deck to support wet concrete loads and the adequacy of the reinforced concrete slab to support the uniform loads when the structure is in service. A critical consideration in the design of non-composite slabs deals with the position of mild reinforcement. Specifically, to maximize the efficiency of thicker slabs, the reinforcement consisting of welded wire mesh is often draped as shown in Fig. 7.23 to maximize the internal resistance moment arm in the regions of both positive and negative reinforcement. In thinner slabs, the reinforcement is typically centered in the solid thickness of the slab and the portion of the slab above the reinforcement in the negative-moment regions, and the portion of the slab below the reinforcement in the positive-moment regions are deemed ineffective. The manufacturers' design tables indicate which reinforcement pattern is required for each particular combination of deck geometry and slab thickness. Manufacturers' tables will indicate the concrete compressive strength f_c' on which the presented design values are based. An f_c' of 4 ksi is a commonly used value. When a reinforced concrete slab is not sufficient in resisting the imposed loads, the engineer may elect to rely on steel deck alone for support of the applied loads. When this is done, web crippling of the deck through a positive evaluation of the support bearing width must be considered.

As an example, assuming that a non-composite Vulcraft 3C deck was selected for the composite joist system from Example 7.3, one must first verify the adequacy of the bare deck under construction loads. As can be seen from Fig. 7.24, for a 6.5-in. thick slab, the Vulcraft 3C Ga. 22 deck is capable of achieving a clear span of 9 ft-4 in. under the

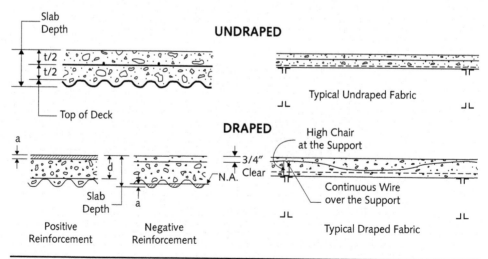

Figure 7.23   Non-composite slab reinforcement. (*Vulcraft.*[34])

weight of normal-weight wet concrete and construction live loads when the deck is continuous over a minimum of three spans. Therefore, this indicates that the same deck profile and thickness are adequate at the construction stage for the 8-ft center-to-center spans supporting a 6.25-in. thick slab, as presented in Example 7.3.

For the slab in the Example 7.3, the clear span between the edges of successive joist top chords equals $8 - (3 + 3 + \frac{3}{4})/12 = 7.44$ ft. Although the manufacturer's design data

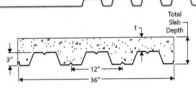

3 C CONFORM

Interlocking side lap is not drawn to show actual detail.

MAXIMUM CONSTRUCTION CLEAR SPANS (S.D.I. CRITERIA)

Total Slab Depth	DECK	WEIGHT PSF	NW CONCRETE N=9 145 PCF			WEIGHT PSF	LW CONCRETE N=14 110 PCF		
			1 SPAN	2 SPAN	3 SPAN		1 SPAN	2 SPAN	3 SPAN
6 (t=3.00)	3C22	56	8-4	8-10	10-1	43	9-3	10-9	11-9
	3C20	57	9-8	11-10	12-3	43	10-9	13-1	13-6
	3C18	57	11-10	14-2	14-2	44	12-11	15-2	15-2
	3C16	58	12-2	14-4	14-10	45	13-7	15-9	16-0
6.5 (t=3.50)	3C22	62	8-0	8-3	9-4	48	8-11	10-0	11-4
	3C20	63	9-3	11-5	11-9	48	10-4	12-7	13-0
	3C18	63	11-4	13-9	13-10	49	12-7	14-9	14-9
	3C16	64	11-7	13-10	14-3	49	13-0	15-2	15-7
7 (t=4.00)	3C22	68	7-9	7-8	8-8	52	8-7	9-4	10-8
	3C20	69	9-0	10-11	11-4	53	9-11	12-2	12-7
	3C18	69	11-0	13-3	13-6	53	12-3	14-5	14-5
	3C16	70	11-4	13-4	13-9	54	12-6	14-9	15-3
7.5 (t=4.50)	3C22	74	7-7	7-2	8-2	57	8-3	8-10	10-0
	3C20	75	8-9	10-2	11-0	57	9-7	11-10	12-2
	3C18	75	10-9	12-10	13-3	58	11-9	14-2	14-2
	3C16	76	11-0	12-11	13-4	59	12-1	14-3	14-9
8 (t=5.00)	3C22	80	7-5	6-9	7-8	61	8-0	8-4	9-5
	3C20	81	8-7	9-7	10-8	62	9-3	11-6	11-10
	3C18	81	10-6	12-5	12-10	62	11-5	13-10	13-11
	3C16	82	10-9	12-6	12-11	63	11-8	13-11	14-4

REINFORCED CONCRETE SLAB ALLOWABLE LOADS

Slab Depth	REINFORCEMENT		Superimposed Uniform Load (psf) - 3 Span Condition										
			Clear Span (ft-in.)										
	W.W.F.	As	6-6	7-0	7-6	8-0	8-6	9-0	9-6	10-0	10-6	11-0	11-6
6 (t=3.00)	6X6-W2.9XW2.9	0.058*	125	108									
	4X4-W2.9XW2.9	0.087	185	160									
	4X4-W4.0XW4.0	0.120	246	212									
6.5 (t=3.50)	6X6-W2.9XW2.9	0.058*	154	133	116	102							
	4X4-W2.9XW2.9	0.087	229	198	172	151							
	4X4-W4.0XW4.0	0.120	306	264	230	202							
7 (t=4.00)	6X6-W2.9XW2.9	0.058*	183	158	138	121	107	96					
	4X4-W2.9XW2.9	0.087	273	235	205	180	159	142					
	4X4-W4.0XW4.0	0.120	366	316	275	242	214	191					
7.5 (t=4.50)	6X6-W2.9XW2.9	0.087*	316	273	238	209	185	165	148	134	121		
	4X4-W4.0XW4.0	0.120	400	368	320	281	249	222	200	180	163		
	4X4-W5.0XW5.0	0.150	400	400	392	345	306	273	245	221	200		
8 (t=5.00)	6X6-W2.9XW2.9	0.087*	360	310	270	238	210	188	168	152	138	126	115
	4X4-W4.0XW4.0	0.120	400	400	365	321	284	254	228	205	186	170	155
	4X4-W5.0XW5.0	0.150	400	400	400	395	350	312	280	253	229	209	191

NOTES:
1. * As does not meet A.C.I. criterion for temperature and shrinkage.
2. Recommended conform types are based upon S.D.I. criteria and normal weight concrete.
3. Superimposed loads are based upon three span conditions and A.C.I. moment coefficients.
4. Load values for single span and double spans are to be reduced.
5. Vulcraft's painted or galvanized form deck can be considered as permanent support in most building applications. See page 23.
   If uncoated form deck is used, deduct the weight of the slab from the allowable superimposed uniform loads.
6. Superimposed load values shown in bold type require that mesh be draped. See page 23.

FIGURE 7.24 Non-composite slab design data. (Vulcraft.[34])

as shown in the figure do not provide design values for a 6-in. slab thickness for clear spans exceeding 7 ft, and the slab thickness of 6.25 in. is not included in the design tables, it is clear that the values for a 6-in. slab with a clear span of 7 ft are very similar to those for a 6.5-in. slab with a clear span of 6.5 ft. To this end, it can be assessed by inspection that for a 6.25-in. thick slab with a 7.44-ft clear span, continuous over at least three spans and reinforced with a draped 4×4–W2.9×W2.9 welded wire mesh will have a superimposed uniform load capacity of approximately 160 psf, far exceeding the required superimposed uniform load of 60 psf, consisting of 40 psf live load and 20 psf. Figure 7.25 shows another option available to the designer in supporting the service load. Specifically, if the contribution of reinforced slab is to be neglected completely, the manufacturer's design table considering only the deck capacity, depicted in Fig. 7.25, shows that the preceding selected Vulcraft 3C Ga. 22 can support a service load of 147 psf with a clear span of 7.5 ft and minimum continuity over two spans. For this example, the total service load supported by the deck consists of the slab weight of 59.38 psf, the deck self-weight of 1.62 psf, the superimposed live load of 40 psf, and the superimposed dead load of 20 psf, for a total of 121 psf, significantly less than the capacity of at least 147 psf. It should be noted that although 2009 IBC, Sec. 1607.9 permits live load reduction for one-way members such as decks, the live load tributary area in this example is not large enough to result in a live load reduction. Finally, the design table in Fig. 7.25 shows that the load of 214 psf will cause a deflection of $L/240$. The 2009 IBC Table 1604.3 stipulates a deflection limit of $L/240$ under the combined dead and live load. Since 214 psf > 147 psf, this criterion is satisfied. For live load, 2009 IBC, Table 1604.3 stipulates a deflection limit of $L/360$. The allowable load can be extracted from Fig. 7.25 simply by proportioning the value provided for the $L/240$ limit, $(240/360)$ 214 = 142.67 psf. Obviously, since 142.67 psf > 40 psf, the live load deflection criterion is satisfied. Lastly, Fig. 7.25 indicates a requirement that a minimum interior bearing of 5 in. and exterior bearing of 2.5 in. be provided. Since the actual interior bearing width equals 3 + 3 + ¾ = 6.75 in., a web crippling check is not necessary at an interior support. Similarly, the deck support detail at the perimeter joist should indicate a minimum bearing width of 2.5 in.

If a composite deck is selected, the design checks also typically are performed using tabular design aids from the manufacturer's literature. The checks consist of verifying a minimum bearing width at support, verifying the recommended slab reinforcement provided for each combination of deck and slab geometry, and a check of permissible clear span for a given combination of loading, deck type, slab thickness, and span continuity. When consideration of web crippling is required, as discussed previously, the check must be performed as provided by ANSI/AISI S100-07, Chap. C. The deck manufacturers' literature typically provides deck web crippling values for various support widths to aid such a check.

The preceding discussion focused on the role of the deck in supporting vertical loads. Steel decks also perform a significant role in providing lateral bracing for supporting members. As noted in preceding sections, the deck attachments to members are mandated at spacings no larger than 36 in. by both SJI-K-1.1-05 and SJI-CJ-1.1-06. Furthermore, specific deck diaphragm attachment patterns further reduce this spacing to a range of 6 to 18 in. Typical deck attachments in most cases will result in a fully braced member.

Lastly, nearly all decks serve an important role in providing floor and roof diaphragms. The role of a diaphragm, the load path thereto, computation of wind and seismic loads for diaphragms and diaphragm elements such as chords and collectors, and types of diaphragms are discussed in Chap. 2 of this book. Diaphragm design consists of verifying

ALLOWABLE UNIFORM LOAD (PSF)

TYPE NO.	NO. OF SPANS	DESIGN CRITERIA	CLEAR SPAN (ft-in)												
			6-6	7-0	7-6	8-0	8-6	9-0	9-6	10-0	10-6	11-0	11-6	12-0	12-6
3C22	1	Fb = 30,000	196	169	147	129	114	102	92	83	75	68	62	57	53
		Defl. = l/240	175	140	114	94	78	66	56	48	41	36	32	28	25
		Defl. = l/180	233	186	151	125	104	88	75	64	55	48	42	37	33
	2	Fb = 30,000	177	155	137	122	109	98	88	80	73	67	62	57	52
		Defl. = l/240	420	336	273	225	188	158	134	115	100	87	76	67	59
		Defl. = l/180	560	448	364	300	250	211	179	154	133	116	101	89	79
	3	Fb = 30,000	212	186	165	147	132	119	108	98	90	82	76	70	65
		Defl. = l/240	329	263	214	176	147	124	105	90	78	68	59	52	46
		Defl. = l/180	438	351	285	235	196	165	140	120	104	90	79	70	62
3C20	1	Fb = 30,000	252	218	189	167	148	132	118	107	97	88	81	74	68
		Defl. = l/240	220	176	143	118	98	83	70	60	52	45	40	35	31
		Defl. = l/180	293	235	191	157	131	110	94	81	70	61	53	47	41
	2	Fb = 30,000	242	211	185	164	146	131	118	107	97	89	81	75	69
		Defl. = l/240	529	424	345	284	237	199	170	145	126	109	96	84	74
		Defl. = l/180	706	565	459	379	316	266	226	194	167	146	127	112	99
	3	Fb = 30,000	294	257	226	201	179	161	145	131	120	109	100	93	85
		Defl. = l/240	414	332	270	222	185	156	133	114	98	85	75	66	58
		Defl. = l/180	552	442	360	296	247	208	177	152	131	114	100	88	78
3C18	1	Fb = 30,000	364	314	273	240	213	190	170	154	139	127	116	107	98
		Defl. = l/240	300	240	195	161	134	113	96	82	71	62	54	48	42
		Defl. = l/180	400	320	260	214	179	151	128	110	95	82	72	64	56
	2	Fb = 30,000	358	311	272	240	214	191	172	156	141	129	118	109	100
		Defl. = l/240	721	577	469	387	323	272	231	198	171	149	130	115	101
		Defl. = l/180	962	770	626	516	430	362	308	264	228	198	174	153	135
	3	Fb = 30,000	439	382	335	296	264	236	213	193	175	160	147	135	125
		Defl. = l/240	564	452	367	303	252	213	181	155	134	116	102	90	79
		Defl. = l/180	753	603	490	404	337	284	241	207	179	155	136	120	106
3C16	1	Fb = 30,000	383	330	288	253	224	200	179	162	147	134	122	112	104
		Defl. = l/240	378	302	246	203	169	142	121	104	90	78	68	60	53
		Defl. = l/180	504	403	328	270	225	190	161	138	119	104	91	80	71
	2	Fb = 30,000	367	319	279	246	218	195	176	159	144	132	121	111	102
		Defl. = l/240	909	728	592	488	407	343	291	250	216	188	164	145	128
		Defl. = l/180	1213	971	789	650	542	457	388	333	288	250	219	193	170
	3	Fb = 30,000	451	392	344	304	270	242	218	197	179	164	150	138	127
		Defl. = l/240	712	570	463	382	318	268	228	195	169	147	129	113	100
		Defl. = l/180	949	760	618	509	424	357	304	261	225	196	171	151	133

Minimum exterior bearing length is 2.5 inches.
Minimum exterior bearing length is 5.0 inches.

Figure 7.25 Non-composite deck design data. (Vulcraft, 2008.[34])

the adequacy of diaphragm shear strength and diaphragm stiffness used in deflection calculations. Section D5 of ANSI/AISI S100-07 provides factors of safety Ω and resistance factors ϕ for steel deck diaphragms subjected to wind and earthquake loads and attached using welds or mechanical fasteners such as screws and power-actuated fasteners. Although welds, in the form of arc seam spot welds, commonly referred to a *puddle welds*, remain a preferred method of attaching deck diaphragms, recent research indicates that the quality-assurance issues pertaining to lapped decks, thicker decks, and deck welding time may result in a significantly underestimated attachment strength.[18] Furthermore, deck connections to the supporting members featuring puddle welds were found to exhibit insignificant ductility compared with the attachments featuring screws and power actuated fasteners.[31] The strength and stiffness of diaphragms can be established by either test or by calculation. The ANSI/AISI S100-07, Sec. D5–provided Ω and ϕ factors apply whether the diaphragm strengths are established by test or by calculation. For steel deck diaphragms, the diaphragm shear strength and shear modulus can be determined using AISI *Test Standard for Cantilever Test Method for Cold-Formed Steel Diaphragms* (AISI S907-08).[5] When the diaphragm strength and stiffness are determined by calculation, methods presented in SDI *Diaphragm Design Manual* (SDI DDM-03)[18a] can be used for either bare deck or concrete top decks. In routine design calculations, direct application of the SDI DDM-03 computation typically is not necessary because the deck manufacturers' literature typically provides tabulated values for determining the strength and stiffness of bare and concrete-topped steel deck diaphragms based on the specific pattern of deck attachments, deck type and thickness, type of attachment, and thickness of the concrete topping, if present.

Joist and joist girders may perform a significant role as a part of a steel deck diaphragm. As can be seen from Fig. 7.16, for diaphragm loading V_x, the joist girders along gridlines B and C will act as diaphragm collectors, subjecting their top chords to additional axial compression or tension depending on the direction of the diaphragm shear. Furthermore, the eccentricity of the diaphragm shear with respect to the geometric centroid of the joist girder top chord also will result in top chord moments. The structural engineer with overall responsibility for the project must indicate in the construction documents the magnitude and source (i.e., seismic or wind) of any collector forces to which joist chords are subjected. Furthermore, by inspection of Fig. 7.16, it is also clear that intermediate K-Series joists between gridlines 1 and 2 will be required to transfer the diaphragm shear accumulating along the perimeter of the diaphragm (i.e., along gridlines A, B, and C) into the supporting lateral force resisting elements or collectors. The transfer mechanism is known as *roll-over resistance of joist seats*. This concept is illustrated in Figure 7.26. Analysis shows that a typical joist seat can transfer a factored force of up to approximately 8 kips through its roll-over resistance.[10] Since the structural engineer with overall responsibility for the project cannot anticipate the precise geometry of the joist seat at the joist selection and framing stage, the required roll-over resistance per joist seat should be included in the construction documents. As shown in Fig. 7.26, when the combined roll-over resistance of joist seats is not sufficient to resist the entire magnitude of the perimeter diaphragm shear, V_D, the transfer mechanism can be augmented by introduction of joist substitute members, such as 2.5K joists, oriented perpendicular to the span of main joist members. In such a scenario, the in-plane shear distortion resistance of the joist substitute web is relied upon to provide additional transfer mechanism for the diaphragm forces. While the joist manufacturer is responsible for proportioning of joist substitute members when used for resistance of diaphragm

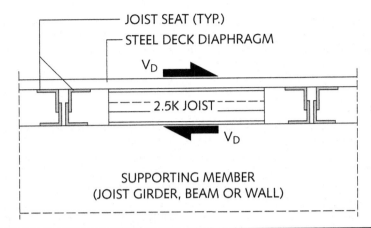

Figure 7.26   Roll-over resistance of joist seats

forces, a collaborative and interactive effort is required between the joist design engineer and the engineer with the overall responsibility for the project in determining the need for and final configuration of the diaphragm force transfer system. The reader is referred to the Fisher, West, and Van de Pas (2002)[10] for further information on use of open-web joists in diaphragm load-path applications.

7.5 International Building Code

The 2009 IBC Sec. 2206 provides the building code framework for the design of open-web steel joist systems. Aside from referencing the four main SJI specifications for the design of open-web steel joists, its chief role is to delineate the responsibilities of the structural engineer with overall responsibility for the project, who specifies the joist, and the joist manufacturer, including its engineer responsible for joist design.

In short, the structural engineer with overall responsibility for the project must indicate in the construction documents the joist member layout, end-support conditions, anchorage, non-SJI standard bridging requirements, bearing connection design for resistance of uplift loads, and any of the non-uniform gravity load conditions, such as concentrated forces, patch loadings, uplift forces, end-member moments, and connection forces resulting from the attachment of other members. The structural engineer also must specifically indicate the requirement for any non-parallel-chord geometric conditions and the deflection criteria, if different from that stipulated by the SJI specifications. It should be noted that fulfillment of the requirement to indicate all building component and cladding loads stipulated by 2009 IBC Sec. 1603.1.4 can be used toward satisfying the requirement to report the joist uplift loads to the joist manufacturer. If the uplift loads are reported per 2009 IBC Sec. 1603.1.4, the countering gravity dead loads must be specifically provided in the construction documents, allowing the joist designer to establish the net uplift effects.

The joist manufacturer is responsible for submitting detailed placement plans indicating connection requirements for joist supports, joist girder supports, field splicing, and bridging attachments to other structural elements. Also, such plans must list all the loading and deflection criteria used in the member design and must be accompanied by a letter of certification stating compliance with the design criteria and construction documents, as stipulated by 2009 IBC Sec. 1704.2.2. Also, at the request of the structural engineer with overall responsibility for the project, the joist manufacturer must submit detailed design calculations, performed by a registered design professional.

References

1. American Concrete Institute (ACI). 2008. *Building Code Requirements for Structural Concrete* (ACI 318-08). ACI, Farmington Hills, MI.
2. American Institute of Steel Construction (AISC). 2005. *Specification for Structural Steel Buildings* (ANSI/AISC 360-05). AISC, Chicago, IL.
3. American Institute of Steel Construction (AISC). 2005. *Manual of Steel Construction* (AISC 325-05), 13th ed. AISC, Chicago, IL.
4. American Iron and Steel Institute (AISI). 2007. *North American Specification for the Design of Cold-Formed Steel Members* (ANSI/AISI S100-07). AISI, Washington, DC.
5. American Iron and Steel Institute (AISI). 2008. *Test Standard for Cantilever Test Method for Cold-Formed Steel Diaphragms* (AISI S907-08), *AISI Cold-Formed Steel Design Manual* (D100-08). AISI, Washington, DC.
6. American Society of Civil Engineers (ASCE). 1991. *Standard for the Structural Design of Composite Slabs* (ANSI/ASCE 3-91). ASCE, New York, NY.
7. Beck, H., and Engelhardt, M. D. 2002. "Net Section Efficiency of Steel Coupons with Power Actuated Fasteners," *American Society of Civil Engineers (ASCE) Journal of Structural Engineering*, Vol. 128, Number 1, pp. 12–21.
8. Chien, E. Y. L. and Ritchie, J. K. 1984. *Design and Construction of Composite Floor Systems.* Canadian Institute of Steel Construction (CISC), Markham, ON, Canada.
9. Englehardt, M. D., Kates, Z., Beck, H., and Stansey, B. 2000. "Experiments on the Effects of Power-Actuated Fasteners on the Strength of Open Web Steel Joists," *American Institute of Steel Construction (AISC) Engineering Journal*, Fourth Quarter, pp. 157–166.
10. Fisher, J.M., West, M.A., and Van de Pas, J.P. 2002. *Designing with Vulcraft Joist, Joist Girders and Steel Deck*, 2nd ed. Vulcraft, Norfolk, NE.
11. Holterman, T., Perry, M., and Green, P. S. 2009. "Specifying Steel Joists, Joist Girders and the International Building Code, in *Proceedings of the American Society of Civil Engineers (ASCE) Structures Congress*, Austin, TX.
12. International Code Council (ICC). 2009. *International Building Code (2009 IBC)*. ICC, Washington, DC.
13. International Code Council (ICC). 2012. *International Building Code (2012 IBC)*. ICC, Washington, DC.
14. Murray, T. M., Allen, D. E., and Ungar, E. E. 1997. *Design Guide 11—Floor Vibrations Due to Human Activity.* American Institute of Steel Instruction (AISC), Chicago, IL.
15. Samuelson, D. 2005. *Estimated Chord Areas and I_{chord} for Composite Joists.* Vulcraft Research and Development, Norfolk, NE.
16. Shultz, W., and Samuelson, D. 2010. Estimating A_{bc}, A_{tc}, d_e, and I_x for K-Series joists., Vulcraft Research and Development, Norfolk, NE.
17. Simpson, Gumpertz & Heger. 2008. *An Experimental Study of the Behavior of Trusses Fabricated with High-Strength Low-Alloy Microalloyed Vanadium Steel.* Simpson, Gumpertz & Heger, Research Report (SGH Project 047079), Waltham, MA.
18. Snow, G. L., and Easterlind, W. S. 2008. "Strength of Arc Spot Welds Made in Single and Multiple Steel Sheets," in *Proceedings of the 19th International Specialty Conference*

on Cold-Formed Steel Structures, Missouri University of Sciences and Technology, Rolla, MO.

18a. Steel Deck Institute (SDI). 2003. *Diaphragm Design Manual* (SDI DDM-03). SDI, Chicago, IL.

19. Steel Deck Institute (SDI). 2006. *Standard for Composite Steel Floor Deck* (ANSI/SDI-C1.0-06). SDI, Fox River Grove, IL.

20. Steel Deck Institute (SDI). 2006. *Standard for Non-composite Steel Floor Deck* (ANSI/SDI-NC1.0-06). SDI, Fox River Grove, IL.

21. Steel Deck Institute (SDI). 2006. *Standard for Steel Roof Deck* (ANSI/SDI-RD1.0-06). SDI, Fox River Grove, IL.

22. Steel Joist Institute (SJI). 2005. *42nd Edition Standard Specifications—Load Tables and Weight Tables for Steel Joists and Joist Girders—K-Series, LH-Series, DLH-Series, and Joist Girders* (2005 SJI Load Tables). SJI, Forest, VA.

23. Steel Joist Institute (SJI). 2005. *First Edition Standard Specifications for Composite Steel Joists—Weight Tables and Bridging Tables, Code of Standard Practice* (2005 SJI Composite Joist Load Tables). SJI, Forest, VA.

24. Steel Joist Institute (SJI). 2006. *Standard Specification for Composite Steel Joists, CJ-Series* (SJI-CJ-1.0-06). SJI, Forest, VA.

25. Steel Joist Institute (SJI). 2005. *Standard Specification for Joist Girders* (SJI-JG-1.1-05). SJI, Forest, VA.

26. Steel Joist Institute (SJI). 2005. *Standard Specification for Open Web Steel Joists, K-Series* (SJI-K-1.1-05). SJI, Forest, VA.

27. Steel Joist Institute (SJI). 2005. *Standard Specification for Longspan Steel Joists, LH-Series, and Deep Longspan Steel Joists, DLH-Series* (SJI-LH/DLH-1.1-05). SJI, Forest, VA.

28. Steel Joist Institute (SJI). 1988. *Technical Digest No. 5—Vibration of Steel Joist–Concrete Slab Floors*. SJI, Myrtle Beach, SC.

29. Structural Engineering Institute (SEI) of the American Society of Civil Engineers (ASCE). 2006. *Minimum Design Loads for Buildings and Other Structures, Including Supplements Nos. 1 and 2* (ASCE/SEI 7-05). ASCE, Reston, VA.

30. Structural Engineering Institute (SEI) of the American Society of Civil Engineers (ASCE). 2002. *Design Loads on Structures during Construction* (ASCE/SEI 37-02). ASCE, Reston, VA.

31. Tremblay, R., and Rogers, C. A. 2003. "Inelastic Seismic Response of Frame Fasteners for Steel Roof Deck Diaphragms," *American Society of Civil Engineers (ASCE) Journal of Structural Engineering* 129(12):1647–1657.

32. Viest, I. M., Colaco, J. P., Furlong, R. W., Griffis, L. G., Leon, R. T., and Wyllie, L. A. 1997. *Composite Construction Design for Buildings*. McGraw-Hill, New York.

33. Vulcraft. 2007. *Vulcraft Steel Joist and Joist Girders*. Vulcraft, Norfolk, NE.

34. Vulcraft. 2008. *Vulcraft Steel Roof and Floor Deck*. Vulcraft, Norfolk, NE.

Index

"f" indicates material in figures. "t" indicates material in tables.

A

Acceptance Criteria (AC) documents, 10–11
accidental torsional moment, 72
across-wind load, 51, 53
adhesives, 314, 316, 330, 350, 351
adhesive joints, 321
air, 54, 57
air temperature, 128, 134, 139–140, 303, 315
air temperature adjustment factor, 302–303, 307, 309, 315
allowable stress design (ASD), 6, 7, 81–86
American Concrete Institute (ACI), 7
American Institute of Steel Construction (AISC), 173
American Institute of Timber Construction (AITC), 7, 354–355
American Iron and Steel Institute (AISI), 173
American Lumber Standard Committee (ALSC), 133
American National Standards Institute (ANSI), 8
American Plywood Association (APA), 7, 338, 348
American Society for Testing and Materials (ASTM), 126
American Society of Agricultural Engineers (ASAE), 7
American Society of Civil Engineers (ASCE), 7, 173
American Welding Society (AWS), 140
amplification factor, 186
arc seam welds, 167

arc spot welds, 167–168, 412
arched roofs, 43, 134
architectural design, 12, 19
awnings, 43

B

balance point, 218
band joists, 345
barrel-vault roofs, 45
base flood elevation (BFE), 60, 61f
base length/bay length, 361
beams
    bond, 90, 191, 269
    in braced-frame systems, 23
    cambered, 303
    in cold-formed-steel construction, 15, 151–152, 154–158, 161–166
    collector, 73f, 91
    deep vs. flexural, 345
    diaphragms as, 72–73, 88, 345
    load path through, 73, 89
    in masonry construction, 191, 194–213, 221, 243, 266, 288
    in moment frames, 13–18
    in open-web joist systems, 362, 367f, 366
    in post-and-beam construction, 23
    in structural timber construction, 15–16, 299–300, 303–307
beam-columns, 164, 308
beam-on-elastic-foundation approach, 299–300, 335, 344–345

bearing area, 192, 192f
bearing area factor, 307
bearing factor, 170, 170t
bearing-wall systems, 14f, 18–21, 31, 65, 68, 75, 77, 86, 118
bed joints, 124f, 125, 280
bending stress, 135, 304, 305, 310–313
blocks, 126, 127f
blow count, 66
boilers, 79
bolts, 168–171, 280–286, 290–294, 308, 314–323, 327, 328, 354, 371–372
braced-frame systems, 14f, 23–24, 31, 48, 72, 77, 89, 119
brackets, 212
breaking wave forces, 62–64
British Standard Institution (BSI), 354
brittle response chart, 28f
buckling coefficient, 153
building codes, 3
building frame systems, 12–13, 14f, 21–23, 31, 65, 69, 75, 86, 115–118
Building Seismic Safety Council (BSSC), 26
buoyancy, 12, 60, 81. See also uplift
buoyant force, 61
butt joints, 167

━━ C ━━

cable-braced frames, 31
California, 3, 26–27, 33
Canada, 149
Canadian Sheet Steel Building Institute (CSSBI) Web site, 173
canopies, 43
cantilever column systems, 14f, 24, 29, 31, 32, 85–86
cantilever walls, 185f, 187, 236, 245–246, 351
capacity-reduction factor, 196, 282–284
certification letter, 413–414
chloride salts, 139
clay masonry units, 125–128, 132, 195, 243, 250
cold-formed steel
    attributes of, 147
    bending strength of, 153
    buckling of, 123, 150–156, 162–165
    compressive stress of, 152
    cracking of, 137

curtain-wall studs, 173–179
ductility of, 121–123
effective yield moment of, 154
elastic buckling stress of, 154–156
fabrication of, 4
fracturing of, 137
IBC on, 4
illustration of members, 148f
manufacturing, 137, 149–150
MOE of, 152, 155, 163
moment of inertia for, 165
inside-radius-to-thickness ratios for, 123
section modulus of, 154–155, 164
shear modulus of, 156
shear strength of, 153, 156–157
stress-strain curves of, 122
tensile strength of, 121–123, 153, 169–171
trusses, 19
width-to-thickness ratio of, 151–152
yield strength of, 121
yield stress of, 122–123
Cold-Formed Steel Engineers Institute (CFSEI) Web site, 173
cold-formed-steel construction
    ASD of, 157, 161, 164, 165–166
    bearing-wall system, 18
    of bearing-wall systems, 18–20
    bending and shear interaction in, 157–158
    bending and web crippling, 161–162
    benefits of, 20
    of building frame systems, 23–26
    connections in, 20, 150, 153, 167–172, 175, 179–180, 314
    deflection amplification factors for, 31
    designing, 149–180
    direct strength method for, 151
    effective-width method for, 150–154
    fuses in, 29
    IBC on, 4
    LRFD of, 150, 157–158, 162, 164, 166
    moment frames for, 14–15, 31
    of non-bearing-wall systems, 21–23
    overstrength factors in, 31, 354
    resistance factors for, 150, 158, 162

response-modification coefficients for, 31
seismic design of, 26
serviceability of, 153, 174
standards for, 6, 8, 9, 147–149
of storage racks, 15
storage racks, 15
strength design of, 154–156
trusses in, 149
uplift in, 168
walls in, 149, 237f
web crippling, 153, 158–161
Wei-Wen Yu Center for, 173
welding requirements for, 4, 167–168
columns. *See also* beam-columns
in braced-frame systems, 23
buckling of, 151–152
classification of, 308
in cold-formed-steel construction, 15, 151–152, 162
diaphragms and, 73
influence area of, 40
joist substitutes and, 365
K-factor, 9, 308
live load on, 92
load path through, 87
load-duration factor for, 309
in masonry construction, 126, 202, 210–231, 236
in moment frames, 13–14
in post-and-beam construction, 15–16, 23
slenderness ratio for, 216–217, 225
in storage racks, 15
in structural timber construction, 308–313
ties for, 212
width-to-thickness ratio of, 151–152
column stability factor, 308, 309
component amplification factor, 78
components and cladding (C&C)
bearing-wall systems as, 18
construction documents on, 388–391, 413
load estimation for, 48, 52–57, 99–100, 112–113, 173–174, 185
load paths of, 87
MWFRS and, 48
composite slabs, 8

compression block factor, 395
compression buckling stress, 372
compressive stress, 152, 153f
concrete, 4, 8, 88, 138, 315, 353
concrete masonry units (CMUs)
ASD of, 276–280
blocks in, 126
classification of, 126
columns, 223–228
compression in, 132, 250
cross web in, 127f
dimensions of, 126
earthquake loads on, 186
face shell of, 127f
grout in, 125
reinforcement for, 126, 254–262, 269
seismic design of, 270–280
strain in, 195
strength design of, 270–276
Young's modulus for, 243
concrete slabs, 7, 88–89, 135, 137–140, 359–361, 380, 394–397, 400, 405–413
construction documents, 12–13
construction loads, 39, 299t, 302, 309, 396–397, 406–409
Council on Tall Buildings and Urban Habitat (CTBUH), 1
cracking moment, 197
crane load, 43, 83, 84, 85
crane loads, 43, 83, 84, 85
creep, 7, 298, 302, 352
cross-grain reinforcement, 134
curtain walls, 48, 173
curtain-wall studs, 173–179
curvature factor, 306
curved roofs, 41, 42, 45–46
cylindrical buildings, 53

D

damping coefficients, 298
dead loads, 38–39
debris impact load, 58, 64
decks
design standards for, 7–8
load paths through, 88, 95–96
open-web joist systems and, 135, 137–141, 357, 360, 361, 371–373, 377–378, 396–397, 405–413

decks (*Cont.*)
    in seismic design, 76
    steel. *See* steel decks
    welding requirements for, 167–168
    wind pressures on, 48
deflection amplification factor, 30–32, 68, 76
deformation capacity, 30
design flood elevation (DFE), 60, 61, 64
design length, 361–362
design stillwater flood depth, 60–61
detailed plain masonry shear walls, 31, 240
diaphragm factor, 315, 325
directionality factor, 55, 57
domes, 43, 45, 56, 134
doors
    in cold-formed-steel construction, 147
    in masonry construction, 187, 213, 246
    in structural timber construction, 339, 350
    wind loads on, 48, 87
dowel connections, 314–316, 318–327, 354
drag, 61–62, 64
drawings, shop, 13
drawings, structural, 12, 389
drift, 15, 18, 23, 67, 72, 76, 77
drift pins, 318t, 319
ductile response chart, 28f

E

earthquake loads
    construction documents on, 412
    in effective seismic weight, 71
    ELFP for, 69–72, 74–75, 115–118
    flood hazard zones and, 84
    on masonry construction, 28, 118–119,
        183–184, 186–187, 236–239, 242–250,
        270–280
    on nonbuilding structures, 80
    on open-web joist systems, 359–360,
        383, 412
    overstrength factor and, 74
    paths for, 11–12, 68, 92
    SDC and, 67–68
    simplified design procedure for, 75–77,
        118–119
    on structural timber construction,
        298–303, 313, 331, 339
earthquake load effect, 37, 38t, 80, 85
earthquakes, 26–27, 32, 64–66, 92

effective ground acceleration, 186
effective length factors, 163, 309
effective length method, 9, 150–154
effective seismic weight, 69, 71, 76, 77,
    116–117
effective shear capacity ratio, 340
effective width method, 174
effective wind area, 99, 100, 105–107, 112
elastic buckling stress, 154–155, 163, 372,
    375, 403
elastic response curve, 27–28, 27f
elastic section modulus, 154–155, 374
electrical nonstructural components, 77–79
elevators, 2, 39, 79, 89
enclosed buildings, 51, 52–56, 57, 95–114
end bearing stress, 310
end joists, 345
end-grain factor, 315, 329
epoxy anchors, 280
epoxy-coated reinforcement bars, 288
equal-displacement seismic design, 30
equivalent lateral force procedure (ELFP),
    65, 69–72, 74–75, 115–118
erosion, 58, 61f
escalators, 79
Euler buckling, 308
Evaluation Service Report (ESR), 10
expansion anchors, 10, 280
expansion joints, 50, 126
expected strength, 30
explosions, 86
exposure categories, 50, 97
exposure factor, 44, 93
exterior nonstructural walls, 78
extreme compression stress, 154

F

fabrication submittals, 13
Federal Emergency Management Agency
    (FEMA), 26–27, 59
fiber reinforcement, 135, 140, 360, 408
fiberboard, 19
fillet welds, 167, 168f
finger joints, 305
fires, 86
fire protection, 5, 39, 77
fire ratings, 20
fire-retardant treatment, 315

flat roofs, 41, 42, 44, 47, 48, 51–53, 58, 71, 82–84, 92–94

flexible buildings, 51, 53–56, 65, 96

floods, 38t, 58–64, 82, 84

flood insurance rate maps (FIRMs), 59

flood vents, 63

floors

    adhesives in, 351

    in light-frame construction, 18

    loads on, 39–41, 42f, 266, 300, 302

    LRFD of, 81

    seismic design of, 71–74, 77

    strength design of, 81

    tongue-and-groove sheathing for, 351–352

    vibration in, 353, 380, 404

floor joists, 19, 21, 25f, 140, 149, 300

flotation, 58

flow rate, rain, 114

fluid load, 38t

flutter, 51, 53

footings, 26, 76, 87, 90, 212

force transfer approach, 338–341, 344f

format adjustment factor, 299, 303

foundations

    costs for, 26

    design standards for, 7, 8

    geotechnical report on, 12

    IBC on, 4

    rotation of, 242–243, 245, 261

    in seismic design, 4, 76

fracture tensile strength, 30, 153

freeboard, 60, 61f

freestanding walls, 53–56

fundamental frequency, 51, 55, 56, 96, 353

fundamental period, 67, 69, 70f, 71, 79, 116, 186

fuse, 12, 29–30

G

gable roofs, 45–46, 48f, 51–53, 56, 94, 107–108, 113

galloping, 51, 53

garages, passenger vehicle, 41

garages, public, 71, 81–82

geometry factor, 315, 325, 326t, 329

geotechnical reports, 12, 187

girders, 190, 212, 213, 270, 271f, 353. *See also* joist girders

glass, 78, 298

global overturning, 330–331

glued-laminated wood, 7, 134, 303, 305–306, 309, 321, 328, 352, 354–355

groove welds, 167

ground motion design values, 115

ground motion parameter calculator, 65

groundwater, 12, 38t, 60

group action factors, 315–316, 321, 324, 325, 354

grout, 7, 13, 123–126, 131–132, 195, 202, 280, 282f, 286

gust-effect factor, 55, 57

gypsum board, 19, 149, 311, 341

H

Hankinson's formula, 313, 324

head joints, 124f, 125f

headers, 6, 148, 333, 336

heating, ventilation, and air conditioning (HVAC) equipment, 39, 79

height-to-width ratio, 51, 56, 108

high-rise buildings, 1, 33, 86

high-strength low-alloy steel (HSLAS), 121–122, 136, 137

hinges, 27f, 29, 30, 250, 289

hip roofs, 45–46, 48f, 51–53, 94

hurricane-prone regions, 49, 50

hydraulic head, 114

hydrodynamic loads, 58, 60, 61–62

hydrostatic forces, 60–61

hydrostatic loads, 58, 60–61

I

ice, 38t, 64

I-joists, 21

impact force, 64

impact loads, 43, 60, 61, 64, 299t, 303, 315

importance factor, 44, 49, 67, 69–70, 77–78, 94, 96, 116, 186

inelastic buckling stress, 162, 403

inertia, 62, 64, 92, 335, 380

influence area, 40

inside-radius-to-thickness ratios, 123

inspections, 4, 10

interaction diagrams, 215–216, 218, 253
interior walls, 43, 128
intermediate reinforced-masonry shear
    walls, 31–32, 241–242, 243t, 249–250, 289
intermediate steel moment frames, 32
International Accreditation Service (IAS), 10
International Building Code (IBC), 3–10
International Code Council (ICC), 4–5
International Code Council Evaluation
    Service (ICC-ES), 10–11

J

jambs, 179, 186–187
JG-Series joists. *See* joist girders
Johansen's mechanics-based methodology
    for timber connections, 321–323
joints, 131, 191, 194, 236–238. *See also specific
    joints*
joists. *See also specific joists*
    anchorage forces and, 190
    design standards for, 7–8
    diaphragms and, 73
    HSLAS for, 136
    IBC on, 4, 7
    load path through, 90, 190
    open-web systems. *See* open-web joist
        systems
    timber beam strength and, 305
    welding of, 137
    wind pressures on, 48
joist extensions, 360f, 363f, 365–366, 367f
joist girders
    design standards for, 7, 135, 359, 367
    designing, 391–392
    IBC on, 4
    illustration of, 93f, 368f, 369f
    in masonry construction, 270, 271f
    steel decks and, 412
    in structural timber construction, 353
    with unequal joist reaction spacings, 382
    welding requirements for, 364
joist hanger, 300
joist substitutes, 363f, 365, 367f, 412–413

K

K-factor, 9, 308
K-Series joists, 7, 135–137, 358, 364,
    370–390, 412

KCS-Series joists, 364–365, 377, 390–391
kinetic energy, 28
knee braces, 16, 17f
Kwik Bolt 3 (Hilti), 10

L

ladder-type joint reinforcement, 131
lag screws, 318–320, 325, 327, 328
landscaped roofs, 43
lap splices, 286–289
*Leadership in Energy and Environmental
    Design* (LEED), 357
ledgers, 134, 272, 284–285, 291f
leeward drift, 46
length adjustment factor, 308
LH-Series joists, 360f, 366
light-frame construction, 18, 26, 30, 31, 33.
    See also stud-wall construction
light-frame steel construction, 6, 18–21,
    24–25, 29, 31, 147–154, 167, 174, 354
light-frame timber construction, 4, 18–21,
    22f, 24, 29, 31, 298–301, 307–308, 331–333,
    350, 353–355
lime, 128–129, 130t
limit-state equation, 196
lintels, 192–210
live loads
    about, 39–43
    in design process, 81–82
    effective seismic weight and, 71
    on open-web joist systems, 359, 376,
        383, 396
    reduction of, 40–41, 92
    on structural timber construction, 298,
        299t, 302, 303, 309, 353
live-load element factor, 40
load. *See also specific types of*
    in bearing-wall systems, 18
    combinations of, 81–86
    in construction documents, 4
    diagrams, 389–391
    ductility and, 28
    fundamental frequency and, 353
    IBC on, 4, 5–6, 37
    maximum, overstrength, 29
    in moment frames, 14
    in non-bearing-wall systems, 21
    notations for, 38t

in post-and-beam construction, 15–16
seismic design forces and, 78
serviceability and, 153
load and resistance factor design (LRFD)
    of cold-formed steel components, 150,
       157–158, 162, 164, 166
    environmental factors in, 302
    IBC on, 6, 81
    load combinations in, 81–82
    of open-web joist systems, 370, 392
    for structural timber construction, 134,
       298–299, 303
load method, 122, 124f
load-duration factor, 84, 134, 298–299, 302,
    303, 307, 309, 312–315
loaded edge condition, 317
load-slip modulus, 316
local overturning, 330–331
long-period site coefficient, 66
long-period transition period, 69
long-span joists, 7, 359
low-rise buildings
    costs for, 26
    definition of, 2, 51
    elastic, 28
    enclosure classification, 51, 96
    height of, 2, 51, 96
    natural periods of, 32
    properties of, 26
    response spectrum range for, 32
    statistics on, 1
    structural systems, 13–25, 31–32
    trends in, 2
    wind load paths through, 88f
lumber, 21, 133–135, 303–305, 308,
    310–313, 352

M

machine-evaluated lumber (MEL), 135
machine-stress-rated (MSR) lumber, 134–135
main wind force-resisting system (MWFRS),
    48–57, 87–91, 95–98, 101–105, 108–111
manuals, design, 8, 9
masonry
    AC documents for, 10
    brittle vs. ductile response of, 239
    components of, 7
    MOE of, 187, 202, 266

    moment of inertia for, 243–244,
      266–267
    plain, 28, 31, 239–240
    prestressed, 31
    reinforced. See reinforced masonry
    unreinforced. See unreinforced
      masonry
masonry construction
    air temperature and, 128, 139–140
    amplification factor for, 186
    area of, 244
    ASD of
        anchors, 285–286, 293–294
        axial loads and flexure on,
          220–223, 228–231, 233–236
        flexure on, 194, 201–207, 209–210
        reinforcement in, 131, 287–290
        shear walls, 239, 240, 253–254,
          262–264
        slender walls, 276–280
    aspect ratio of, 238, 250
    axial loads and flexure on, 210–236
    axial loads on, 194
    balance point of, 218
    bearing loads on, 191–192, 193f
    bearing-wall systems, 18, 19f, 31
    braced frames in, 23–24
    buckling of, 239
    in building-frame systems, 31, 116
    capacity-reduction factor for, 196
    clay for, 128
    concentrated loads on, 191, 192f
    connections in, 68, 74–75, 118, 184,
      186–191, 240, 280–294
    cracking of, 28–29, 184, 187, 197,
      200–201, 204–208, 214, 221, 239, 245
    creep in, 7
    deflection amplification factors for,
      31–32
    deflection of, 242–247
    deformation modes and failure
      patterns for, 236–238
    depth of, 194
    design standards on, 6–10
    designing, 194–280
    drag forces on, 210
    ductility of, 29, 199, 238–239, 242,
      249–250

masonry construction (*Cont.*)
    earthquake loads on, 28, 118–119,
        183–184, 186, 213, 270–280
    in ELFP, 115–118
    energy dissipation with, 29
    external pressure coefficient for, 105–106
    fabrication submittals for, 13
    failure of, 28, 29
    flexibility of, 242–244, 246
    flexural loads on, 194–210
    flow chart for design of, 268f
    girders in, 190, 212, 213, 270, 271f
    grout in. *See* grout
    height of, 184, 185f, 187, 192f
    hinges in, 250
    IBC on, 4, 7
    illustration of, 124f, 125f
    interaction diagram for, 215–216,
        218, 249
    joist substitutes and, 365
    lintels in, 192–194
    load paths through, 184, 191
    materials for, 123–133
    moment frames and, 14, 18, 25, 32
    non-bearing-wall systems, 22, 23f
    nonstructural components anchored
        to, 189
    open-web joist systems and, 18
    out-of-plane forces on, 118, 183–191,
        213–214, 250t, 268f
    overstrength factor for, 31–32, 186
    radius of gyration for, 220, 277
    reinforcement in. *See* reinforcement
    reinforcement ratio for, 198f, 199,
        250–251, 253, 261
    response-modification coefficients for,
        29, 31–32, 116, 186, 236, 250
    rigidity of, 242–248
    section modulus of, 197, 221, 266
    seismic design for, 25, 26, 28, 68, 74–75,
        115–119, 191, 212, 236–239, 242–250,
        254–262
    seismic performance of, 33
    serviceability of, 7
    shear-span ratio of, 228, 253
    shrinkage in, 7, 126, 129, 132,
        139–140, 221
    slenderness ratio for, 216–217, 225, 278

    soil and, 187
    stiffness of, 242, 245, 267f
    strain in, 198f
    strength design of, 194–201, 207–209,
        214–220, 224–228, 231–233, 236,
        239–240, 249–253, 255–262, 265–276,
        287–290
    strength-reduction factor for, 196, 200,
        216, 219
    stress and strain distribution in,
        195–196, 198–199, 203–205, 215,
        221–222, 251–252
    struts in, 189–190
    tensile strength of, 195, 202, 239
    tension reinforcement strain factor
        and, 249–250
    ties for, 212, 213f
    trusses in, 213
    uplift in, 18, 243
    walls in. *See* masonry walls
    weight of, 115, 117, 186, 188
    wind loads on, 95, 99, 105–106,
        111–113, 183–188, 213
    windows and, 246
Masonry Standards Joint Committee
    (MSJC), 7
masonry units, 7, 124f, 125f, 126, 128t, 202,
    241, 250, 276–280. *See also specific types of*
masonry walls
    anchorage of, 74–75
    blocks in, 126
    cantilever vs. fixed walls, 245
    deflection amplification factors for,
        31–32
    designing, 183–194, 199, 202, 210–214
    failure of, 28
    grout in, 131
    illustration of, 124f, 125f
    joist substitutes as, 365
    out-of-plane forces on, 74, 115–119,
        265–280
    overstrength factor for, 31–32
    reinforcement in. *See* reinforcement
    response-modification coefficients for,
        31–32
    shear, 236–265
    slender, 265–280
    wind pressure on, 99, 105–106, 111–113

mass, serviceability, and stiffness, 11
material specifications, 12
maximum considered earthquake (MCE), 26–27, 65–66
maximum earthquake load effect, 38t
mechanical nonstructural components, 77–79
mechanical splices, 289
Metal Building Manufacturers Association (MBMA) Web site, 173
metal buildings, 23, 24, 31, 50–51, 357
Metal Construction Association Web site, 173
metal plate connectors (MPCs), 314
metal side-plate factor, 315, 329
microwaves, 135
mid-rise buildings, 2, 32, 33f
modification factor, 170
modular ratio, 203, 267
modulus of elasticity (MOE)
    of cold-formed steel, 152, 155, 163
    fundamental frequency and, 353
    group action factor and, 316
    of K-Series joist steel, 373
    load-duration factor and, 299
    of lumber, 134–135, 349
    of masonry, 187, 202, 266
    serviceability and, 11
    of steel reinforcement, 195
    time effect factor and, 299
modulus of inertia, 335, 380
modulus of rupture, 197, 266
moisture content adjustment factor, 302–303, 306, 309. *See also* wet service factor
moment capacity, 197
moment diagram shape factor, 374
moment frames
    about, 13–18
    cantilever column systems and, 24, 29
    cold-formed-steel special bolted, 29, 31
    drift and, 77
    hinges and, 29
    masonry construction and, 25, 32
    open-web joist systems and, 357, 364
    steel, 32
    structural timber construction and, 354
    wind loads and, 48, 50, 87, 89

moment joints, 17, 29
moment magnifier procedure, 187
mortar, 7, 123, 124f, 125, 128–132, 202
multiple-folded-plate roofs, 45

N

nails, 318–321, 324–325, 338, 340–341
National Earthquake Hazards Reduction Program (NEHRP), 26
National Flood Insurance Program (NFIP), 59, 60
National Institute of Building Sciences (NIBS), 26
National Lumber Grades Authority, 133
NDS Supplement for Glued-Laminated Softwood Timber, 309
Newton's second law of motion, 32
non-bearing-wall systems, 21–23
normal impact loads, 64
Northern Softwood Lumber Bureau, 133

O

occupancy
    change of, in existing buildings, 5
    live load from, 39–41
    after MCE event, 33
    in seismic design, 67, 115
    in snow load calculations, 44, 93–94
occupancy categories
    importance factor and, 49, 67, 70, 78, 116
    in seismic design, 67, 70, 75t, 78, 116, 118
    structural integrity and, 86
    in wind load calculations, 52–53, 95, 96
occupancy groups, 39, 41
offset method, 122, 124f
offset walls, 339
one-way slabs, 40
open buildings, 51, 53–56
open-web joist systems
    about, 357–360
    composite, 357–360, 392–410
    designations for, 364
    designing, 370–381
    external pressure coefficient, 107

open-web joist systems (*Cont.*)
    flexural model of, 370–377
    IBC on, 4
    joist extensions, 360f, 363f, 365
    joist girders. *See* joist girders
    joist substitutes for, 365
    materials for, 135–143, 370
    in moment frames, 357, 364
    in non-bearing-wall systems, 21
    non-composite, 357, 366, 363f, 380,
        390–391, 408, 412–413
    pre-composite stage, 396–397, 401–403
    reinforcement for, 376–377, 408
    standards for, 7
    uplift on, 361, 371, 379, 381, 382f, 389, 413
    vibration in, 380–381, 404
ordinary plain masonry shear walls, 31, 240
ordinary reinforced-masonry shear walls,
    31–32, 240f, 241, 242, 243t
ordinary steel concentrically braced frames,
    31
ordinary steel moment frames, 32
overstrength factors, 29–32, 68, 74, 76, 81,
    85–86, 186, 330, 336, 354
overstrength of the structure, 27f, 29–31
overturning, 20–21, 72, 80, 81, 85, 87, 301,
    330–336, 351

P

panel points, 361, 366, 372–377
parallel-chord configuration, 364
parapets, 46, 58f, 131, 186, 214f, 270, 373
partially enclosed buildings, 51, 53–56, 57
partitions, 16, 18, 39, 40, 43, 71, 78, 128,
    186, 348
P-delta effects, 72, 265–266
pea gravel, 132
penetration-depth factor, 315
perforated shear walls, 339–342, 344f
performance specifications, 13
permanent loads, 39
permit documents, 12
piers, 210–223, 236, 245–248, 338, 341
pilasters, 184, 212, 213, 214f, 243
pitched roofs, 41, 42, 55, 56
plain masonry shear walls, 28, 31, 239–240
plastic hinge, 27f, 29, 250, 289
plate anchor bolts, 280–281

plate buckling coefficient, 152
plate elastic buckling stress, 152
plywood, 17–20, 340–342
ponding instability, 47, 58, 95
portland cement, 126, 128–129, 130t, 132
post-and-beam construction, 15–16, 17f, 23,
    24, 31
potential energy, 28
power-actuated fasteners (PAFs), 141, 142f,
    143, 360, 372–373, 412
precision units, 126
press brakes, 150
pressure coefficients, 52, 54–57, 63, 102–103,
    105–107, 109
pressure vessels, 79
prestressed masonry shear walls, 31
prism test methods, 132
Product Technical Information catalogue
    (SSMA), 19
project specifications, 13
puddle welds, 140–141, 167–168, 373,
    397, 412
Puerto Rico, 65
purlins, 48, 312

R

R-Type joist extensions, 365–366
Rack Manufacturers Institute (RMI), 15, 173
racking, 149, 301, 331–333, 336
radius of gyration, 220, 277, 372–373,
    376, 402
rafters, 149
rain loads, 38t, 47, 57–58, 114–115
rain-on-snow surcharge, 47, 94
rainwater depth, 57–58
redundancy factor (ρ), 152
reduction factor (R), 41
redundancy factor, 80
regularly-shaped buildings, 51–57, 75, 96
reinforced masonry
    allowable loads, 409f
    axial loads and flexure on, 210–236
    with bearing-wall systems, 18, 19f
    buckling of, 187
    components of, 123
    compression face of, 194
    compression in, 132, 138
    cracking of, 28–29, 184, 187, 221

depth of, 194
design standards for, 8, 140
drag forces on, 210
energy dissipation with, 29
flexural loads on, 194–210
IBC on, 4
load path through, 90, 191
moment frames and, 18, 25
with non-bearing-wall systems, 22, 23
in seismic design, 25, 33, 115–118
splices for, 286–289
steel decks and, 19, 406, 408
reinforcement
  cross-grain, 134
  fiber, 135, 140, 360, 408
  ladder-type joint, 131
  steel. *See* steel reinforcement
  welded-wire, 139–140, 408
reinforcement ratio, 197–199, 204, 208, 222, 250, 251f, 253, 261, 270
reinforcement size factor, 288
resistance factors, 11, 150, 158, 162, 303, 396
resistance welds, 167
response spectrum, 32–33
response-modification coefficients
  about, 29–32
  for building frame systems, 116, 119
  for masonry construction, 186, 236, 250
  for seismic force-resisting systems, 68, 70, 76, 78
  for structural timber construction, 330
retaining walls, 12, 187, 231–236, 237f
rigid buildings, 51–57
rim joists, 300, 335, 337f
rivets, 314, 316, 328–329, 354
Rockwell C hardness (HRC), 141, 143
rod-braced frames, 31
roll-forming, 149
roofs. *See also specific types of*
  adhesives in, 351
  angle of, 55, 96, 97, 102, 106
  architectural design and, 19
  ASTM A653 steel for, 121
  awnings and canopies on, 43
  with bearing-wall system, 18, 19, 21, 22f
  drainage systems on, 57–58
  existing, 47

exposure factor for, 44
external pressure coefficient for, 55, 102, 106
gardens on, 39m42
height of, 50–57, 78, 96, 97, 102, 105, 186
importance factor for, 44
landscaped, 43
in light-frame construction, 18
load on, 38t, 39–58, 81–85, 92, 266, 272, 309
materials for, 44, 93–94
multiple-folded-plate, 45
net pressure coefficient for, 57
in non-bearing-wall systems, 21, 23f
obstructions on, 44, 94
occupancy of, 39, 42, 44
open-web joist systems for framing, 135, 357, 360, 361, 389–392, 406–408
parapets. *See* parapets
pilasters and, 214f
projections from, 46
rafters, 149
in seismic design, 72, 187–188
slipperiness of, 44, 94
slope of, 42–48, 51, 53, 56, 58
special-purpose, 41
temperature of, 44, 315
thermal factor for, 44
timber for framing, 21, 22f, 24
trusses for framing, 19, 21, 22f
walls anchored to, 74
weight of, 115, 116
wind pressures on, 47–48
roof decks, 7, 18, 48, 135, 137, 138f, 360, 406–408
roof joists, 19, 346, 379

S

S-Type joist extensions, 365
SAE, 143
St. Venant torsion, 156
sand, 128–129, 132
sawtooth roofs, 45–46, 56, 81
scour, 62
screws. *See also specific types of*
  in cold-formed-steel construction, 19, 20, 171–172, 179–180

screws (*Cont.*)
    design standards for, 11
    in open-web joist systems, 141–143,
        360, 372–373, 412
    in structural timber construction,
        318–321, 324–325, 341
scuppers, 58
section modulus, 154–155, 164, 197, 221, 266,
    304, 312, 386
segmented shear walls, 336–342
seismic design
    architectural design and, 12
    of bearing-wall systems, 65, 68, 75, 118
    of braced-frame systems, 72, 119
    of building frame systems, 65, 69, 75,
        115–118
    calculations for, 27–29
    of cold-formed-steel construction, 26
    deflection amplification factors for, 68
    deformations in, 11
    diaphragms in, 72–74
    direction of, 117
    distribution of, 68f
    ELFP for, 65, 69–72, 74–75, 115–118
    equal-displacement, 30
    geotechnical report for, 12
    goal of, 11
    key aspects of, 27
    of light-frame construction, 18, 30, 33
    load in. *See* earthquake loads
    load paths through, 92
    of masonry construction, 25, 26, 33, 68,
        74–75, 115–119, 191, 244, 247–249,
        254–262, 270–280
    MCE contour maps for, 65
    moment frames in, 14, 25
    of nonbuilding structures, 14–15, 79–80
    on nonstructural components, 78, 186
    nonstructural components in, 30,
        77–79, 186
    objectives of, 28, 30
    offset of vertical components in, 74f
    P-delta effects in, 72, 265–266
    philosophy of, 26, 30
    role of, 11, 33
    simplified design procedure for, 75–77,
        118–119
    of structural timber construction, 26,
        298, 321–322, 336, 354

seismic design category (SDC)
    for building frame systems, 69
    building irregularities and, 72
    determination of, 12, 67
    ELFP and, 67–69
    ground motion value and, 115
    load paths and, 92
    for masonry construction, 74–75, 191,
        212, 242, 243t
    moment frames and, 14
    of nonstructural components, 77
    in simplified procedure, 76
    for structural timber construction, 330
seismic response coefficient, 67, 69–70, 116
serviceability, 7, 11, 153, 174, 351–354
shape factor, 61, 374
sharp-yielding steel, 122
shear buckling coefficient, 156
shear capacity ratio, 340, 342
shear modulus, 156, 243, 246
shear-lag effect, 315–316, 321, 354
shell effects, 330
short-period site coefficient, 66, 76
siding, 121, 340–341
signs, 53–56
sill track to jamb stud connection, 179–180
simple diaphragm buildings, 50, 56, 95–96, 108
site, project, 12, 44, 51–54, 56, 108
site classes, 65, 66, 75t, 115, 118
site coefficient, 66, 76–77
slabs, 7–8, 26, 88, 212. *See also specific*
    types of
slenderness K-factor, 9
slenderness λ-factor, 152
slenderness Q-factor, 372, 375
slenderness ratio, 216–217, 225, 278, 308, 321,
    372–373, 379, 378, 397–398, 402
slenderness-reduction factor, 372, 375
SLH-Series joists, 366
sliding, 81, 85, 87
sliding snow, 44, 45, 47
slumped units, 126
snow density, 46, 94
snow drifts, 44–47, 82, 84, 94
snow loads
    balanced vs. unbalanced, 95f
    calculating, 45–47, 82, 84, 94
    ground, 44, 93

on open-web joist systems, 359, 382, 406

in structural timber construction, 299t, 302, 311

Society of Automotive Engineers (SAE), 143

soft-mud process, 127

soils, 4, 43, 65, 66, 85, 183, 187

Southern Pine Inspection Bureau, 133

special impact loads, 64

special reinforced-masonry shear walls, 31–32, 115–119, 214, 241–242, 243t, 249–250, 253, 254–262, 289, 290

special steel moment frames, 32

special-purpose roofs, 41

specifications, 6, 8–9, 12–13. *See also individual specifications*

specifications, 6, 8–9

spikes, 318t, 320, 324–325

splices, 18f, 30, 74, 85, 190–191, 280, 286–289, 349

split-face units, 126

split-ring connectors, 315–316, 319, 327–328, 354

stability factor, 305

stainless steel, 4, 315

stairs, 89

standards, design, 6, 8–9, 11

staples, 324

steel, 32, 121–123, 137, 149, 357. *See also specific types of*

steel concentrically braced frames, 31

Steel Deck Institute (SDI), 173

steel decks
    design standards on, 7–8
    IBC on, 4
    in light-frame construction, 19, 21
    materials for, 149
    with open-web joist systems, 18, 135, 137–138, 140, 360, 361, 373, 396–397, 405–413

Steel Framing Alliance Web site, 173

Steel Joist Institute (SJI), 357, 358, 369

steel joists, 135–137, 149

steel reinforcement
    about, 131
    axial loads and flexure on, 214, 216–223, 226
    blocks and, 126

design standards on, 7–8

failure of, 28, 29

flexural loads on, 195–210

grout and, 131

IBC on, 4

in shear wall design, 240–242, 249–262

shop drawings for, 13

in slender wall design, 267, 270

splices for, 286–289

ties with, 213f

steel side plates, 354

Steel Stud Manufacturers Association (SSMA), 19, 173

stepped roofs, 56

stepped walls, 339

stiff-mud process, 127

stirrup hooks, 289

storage racks, 15, 16f, 24

story drift, 67, 72

strain hardening, 123, 123f

strap brace, 20, 21f, 29–31

strength design, 37, 81–82, 85–86

strength-reduction factor, 196, 200, 216, 219, 370–372, 374–375, 395, 405

stress-strain curves, 122, 123f, 124f

structural design, 11–12, 87

structural drawings, 389

Structural Engineers Association of California (SEAOC), 267

structural specifications, 12–13

Structural Stability Research Council Web site, 173

structural steel frames, 21–23, 357

structural timber construction
    ASD of, 84, 134, 299, 302, 303, 309, 311, 313
    aspect ratio in, 336, 338, 346, 348
    bearing-wall systems in, 18–20
    braced frames in, 24
    cantilever column systems in, 24
    compression members in, 308–313
    connections in, 298, 305, 308, 312, 313–329, 338–342, 344f, 354
    creep in, 298, 302, 352
    damping coefficients for, 298
    design manuals for, 9
    design philosophy for, 298–299
    design standards for, 7, 8, 354

structural timber construction (*Cont.*)
    ductility in, 298, 321–322
    earthquake loads on, 298–303, 313, 331, 339
    environmental factors in, 302–304
    flexural members in, 303–307
    fundamental frequency of, 353
    fuse in, 29
    girders in, 353
    IBC on, 4, 7, 342, 354–355
    live loads on, 298, 299t, 302, 303, 353
    load and grain direction in, 313, 314f, 329
    load paths in, 87, 298–301
    load-duration factor for. *See* load-duration factor
    LRFD for, 134
    masonry and, 1–2
    moisture content adjustment factor for, 302–303, 306, 309
    moment frames in, 14, 15–18, 354
    non-bearing-wall systems in, 23
    objectives of, 297–298
    overstrength factor for, 330
    overturning in, 301, 330–336
    prescriptive, 342–345
    racking load on, 301, 331–333
    response-modification coefficient for, 330
    seismic design of, 26, 298, 321–322, 330, 354
    serviceability of, 351–354
    shrinkage in, 314, 325, 351
    tension and bending in, 310–312
    tension members in, 307–308
    time-effect factor for, 134, 298–299, 303
    in United States, 1
    uplift in, 301–302, 307, 330–336, 339–345
    vibration in, 353
    walls in, 7, 301, 330–342, 344f
    wind loads on, 298–303, 313, 330, 331, 339–342
struts, 74, 189–190, 338, 345–349
studs
    for cold-formed-steel applications, 19–20, 149
    curtain-wall, 173–179
    jamb, 178f
    in in-line framing, 334
    load path through, 90
    in open-web joist systems, 135, 140, 360
    serviceability of, 174
    for structural timber applications, 20–21, 305
stud tracks, 19
stud-wall construction, 6, 19–21. *See also* light-frame construction
suction, 48, 51, 53, 61, 101f
supertall buildings, 1
surface-roughness categories, 49–50, 95. *See also* terrain categories
suspended ceilings, 78
system overstrength factor, 30, 68

T

Tennessee, 3
tension cracks, 200–201
tension reinforcement strain factor, 249–250
tension rupture, 153, 169
terrain categories, 93. *See also* surface-roughness categories
test coupons, 136
Texas, 3, 138f
The Masonry Society (TMS), 7
thermal factor, 44, 93–94
third-party design submissions, 13
tie anchorage hooks, 289
tilt-up walls, 357
timber, 133–134, 297, 298, 302–309, 314–315, 320
time-effect factor, 134, 298–299, 303
toe nail factor, 315, 318–319
topographic factor, 52, 54–55, 57, 97
tornadoes, 49
torsional moments, 72, 76
tributary area, 40–43, 76, 108, 114, 388
tripartite log chart, 32–33
troughed roofs, 55, 56
Truss Plate Institute (TPI), 7, 354–355
trusses
    in braced-frame systems, 23
    in cold-formed-steel construction, 149
    design standards for, 7
    in light-frame construction, 18

in masonry construction, 213
open-web joist systems and, 357–358
performance specification for, 13
for roof framing, 19, 21, 22f
for structural timber construction, 134,
299, 307–308, 311, 314, 354–355
truss-type joint reinforcement, 131
turned-down slabs, 26

U

Uniform Building Code (UBC), 26, 41, 84
United States. *See also specific states*
long-period transition period for, 69
low-rise building construction in, 1
MCE in, 26, 65
United States Department of Agriculture
(USDA), 319
United States Department of Commerce, 133
United States Geological Survey (USGS),
ground motion parameter calculator, 65
United States Green Building Council
(USGBC), 357
unloaded edge condition, 317
unreinforced masonry, 28–29, 33, 186,
239–240, 242
uplift. *See also* buoyancy
in cold-formed-steel construction, 168
of foundations, 242
from hydrostatic forces, 60
in masonry construction, 18, 243
on open-web joist systems, 361, 371,
379, 381, 382f, 389, 413
from shoaling waves, 62
in structural timber construction,
301–302, 307, 330–336, 339–345
U-stirrups, 207
utility lines, 79

V

vehicular impact, 86
velocity exposure coefficient, 57, 108
velocity pressure exposure coefficient,
54, 102
vertical distribution factor, 71
vibration, 64–65, 69, 353, 380–381, 404
vitrification, 127
Voluntary Product Standards, 133
vortex shedding, 51, 53

W

walls
adhesives in, 351
in cold-formed-steel construction, 237f
external pressure coefficient for, 55
in light-frame construction, 18
load classification of, 39
in masonry construction. *See* masonry
walls
net pressure coefficient for, 57
nonbearing, 21–23
offset, 339
parapets. *See* parapets
retaining, 12, 187, 231–236, 237f
in seismic design, 74–76, 118, 183–191,
242, 243t
stepped, 339
in structural timber construction, 7,
301, 330–342, 344f
support point of, 361
tilt-up, 357
wind pressures on, 47–48
water
fresh, 60
groundwater, 12, 38t, 60
mass density of, 62
in mortar mix, 129
pressure of, 12
salt, 60
velocity of, 60, 61–62
waves, 58–64
web slenderness coefficient, 158
Wei-Wen Yu Center for Cold-Formed Steel
Structures Web site, 173
welds, 4, 137, 141, 167, 364, 377. *See also*
specific types of
welded splices, 289
welded-wire reinforcement, 139–140, 408
Western Wood Products Association, 133
wet service factor, 307, 315. *See also* moisture
content adjustment factor
wind
bearing-wall systems and, 18
load from. *See* wind loads
snow loads and, 45
wind directionality factor, 85, 102
wind exposure categories, 47, 49–50, 52
wind loads

alternate all-heights method, 6, 47, 49, 56–57, 108–114

in ASD, 83, 84–85

on awnings and canopies, 43

building surface area and, 92

calculating, 6, 47–57, 95–114

cladding and, 48

construction documents on, 388, 412

on curtain wall studs, 173–174, 177

for flexible buildings, 53–56

on ice, 38t

importance factor for, 49

on jambs, 187

on joist girders, 412

on masonry construction, 95, 99, 105–106, 111–113, 183–188

on multi-story buildings, 331f

on open-web joist systems, 95, 100, 101f, 107–108, 359–360, 376, 382, 390

paths for, 50, 87–91

for rigid buildings, 51–57

on structural timber construction, 298–303, 313, 330, 331, 339–342

wind pressure, 47–57, 95–114, 185

wind pressure adjustment factor, 52, 53, 97, 185

wind speed, 47, 49, 57, 95

wind stagnation pressure, 57, 108

wind-on-ice load, 38t

windows

in cold-formed-steel construction, 147

in masonry construction, 126, 186–187, 192–194, 213, 246

in structural timber construction, 339, 341, 350

wind loads on, 48, 87

wind-tunnel tests, 47, 49

windward drift, 46, 46f

Wire Reinforcement Institute (WRI), 139

wood

definition of, 298

designing with, objectives of, 135

environment and, 302–303

failure of, 298, 329

for framing, 4, 7, 21, 22f, 24, 32, 87

glued-laminated. *See* glued-laminated wood

granular structure of, 17

lumber. *See* lumber

plywood, 17–20, 340–342

specific gravity of, 319, 327–328

timber. *See* timber

wood decks, 360, 361

wood screws, 318t, 320, 324–325

working stress design. *See* allowable stress design

X

X-bracing, 149

Y

yield displacement, 238–239, 249

yield modes, 321–323, 324t, 354

yield point, 122, 124f

Young's modulus, 243. *See also* modulus of elasticity

International Green Construction Code (IGCC)

The International Code Council (ICC) published the second public version of the *International Green Construction Code* (IGCC) in November of 2010. This Public Version (PV 2.0) is the latest of a series of documents in the code development process leading to the publication of the 2012 IGCC. Partnering in this effort are cooperating sponsors the American Institute of Architects (AIA) and ASTM International, which have supported the development of an adoptable and enforceable green building code.

The IGCC is also augmented with *ANSI/ASHRAE/USGBC/IES Standard 189.1, Standard for the Design of High Performance, Green Buildings Except Low-Rise Residential Buildings,* as an alternate path of compliance. American Society of Heating, Refrigeration and Air-conditioning Engineers (ASHRAE), U.S. Green Building Council (USGBC) and the Illuminating Engineering Society (IES) have joined with ICC as partners in the development of the IGCC.

The IGCC provides the building industry with language that both broadens and strengthens building codes in a way that will accelerate the construction of high-performance, green buildings. It provides a vehicle for jurisdictions to regulate green for the design and performance of new and renovated buildings in a manner that is integrated with existing codes, allowing communities to reap the rewards of improved design and construction practices.

The code includes technical requirements addressing water use efficiency, indoor environmental quality, energy efficiency, materials and resource use, as well as the impact of a building on its building site and its community.

For the latest code development activities related to the IGCC, interested individuals can visit the ICC website at www.iccsafe.org.

Sustainable Attributes Verification and Evaluation™ (SAVE™)

The number of new products and innovative approaches that are introduced into the building construction industry on a continuing basis has been phenomenal. A large number of these new products or practices claim to be "green" or consistent with and promoting the goals of sustainable construction. The question of sustainability becomes more complicated considering the global market and the fact that many products are sold internationally. Because it is difficult for designers, contractors or code officials to verify the credibility of each and every claim regarding sustainable attributes, it is important that there is a reliable source to verify such claims legitimate. Verification requires a methodology and a standardized process by which to evaluate the degree of "greenness" and the sustainable attributes of construction materials, elements and assemblies that establish a building's green performance.

Because of the large number of innovative green products that are being introduced, ICC Evaluation Service (ICC-ES) has created a program to address the issue of green evaluation called the Sustainable Attributes Verification and Evaluation™ (SAVE™) Program.

The SAVE™ Program from ICC-ES® provides the most trusted third-party verification available today for sustainable construction products. Under this program, ICC-ES® evaluates a product's conformance to the requirements contained in current green standards. The SAVE™ Program may assist in identifying products that have been evaluated to multiple SAVE™ guidelines and multiple green building rating systems, standards and codes, such as US Green Building Council's LEED, Green Building Initiative's Green Globes, National Association of Home Builders and International Code Council's National Green Building Standard (ICC 700-2008) and the California Green Building Standard Code.

The ICC-ES Save Verification of Attributes Reports (VAR) are easily accessible online at www.icc-es.org/save. A sample VAR is shown.

ICC-ES SAVE: Verification of Attributes Report™   VAR-3045

Issued January 1, 2011
This report is subject to re-examination in one year.

www.icc-es.org/save | 1-800-423-6587 | (562) 699-0543   *A Subsidiary of the International Code Council®*

DIVISION: 04 00 00—MASONRY
Section: 04 22 00—Unit Masonry

REPORT HOLDER:

BUILD SOLID BLOCK, CO.
1824 BROAD STREET
LONG VIEW, CALIFORNIA 91450
(169) 640-2750
www.buildsolidblock.com

EVALUATION SUBJECT:

BUILD SOLID BLOCK CONCRETE MASONRY UNITS

1.0 EVALUATION SCOPE

Compliance with the following evaluation guideline:

ICC-ES Evaluation Guideline for Determination of Recycled Content of Materials (EG101), dated October 2008.

Compliance eligibility with the applicable sections of the following green building rating systems, standards and codes:

■ International Green Construction Code – Public Version 2.0 (IGCC PV2.0) (see Table 2 for details)

■ 2010 California Green Building Standards Code (CALGreen), Title 24, Part 11 (see Table 3 for details)

■ ICC 700-2008 National Green Building Standard (see Table 4 for details)

■ LEED 2009 for New Construction and Major Renovations (see Table 5 for details)

■ LEED for Homes 2008 (see Table 6 for details)

2.0 USES

Build Solid Block concrete masonry units are used for load-bearing and nonload-bearing wall construction and also for masonry veneer applications.

3.0 DESCRIPTION

Build Solid Block concrete masonry units are manufactured in hollow and solid, structural and nonstructural units with lightweight or normal-weight aggregates. Typical unit widths are 4, 6, 8, 12 and 16 inches (102, 152, 203, 305 and 406 mm), while typical heights are 4, 6 and 8 inches (102, 152 and 203 mm).

The Build Solid Block concrete masonry units contain the minimum type and percentage of recycled content, by weight, as set forth in Table 1.

4.0 CONDITIONS

4.1 Code Compliance:

Evaluation of Build Solid Block concrete masonry units for compliance with the requirements of the International Codes is outside the scope of this evaluation report. Evidence of code compliance must be submitted by the permit applicant to the Authority Having Jurisdiction (AHJ) for approval.

4.2 Green Rating Systems and Standards Eligibility:

The information presented in Tables 2 through 6 of this report provides a matrix of areas of evaluation and corresponding limitations and/or additional project-specific requirements and offer benefit to individuals who are assessing eligibility for credits or points.

The final interpretation of the specific requirements of the respective green building rating system and/or standard rests with the developer of that specific rating system or standard or the AHJ, as applicable.

Compliance for items noted as 'Verified attribute' are subject to any conditions noted in Tables 2 through 6. The user is advised of the project-specific provisions that may be contingent upon meeting specific conditions and the verification of those conditions is outside the scope of this report. Rating systems or standards often provide supplemental information as guidance.

5.0 BASIS OF EVALUATION

The information in this report, including the "Verified Attribute", is based upon the following supporting documentation:

■ ICC-ES Evaluation Guideline for Determination of Recycled Content of Materials (EG101). [IGCC PV2.0 Section 503.2.2 (2); CALGreen Sections A.4.405.3 and A5.405.4; ICC 700 Section 604.1; LEED NC MR Credit 4; LEED Homes MR Credit 2.2]

6.0 IDENTIFICATION

Build Solid Block concrete masonry units are identified with the manufacturer's name and address; the model name; and the VAR number (VAR-3045).

TABLE 1—MINIMUM RECYCLED CONTENT BY WEIGHT SUMMARY

PRODUCT	% PRE-CONSUMER RECYCLED CONTENT	% POST-CONSUMER RECYCLED CONTENT	% MINIMUM RECYCLED CONTENT
Build Solid Block concrete masonry units (all sizes)	10[1]	45	55

[1] Actual per-consumer recycled content amount is 20%; however, pre-consumer content calculations for IGCC and LEED stipulate pre-consumer recycled content that only $1/2$ of the actual value is counted.

TABLES 2 THROUGH 7

SECTION NUMBER	SECTION INTENT	POSSIBLE POINTS	CONDITIONS OF USE TO QUALIFY FOR POINTS	FINDING
TABLE 2—SUMMARY OF AREA OF ELEGIBILITY WITH INTERNATIONAL GREEN CONSTRUCTION CODE – PUBLIC VERSION 2.0				
503.2.2	Recycled content building materials	——	Not less than 55 percent of the total *building* materials used in the project, based on mass or cost, shall comply with Section 503.2.1, 503.2.2, 503.2.3, 503.2.4 or 503.2.5. Compliance shall be demonstrated in accordance with those sections singularly or in combination. Materials regulated by Sections 503.2.2, 503.2.3, 503.2.4 and 503.2.5 shall have a *design life* that is equal to or greater than that indicated in the *building service life* plan in accordance with Section 505.1. **503.2.2 Recycled content building materials.** *Recycled content building* materials shall comply with one of the following: 1. Contain not less than 25 percent combined *post-consumer* and *pre-consumer* recovered material, and shall comply with Section 503.2.3. 2. Contain not less than 50 percent combined *post-consumer* and *pre-consumer* recovered material. The *pre-consumer* recycled content shall be counted as one-half of its actual content in the material.	•
TABLE 3—SUMMARY OF AREA OF ELEGIBILITY WITH 2010 CALIFORNIA GREEN BUILDING STANDARDS CODE (CALGREEN)				
A4.405.3 A5.405.4	Recycled content	——	To achieve Tier 1 - Use materials, equivalent in performance to virgin materials, with a postconsumer or preconsumer recycled content value (RCV) for a minimum of 10% of the total value, based on estimated cost of materials on the project. To achieve Tier 2 - Use materials with a postconsumer or preconsumer RCV for a minimum of 15% of the total value, based on estimated cost of materials on the project. RCV shall be determined as follows: RCV = (% PC X material cost) + 0.5 (% PI X material cost) **Notes:** 1. PC means post consumer waste. 2. PI means post industrial waste.	•
TABLE 4—SUMMARY OF AREA OF ELEGIBILITY WITH ICC-700 NATIONAL GREEN BUILDING STANDARD				
604.1	Use two or more major and/or minor building materials containing recycled content	2 6 max	2, 4 or 6 points may be earned when products are used with another major building component with recycled content of 25% < 50%; 50% < 75%; ≥ 75%, respectively.	•
TABLE 5—SUMMARY OF AREA OF ELEGIBILITY WITH USGBC'S LEED 2009 FOR NEW CONSTRUCTION AND MAJOR RENOVATIONS				
MR 4	Recycled content	1 2 max	To earn 1 point use materials with recycled content such that the sum of postconsumer recycled content plus 1/2 of the preconsumer content constitutes at least 10%, based on cost, of the total value of the materials in the project. To earn 2 points use 20% or more.	•
TABLE 6—SUMMARY OF AREA OF ELEGIBILITY WITH USGBC'S LEED FOR HOMES 2008				
MR 2.2	Recycled content	0.5	To earn 0.5 point use masonry for exterior walls where the masonry with recycled content such that the sum of postconsumer recycled content plus 1/2 of the postindustrial (preconsumer) content constitutes a minimum total recycled content of 25%.	•
•	= Verified attribute			

CPSIA information can be obtained
at www.ICGtesting.com
Printed in the USA
LVOW04*0736010917

547143LV00009B/161/P